AF598558

2405828638

Large-Scale Transport Processes in Oceans and Atmosphere

NATO ASI Series

Advanced Science Institutes Series

A series presenting the results of activities sponsored by the NATO Science Committee, which aims at the dissemination of advanced scientific and technological knowledge, with a view to strengthening links between scientific communities.

The series is published by an international board of publishers in conjunction with the NATO Scientific Affairs Division

A Life Sciences B Physics	Plenum Publishing Corporation London and New York
C Mathematical and Physical Sciences	D. Reidel Publishing Company Dordrecht, Boston, Lancaster and Tokyo
D Behavioural and Social Sciences E Engineering and Materials Sciences	Martinus Nijhoff Publishers The Hague, Boston and Lancaster
F Computer and Systems Sciences G Ecological Sciences	Springer-Verlag Berlin, Heidelberg, New York and Tokyo

Series C: Mathematical and Physical Sciences Vol. 190

Large-Scale Transport Processes in Oceans and Atmosphere

edited by

J. Willebrand
Institut für Meereskunde,
Universität Kiel, F.R.G.

and

D. L. T. Anderson
Department of Atmospheric Physics,
Oxford University, U.K.

D. Reidel Publishing Company

Dordrecht / Boston / Lancaster / Tokyo

Published in cooperation with NATO Scientific Affairs Division

Proceedings of the NATO Advanced Study Institute on
Large-Scale Transport Processes in Oceans and Atmosphere
Les Houches, France
February 11-22, 1985

Library of Congress Cataloging in Publication Data

NATO Advanced Study Institute on Large-Scale Transport Processes in Oceans and Atmosphere (1985 : Les Houches, Haute-Savoie, France)
Large-Scale transport processes in oceans and atmosphere.

(NATO ASI series. Series C, Mathematical and physical sciences; vol. 190)
"Proceedings of the NATO Advanced Study Institute on Large-Scale Transport Processes in Oceans and Atmosphere, Les Houches, France, February 11–22, 1985"—Verso t.p.
"Published in cooperation with NATO Scientific Affairs Division."
Includes bibliographies.
1. Ocean circulation—Congresses. 2. Ocean-atmosphere interaction—Congresses. I. Willebrand, J. (Jürgen), 1941– . II. Anderson, D. L. T. (David L. T.) III. North Atlantic Treaty Organization. Scientific Affairs Division. IV. Title. V. Series: NATO ASI series. Series C, Mathematical and physical sciences ; no. 190)
GC228.5.N37 1985 551.47 86–20426
ISBN 90–277–2353–2

Published by D. Reidel Publishing Company
P.O. Box 17, 3300 AA Dordrecht, Holland

Sold and distributed in the U.S.A. and Canada
by Kluwer Academic Publishers,
101 Philip Drive, Assinippi Park, Norwell, MA 02061, U.S.A.

In all other countries, sold and distributed
by Kluwer Academic Publishers Group,
P.O. Box 322, 3300 AH Dordrecht, Holland

D. Reidel Publishing Company is a member of the Kluwer Academic Publishers Group

Printed in The Netherlands

In memoriam

ADRIAN E. GILL

CONTENTS

PREFACE

One of the major experiments in earth science at the present time is about to begin: the World Climate Research Program (WCRP). The objectives of WCRP are to determine the extent to which climate change can be predicted, and the extent to which human activities (such as increasing the level of CO_2) can influence our climate.

To understand and possibly to predict climate change, one needs a good understanding of the dynamics of the ocean, the atmosphere, and the processes by which they are coupled. Two major programs are being developed within WCRP: TOGA (Tropical Oceans, Global Atmosphere) and WOCE (World Ocean Circulation Experiment). The success of these programs will depend on many things, not least of which is the existence of a pool of active young researchers. This NATO Advanced Study Institute brought together students and young scientists from 13 countries, most of them from Europe and North America. The objective was to provide them with a background in the perceived state of knowledge of atmosphere and ocean dynamics, and to mediate a flavour of the problems presently concerning scientists active in climate related dynamics. In the past, the two disciplines of oceanography and meteorology have largely been carried out separately. But for climate research both disciplines must interact strongly, and another objective of this school was to bring together both oceanographers and meteorologists.

To promote an integrated approach, the lecture presentations were divided into two formats. The principal lectures were given by five eminent scientists, each of them contributing a series of four to six lectures which formed the backbone of the course. Two lecturers concentrated on the atmosphere (M. Blackmon on atmospheric models, B. Hoskins on interpretation of atmospheric observations), another two concentrated on the ocean (P.Rhines on ocean circulation dynamics, P. Welander on specific thermohaline effects and oscillations relevant for climate problems), while one lecturer specifically addressed the coupled problem (A. Gill on tropical atmosphere - ocean interactions; lecture notes are not included in this volume). All lecturers drew parallels between atmosphere and ocean wherever possible.

In support of the major lecture themes we had a number of individual lectures by active researchers. Although the range of topics covered was quite broad, most lecturers took care to point out the relevance of their work to both fluids. The feeling of the organisers was that a useful step towards bridging the gap between oceanographers and meteorologists was achieved. In addition to the organised lectures, we had presentations from many student participants. Regrettably, we have not been able to include these here but we all enjoyed the enthusiasm displayed by the students.

It is a pleasure to acknowledge the generous support of the NATO Scientific Affairs Divison which enabled us to organise this study institute. Also, the support given by the Les Houches School of Physics Winter Program, and by the US National Science Foundation, is gratefully acknowledged. The support provided by these bodies, and the pleasant environment at the Les Houches School of Physics, helped to create a lively and enjoyable spirit among the participants.

We would like to thank Mrs. S. Trier for her never-ending willingness to cope with the technical production of this volume. The assistance of Mrs. H. Bonnes, A. Schurbohm and H. Schomann is gratefully acknowledged.

On a sadder note, we regret to report the untimely death of Adrian Gill, one of our principal lecturers. Leading light in many fields of atmosphere and ocean dynamics, he gave an excellent series of lectures on the coupled atmosphere-ocean system. Students will remember him, not only for his lectures but also for the interest he took in their own research topics during his stay in Les Houches. We dedicate this book to him.

Kiel, July 1986

Jürgen Willebrand David Anderson

BUILDING, TESTING AND USING A GENERAL CIRCULATION MODEL

Maurice L. Blackmon

National Center for Atmospheric Research [1] Boulder, Colorado 80307

INTRODUCTION

The following notes are a summary of four lectures given on modeling the general circulation of the atmosphere. Since five or six hours are hardly sufficient to cover everything in general circulation modeling, and since in addition, these notes are a condensation of what was discussed, the reader will have to consult the references at the end of each lecture to find the details worked out for many of the topics which have been necessarily treated very briefly here. I have not tried to present a complete reference list. However, the references given, plus the references within the references, will give the reader a good working knowledge of the subject.

Let me mention one reference at the outset and that is Volume 17 of *Methods in Computational Physics*. This book contains five review articles, one on computational aspects of numerical models and four other articles, each of which describes a different general circulation model. The reader will notice that alternative points of view are taken in formulating the various models; and that is good, because none of the models is anywhere near perfect. It is my impression that this book is beginning to be out of date. Much of the numerical analysis which is discussed there will continue to be used, but when a comparable, updated volume is written, I expect the articles in it will be much more focussed on improvements in the physical parameterizations used in models. This is, I think, where current models are weakest, particularly in the treatment of clouds. The quality of simulations by general circulation models is improving, due in part to increased understanding of

[1] The National Center for Atmospheric Research is sponsored by the National Science Foundation.

J. Willebrand and D. L. T. Anderson (eds.), Large-Scale Transport Processes in Oceans and Atmosphere, 1–70.

the "physics" which goes into the models. We are also learning more about the atmosphere by using these models. This is, after all, the reward for all the work in model development. It is truly an exciting time to be building, testing and using general circulation models.

I. BUILDING A GENERAL CIRCULATION MODEL

The particular general circulation model (GCM) that I will discuss is NCAR's Community Climate Model, sometimes denoted CCM-Ø. The zero denotes the historical fact that this model, CCM-Ø, was expected to be a forerunner to the "real" model CCM-1. Therefore, a decision was made to keep many of the physical parameterizations in CCM-Ø as simple as possible in order to get the model to run quickly on a CRAY-1 computer, and also in order that later, we would be able to understand better what improvements result from more complicated and, hopefully, more realistic parameterizations. Therefore, several aspects of the model I will outline are not considered to be "state of the art." Much to our amazement when we began using CCM-Ø, we found many aspects of the model's simulations were quite realistic,more so than anticipated. This has resulted in many more studies being done using CCM-Ø than originally envisioned, and consequently a delay in the development of CCM-1.

The journal articles which describe CCM-Ø are Pitcher et al. (1983) and Ramanathan et al. (1983). This model was originally developed by W. Bourke and his collaborators, and the best description of the original model is given in Bourke et al. (1977).

Let me start with a list of model characteristics. I will then discuss these properties in varying detail.

Community Climate Model (CCM-∅)

- primitive equations
- spectral transform dynamics, rhomboidal 15 truncation, 48×40 grid
- 9 vertical levels, σ coordinates
- horizontal and vertical diffusion
- bulk parameterization of surface stress, sensible and latent heat flux
- realistic orography
- realistic land-sea-ice distribution
- surface temperatures calculated over land from energy balance, surface temperature of oceans fixed
- no surface hydrology, fixed snow line, prescribed albedo
- NCAR radiation package, solar and IR radiative transfer, calculated cloudiness
- convective adjustment and condensation

The basic equations we start with are the primitive equations written in spherical coordinates. Gill (1982, pages 91–94) has a nice discussion on the use of spherical coordinates and the accuracy of the approximations involved. Lorenz (1967, pages 16–19) has an excellent discussion of the approximations used to derive the primitive equations. The important point to mention here is that we assume the motions are in hydrostatic balance. The hydrostatic approximation eliminates vertically propagating sound waves and allows us to consider only the lower frequency fluctuations which have more importance for meteorological purposes.

The primitive equations are

$$\frac{D\vec{V}}{Dt} = -f\hat{k} \times \vec{V} - \vec{\nabla}\Phi + \vec{\Im}$$

$$\frac{DT}{Dt} = \frac{\kappa T\omega}{p} + \frac{Q}{c_p}$$

$$\vec{\nabla} \cdot \vec{V} + \frac{\partial\omega}{\partial p} = 0$$

$$\frac{\partial\Phi}{\partial p} = -\alpha$$

$$p\alpha = RT$$

$$\frac{Dq}{Dt} = E - P$$

Here $\vec{V}$ is the horizontal velocity; $\omega = \dot{p}$ is the vertical velocity; $f = 2\Omega\cos\phi$, the Coriolis parameter; Φ, the geopotential; α the specific volumes; T, the temperature and q, the moisture mixing ratio. The thermodynamic constants are R, the gas constant; c_p, the specific heat capacity at constant pressure and $\kappa \equiv R/c_p$. Q is the net rate of heating per unit mass and $\vec{\Im}$ is the net frictional force per unit mass. E and P are the evaporation and precipitation rates, respectively. The operator D/Dt is the time rate of change following a material element. These equations have been written in pressure coordinates because that is the common coordinate system for meteorological analysis. The model actually uses the Phillips σ coordinate for its calculations, where $\sigma = p/p*$, with $p*$ the surface pressure. Rather than discuss the coordinate transformation here, I will refer the reader to the discussion of σ coordinates in Haltiner and Williams (1980, pages 17–19, and pages 215–217).

The momentum equations are further manipulated by taking the curl and divergence to give two equations for scalar quantities $\varsigma = \vec{\nabla} \cdot \vec{V} \times \hat{k}$, the vorticity and $\delta = \vec{\nabla} \cdot \vec{V}$, the divergence. The equations for these quantities are

$$\frac{D}{Dt}(\varsigma + f) = -\delta(\varsigma + f) - \vec{\nabla}\omega \cdot \frac{\partial \vec{V}}{\partial p} \times \hat{k} + \vec{\nabla} \cdot (\vec{\Im} \times \hat{k})$$

$$\frac{\partial}{\partial t}\delta = -\vec{V} \cdot \vec{\nabla}(\varsigma + f) \times \hat{k} - \omega \frac{\partial \delta}{\partial p} + (\varsigma + f)\varsigma - \vec{\nabla}\omega \cdot \frac{\partial \vec{V}}{\partial p} - \nabla^2(\Phi + \frac{\vec{V}^2}{2}) + \vec{\nabla} \cdot \vec{\Im}$$

In this model scalar fields such as $\varsigma, \delta, T \cdots$ are represented as truncated series of spherical harmonics. For example,

$$\varsigma(\vec{x}, p, t) = \sum_{n,m} \varsigma_{n,m}^{(p,t)} Y_{n,m}(\lambda, \phi)$$

The position on the earth's surface is contained in the arguments of the spherical coordinates while the pressure coordinate and time dependence are contained in the expansion coefficient. The series is truncated at wave number 15 using the rhomboidal scheme (R15).

$$0 \leq |m| \leq 15$$

$$|m| \leq n \leq |m| + 15$$

where n is the two-dimensional wave number and m is the zonal wave number. $n - |m|$ gives the number of nodes in the spherical harmonic from pole to pole. With R15 truncation, we are hopefully able to resolve the baroclinic eddies which have zonal scales on the order of $m \simeq 6 - 10$ and larger fluctuations. (We now have some evidence that higher resolution improves the simulation of the Southern Hemisphere's circulation, but higher resolution also slows down the computational speed of the model. This would have made impossible some of the lengthy experiments I will describe later.)

The model has nine levels in the vertical at the values of $\sigma = 0.991$, 0.926, 0.811, 0.664, 0.500, 0.336, 0.189, 0.074 and 0.009. (Roughly equivalent values for pressure levels are obtained by multiplying by 1000 mb, the approximate mean surface pressure.) The large horizontal fluctuations in which we are interested have fairly smooth vertical structure with, for example, maxima in sensible heat flux or momentum flux near the surface and the tropopause, respectively. Greatly increased vertical resolution does not seem warranted. (We do have, however, recent evidence that increased resolution near the tropopause improves the simulation of some features. This entails adding two or three levels to the model.)

Since we have truncated the fields in the model both horizontally and vertically, we need to parametrize the effects of the unresolved eddies. This is commonly done by including dissipation terms in the prognostic equations. Simple forms are usually taken. For example,

$$\partial_t \varsigma = \cdots - \kappa_H \nabla^4 \varsigma$$

is a typical horizontal eddy diffusion approximation, with κ_H the horizontal eddy diffusion coefficient. For the vertical

$$\partial_t \varsigma = \cdots \cdot \vec{\nabla} \cdot [g^2 \partial_p (\rho^2 \kappa_v \frac{\partial \vec{V}}{\partial p}) \times \hat{k}]$$

is used. (The constants and variables are such that for constant density and constant vertical eddy diffusion coefficient, the term would be $\kappa_v(\partial^2 \varsigma / \partial z^2)$. In this model, κ_v is determined by a mixing length approximation

$$\kappa_v = \mu^2 \rho g \left| \frac{\partial \vec{V}}{\partial p} \right|$$

where

$$\mu = \begin{matrix} 30m & p > 500\ mb \\ 0 & p < 500\ mb \end{matrix}$$

These parameterizations are likely to be inadequate for some purposes. Since they operate most strongly on the smallest resolved scales, it seems plausible that these smallest resolved scales are the least trustworthy in the simulations. What we hope is that these parameterizations allow for the dissipation of small-scale fluctuations but do not have any major effect on large-scale features of the circulation.

This model is called a spectral transform model because of the way it manipulates various terms in the prognostic equations. These equations are of the general form

$$\frac{dX_{n,m}}{dt} = A_{n,m} + B_{n,m}(t)X_{n,m}(t) + N(X,X)$$

where $X_{n,m}$ is a typical expansion coefficient for some variable. A and B are independent of the variable X and N represents a nonlinear term such as $\vec{\nabla} \cdot (\vec{V}\varsigma)$, which is quadratic in the stream function and potential function. The obvious solution involves expanding each of the X's in the term $N(X,X)$ in spherical harmonics which reduces N to a multiple summation over expansion coefficients. Bourke et al. (1977) have a good discussion of the resulting problems, and they show that the time required to integrate a high resolution model using this method increases so rapidly as to make the technique unfeasible. The necessary trick is to calculate the nonlinear terms on a grid and to project the full term N into spherical harmonics $N = \Sigma N_{n,m}Y_{n,m}$. Then the time integration can be done efficiently. The reader is strongly advised to go through the discussion of this point in Bourke et al. (1977). My main objective here is just to acquaint the reader with the notion of transforming between spectral and physical space and thereby get to the point that there is a latitude-longitude grid associated with the spherical harmonic truncation.

Bourke et al. (1977) discuss the resolution necessary in the latitude-longitude grid so that no information is lost in the transformation between spectral and physical space. For rhomboidal truncation at wave number N, the number of points needed along a latitude circle is $I \geq 3N + 1$ and the number of latitudes needed is $J \geq 5N/2$. Since $N = 15$ in this model, a grid of dimension $(I,J) = (48,40)$ suffices.

Given the 48×40 grid, we are now in a position to specify the orography which goes into the model and also to determine the distribution of land, sea and ice points. We take a high resolution data set, say $1^\circ \times 1^\circ$ or finer, and average the heights found in the high resolution data over an area surrounding one of the points in the model grid. The distribution of orography which results is shown in Fig. 1a. This distribution is projected into spherical harmonics. If the orography is reconstructed at this point, negative heights occur in several regions, over the oceans near the Andes, Greenland, Antarctica, etc. We smooth out part of these negative heights by applying a short wave smoother to the orographic spherical harmonic coefficients. One smoother is

$$\tilde{T}_{n,m} = (1 - \frac{n^2}{N_0^2})T_{n,m}$$

where $T_{n,m}$ is the expansion coefficient of the orography, N_0 is an adjusted constant and $\tilde{T}_{n,m}$ is the resulting smoothed orography spectral coefficient. Fig. 1b shows the reconstructed orography for this smoothing procedure using $N_0 = 36$. If stronger smoothing is used, the negative heights are reduced but so are the mountain peak values. Bourke et al. (1977) used considerably stronger smoothing than we have (see their Fig. 5). We have found through a comparison of model statistics, such as discussed below in Lecture II, that too much smoothing in the orography spoils the regional differences in eddy statistics.

We also use the high resolution data to determine whether model grid points are land, sea or ice. If the majority of the high resolution points in an area are of one type, say land, then that model grid point is made a land point. For areas with a mixture of land, sea and ice, the model grid point assumes the type of the highest fraction of points in the high resolution data. In Fig. 2a and 2b we show the distribution of land, sea and ice for January and July, respectively.

We also need to specify other properties at the surface. The surface stress is given by a simple bulk aerodynamic formula

$$\vec{\tau}_* = \rho_1 C_d |\vec{V}_1| \vec{V}_1$$

where * denotes a quantity at the surface and subscript 1 denotes a quantity at the first model level. C_d is a drag coefficient which is set equal to 0.004 or 0.001

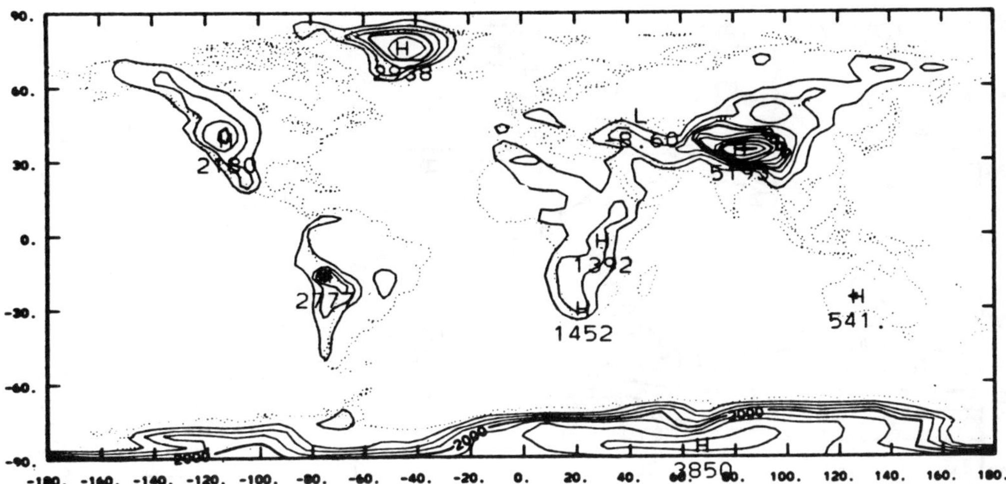

1. (a) Mean orography from high resolution data averaged over each box (approximately 4.5° × 7.5°) on the model grid.

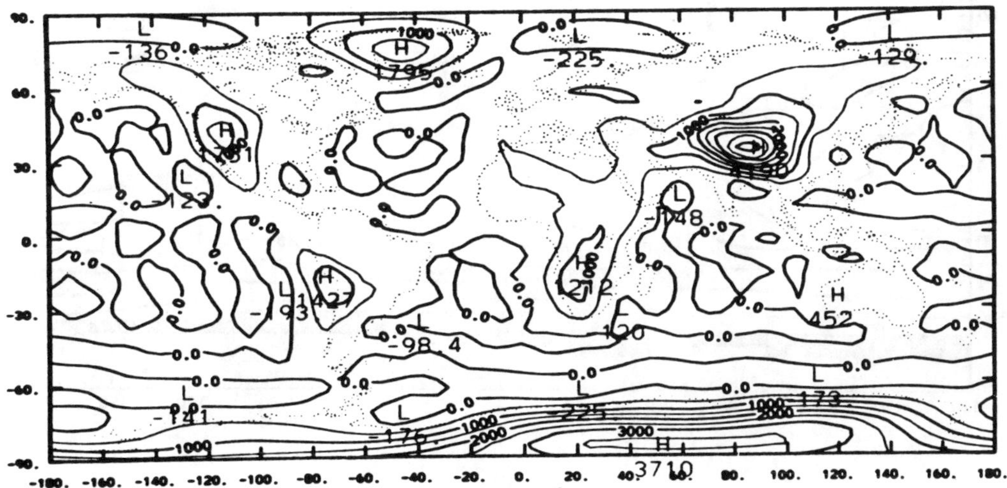

(b) Orography shown in (a) after projection onto R15 spherical harmonics and smoothed as described in text.

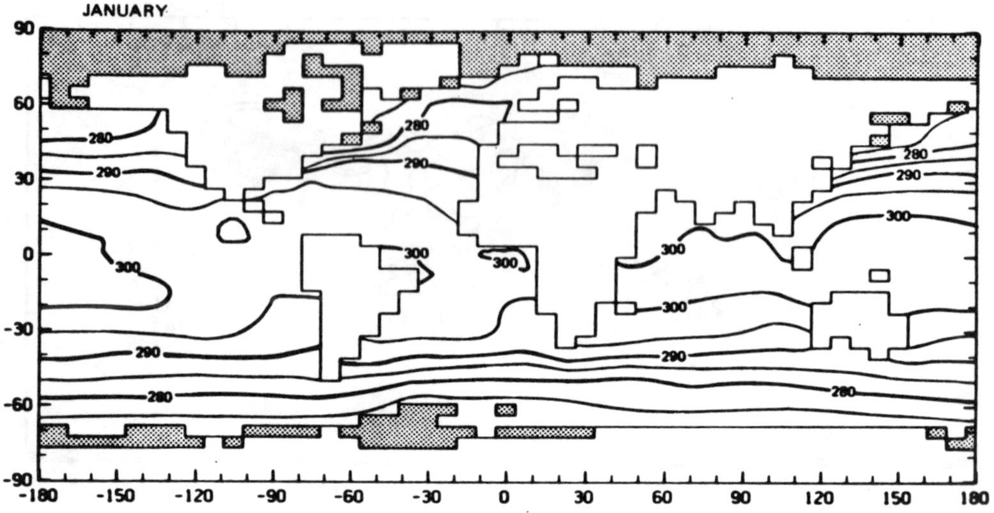

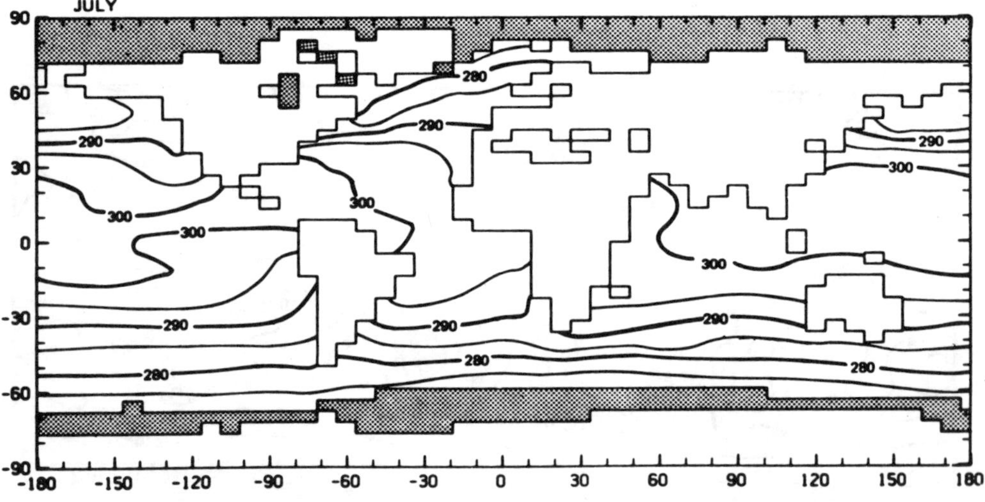

2. The sea surface temperature (K) and land-sea-ice distribution used in the model for (a) January and (b) July. The contour interval is 5K. Stippled areas denote regions of sea ice.

over land or sea, respectively. Similarly, surface sensible and latent heat fluxes are parameterized by the formulas

$$\eta_* = \delta\rho_1 C_d |\vec{V}_1| (\theta_* - \theta_1)$$

$$\beta_* = \rho_1 C_w C_d |\vec{V}_1| (q_s(T_*) - q_1)$$

where $\delta = (p/p_0)(R/c_p)$, θ is the potential temperature, q_s is the saturation mixing ratio and C_w is a fixed "wetness" factor which is taken to be 1 for sea and 0.25 for land. (You will note that this simple, and obviously crude, parameterization makes no distinction between rain forest, farm land and desert.)

Because of the large heat capacity of water, the sea surface temperature is prescribed, and in the simulations I will discuss, it is a constant. The surface temperature over land is calculated by simple energy balance.

$$S_\downarrow + D_\downarrow = \epsilon\,\sigma_s T_*^4 + c_p\eta_* + L\beta_*$$

where $S_\downarrow$ is the downward solar flux absorbed at the surface, $D_\downarrow$ is the downward IR flux at the surface, ϵ is the surface emissivity taken to be 0.75 for deserts and 0.95 for other land surfaces, σ_s is the Stefan-Boltzmann constant, c_p is the specific heat capacity and L is the latent heat of water vapor. (No heat storage is allowed at the ground, although it could easily be incorporated at the cost of another parameter.)

One other surface property is specified in the model and that is albedo. Ideally, albedo is a function of the wavelength of incident radiation, the zenith angle of the radiation beam and the nature of the surface (sea, ice, snow, soil or vegetation, etc.) In this model, the geographical extent of snow is prescribed by season and greatly simplified albedos are specified. Again, we make as few distinctions among the various surface types as seems reasonable. Table 1 lists the various surface types and albedòs used.

The only part of the system we have not discussed is Q, the heating rate. This is not because it is unimportant. It is, after all, what makes the system go. The

Table 1. Albedo and Surface Types

Land (except desert)	0.13	
Desert	0.25	
Ocean	$0.05/(0.15 + \cos\theta)$	θ - solar zenith angle
Ice	0.50	
Snow	0.85	$0.0 \leq \lambda \simeq 0.9\mu$ m
	0.55	$0.9 \leq \lambda \simeq 5.0\mu$ m

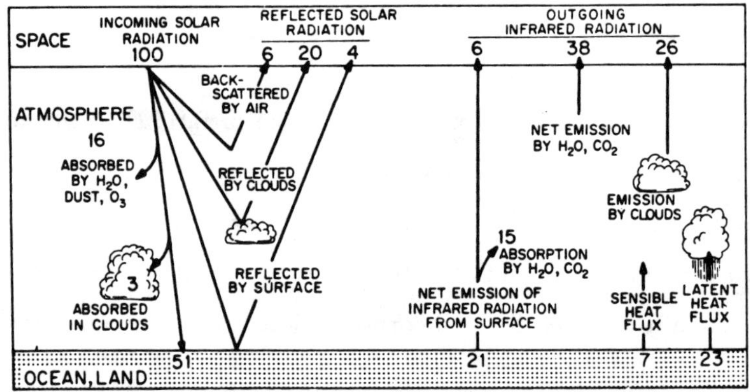

3. Radiation balance for the atmosphere. [Adapted from "Understanding Climatic Change," U.S. National Academy of Sciences, Washington, D.C., 1975.]

problem is that the calculation of Q is difficult. Fig. 3 shows schematically what is involved. Of the incoming solar beam, 6% is scattered by the atmosphere, 20% is scattered by clouds and 4% is reflected by the surface. In addition, 16% is absorbed by various atmospheric components, including dust, and 3% is absorbed in clouds. The solar absorption by dust and by clouds is presently not included in the model, but the other components are. The cloud cover is calculated in the model. Cloud albedo is prescribed to try to get the 20% solar reflection by clouds. This also requires that the total cloud cover be calculated correctly.

We have already discussed the emission of IR from the surface and the sensible and latent heat flux parameterizations. We also take into account emission by clouds and clear sky emission from H_2O, CO_2 and O_3. Again, getting the IR emission from clouds correct means we not only need to have the properties of an individual cloud properly specified but are also calculating approximately correctly the cloud amount and distribution of clouds. The distribution of O_3 is prescribed and the mixing ratio of CO_2 is fixed. The mixing ratio of H_2O is calculated.

The heating rate per unit mass is related to the flux of radiant energy by

$$Q \sim -\frac{\vec{\nabla} \cdot \vec{F}}{\rho}$$

In the model, we are interested in the upward and downward fluxes, so

$$Q = g\frac{\partial F}{\partial p}$$

Equations for the upward and downward flux are found in Ramanathan et al. (1983). The equations depend on the well-known Planck function and also properties of the gases, emissivities and absorptivities which can be determined from data. For infrared radiation, the important gases in the atmosphere are H_2O, CO_2 and O_3. These gases have bands with wavelengths in the range where most of the IR is emitted by a body with the temperature of the earth. Table 2 summarizes

TABLE 2. Clear-sky radiation processes.

Constituent	Comments
I. Longwave	
1. H_2O	Rotational (12–∞ μm), vibration–rotation (4–8 μm), and continuum band (8–12 μm). The *e*-type absorption is neglected.
2. CO_2	Fundamental, 1st and 2nd hot bands in the 12–18 μm region. Includes four isotopes of CO_2: $C^{12}O_2{}^{16}$; $C^{13}O_2{}^{16}$; $C^{12}O^{16}O^{17}$; and $C^{12}O^{16}O^{18}$.
3. O_3	9.6 μm vibration–rotation band.
4. Surface emissivity	Non-black (i.e., emissivity < 1) in the 8–12 μm region.
II. Solar	
Scattering	Rayleigh scattering
Absorption	Includes O_3, H_2O, CO_2 and O_2.

TABLE 3. Clouds and cloud–radiation interactions.

Item	Comments
I. Cloud prescription:	• Clouds are formed if the relative humidity exceeds 80%, the value assumed for saturation. • Cloud fraction is determined by whether the clouds are formed by convective or non-convective processes. • Clouds are not formed in the thin surface layer (σ = 0.991).
II. Solar Albedo	• Depends on zenith angle and cloud type • Accounts for multiple reflection between different cloud layers and between clouds and surface.
Absorption	• Gaseous absorption within clouds included.
III. Longwave emissivity	• For levels with $\sigma \geqslant 0.5$, clouds are assumed to be black. • High level clouds, i.e., $\sigma < 0.5$, are assumed to be black only if the cloud water content exceeds $10\ g\ m^{-2}$.

the various gases and the bands that are parameterized and included in the model. Other gases have bands in the appropriate range of wavelengths, but their concentration in the atmosphere is thought to be small enough so that their effects are generally small compared to the effects presently included in the model.

Figure 4 shows the absorption of radiation integrated through the depth of the atmosphere, as a function of wavelength for various atmospheric constituents. The temperature profile used is typical for midlatitude winter. The absorption ranges between zero (no absorption) and one (complete absorption). The top three panels show the absorption for H_2O, CO_2 and O_3, which are treated in the model. The next two panels are for CH_4 and N_2O which are presently not included. The total absorption is shown at the bottom. For water vapor, the absorption is near total for frequencies ν less than 500 cm^{-1}. As ν increases we pass through the "atmospheric window" where the absorption decreases to about 30% at $\nu \sim$ 1000 cm^{-1}. The absorption then increases again nearly to unity for $1300 \simeq \nu \simeq 2000$ cm. The CO_2 absorption is unity around ν = 700 cm^{-1} which is in the 12–18 μm band. (Fig. 4 was kindly supplied by my colleague J. Kiehl, NCAR.) Remember that what goes into the heating rate is the derivative of the flux with respect to pressure, so we need to be able to calculate similar curves to those shown in Fig. 4 as a function of pressure and temperature. That this can be done in an efficient fashion is quite impressive. Refer to Ramanathan et al. (1983) for some of the parametric forms used. Recent intercomparison of clear-sky radiation codes for several GCMs has shown that all are in reasonable agreement with observations, say within 5–10%.

In this model, we allow precipitation when the relative humidity in some layer at some gridpoint exceeds 80%. Excess moisture precipitates until the relative humidity is reduced to 80%. Precipitation falls out without evaporation into lower layers. When the criterion for precipitation is reached, the model forms clouds. Fig. 5 schematically depicts the various cloud types formed in the model. The model cloud properties are also summarized in Table 3.

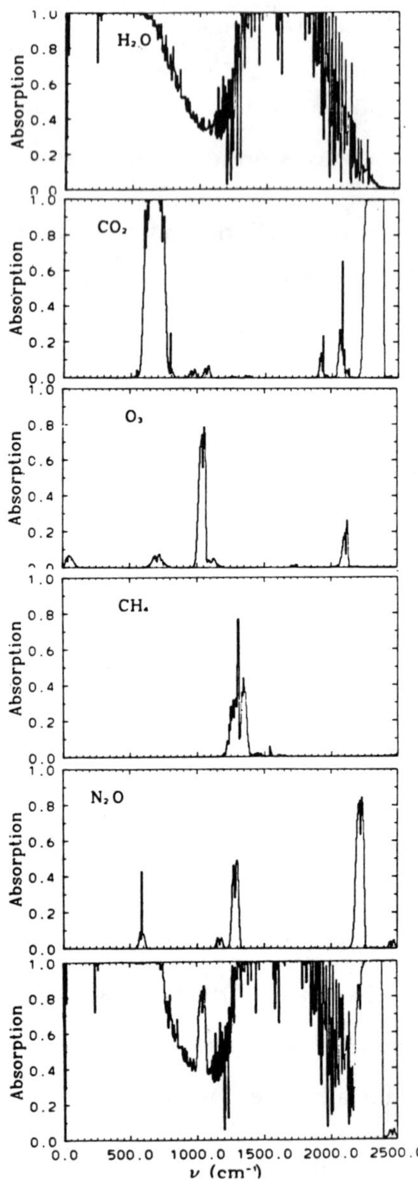

4. Vertically integrated absorption for $H_2O, CO_2, O_3, CH_4, N_2O$ and the sum of these gases. [Figure courtesy of J. Kiehl, NCAR.]

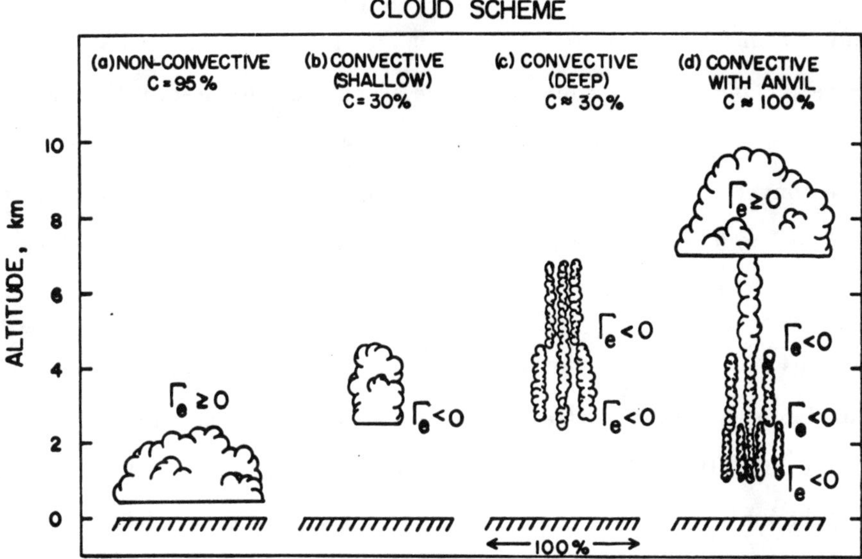

5. Schematic illustration of the nonconvective and convective cloud configurations in the model. Nonconvective clouds that occur in two or more model layers are assumed to be randomly overlapped. The total cloud amount in percent of the area in the model grid is C. The vertical gradient of equivalent potential temperature in a moist atmosphere is Γ_e.

Unlike the clear-sky radiation parameterizations, there are large uncertainties about cloud parameterizations. Ramanthan et al. (1983) give rationalizations for the choices used. One parameter in our scheme which needs to be set is the fraction of cloud cover for nonconvective cloud conditions. Another is the fractional cloud coverage for convective conditions. While an appeal is made to obervations to determine these numbers, it must be admitted that the observational data base is inadequate.

Another area of uncertainty is the relationship between the emissivity and liquid water in clouds. Most models treat all clouds as black-body radiators. However, Ramanathan et al. (1983) discuss evidence that for clouds with small liquid water content, the emission is reduced (emissivity < 1). In the next lecture, I will show evidence that treating all clouds as black bodies biases the simulation toward stronger than observed jets. In CCM-Ø we have chosen not to let clouds form above 300 mb. Clouds below 300 mb typically have large enough liquid water content that they can be considered to be black. A discussion of a first attempt to let emissivity vary with liquid water content is given in Ramanathan et al. (1983).

One final parameterization needs to be included to make the model run for extended periods. If the model were allowed to run as I have described so far, quite frequently regions would develop, due to strong heating at the surface, where $(\partial\theta/\partial p) > 0$ or $(\partial\theta_e/\partial p) > 0$, i.e., the local lapse rate would be unstable. If such instabilities develop in the atmosphere, convection results, although on much smaller scales than the smallest resolved scales in the model. We need, therefore, a parameterization scheme to remove these instabilities. The scheme we use is convective adjustment which was developed by Manabe et al. (1965). Consider first the case of dry convective adjustment. If $(\partial\theta/\partial p) > 0$, and if the layers to be adjusted are not saturated, then the dry static energy is conserved. We have

$$\int_{p_B}^{p_t} dp(c_p T + gz) = \text{constant}$$

subject to two constraints

$$1. \quad \frac{\partial \theta}{\partial p} = 0$$

$$2. \quad \frac{c_p}{g} \int_{p_B}^{p_T} dp \delta T(p) = 0$$

where $\delta T = T_{after} - T_{before}$ is the variation of temperature with pressure. Since the vertical integral of the height is proportional to the vertical integral of temperature in a hydrostatic atmosphere, condition two is a statement of conservation of the dry static energy.

For saturated conditions, the moist static energy is the appropriate quantity to conserve.

$$\int dp(c_p T + gz + Lq) = \text{constant}$$

subject to the conditions

(1) $$\frac{\partial \theta_e}{\partial p}(T + \delta T, q + \delta q, p) = 0$$

(2) $$q + \delta q = q_s(T + \delta T, p)$$

(3) $$\frac{1}{g} \int dp(c_p \delta T + L \delta q) = 0$$

Haltiner and Williams (1980, pages 312–314) have a discussion of this scheme with a simple example.

This has necessarily been a brief description of the model. The code is approximately 10,000 lines long and the most efficient version of the code presently takes about 80 seconds to integrate one model day on a CRAY-1 computer. The data output for one model day consists of approximately 2×10^5 numbers, if the output is sampled at 12-hour intervals. Some of the experiments we have run have been for 1200 model days. What one does with part of this mass of data will be discussed in the next lectures.

II. TESTING A GCM

The quality of the simulation of GCM depends on the dynamics, physics and numerics that go into it—in short, everything. In this lecture summary, I will skip everything to do with numerics, even though there are some interesting examples that could be shown, and present only two examples of the effects of changing physics parameterizations on various aspects of the model simulations. What I will do most extensively is to compare the simulation of CCM-Ø with observations. In some cases, various experiments have shown why model inadequacies are as they are, and I will mention what is likely to be done in CCM-1 to improve the model's performance. I will not be able to go through an elaborate documentation, however.

Pitcher et al. (1983) presents documentation for the mean fields simulated by the model. Ramanathan et al. (1983) discusses in detail the model sensitivity to changes in the radiation and cloud parameterizations in the model. Malone et al. (1984) discusses in detail the stationary and transient eddy characteristics of the model for January and July and for both hemispheres.

The most important modification of the Bourke model which was done at NCAR was the change to the NCAR radiation package as discussed in Ramanathan et al. (1983). Therefore, I want to give two examples of effects of improving physical parameterizations.

Fig. 1 shows the zonal average of the zonal wind for the model (Fig. 1a) and observations (Fig. lb) and the zonal average of the temperature for model (Fig. 1c) and observations (Fig. 1d). The simulations and observations are for Northern Hemisphere winter. There is good qualitative agreement between the model and observed winds, although the model's summer hemisphere winds are too weak by about 5 m s^{-1}. The agreement between model and observed temperatures also

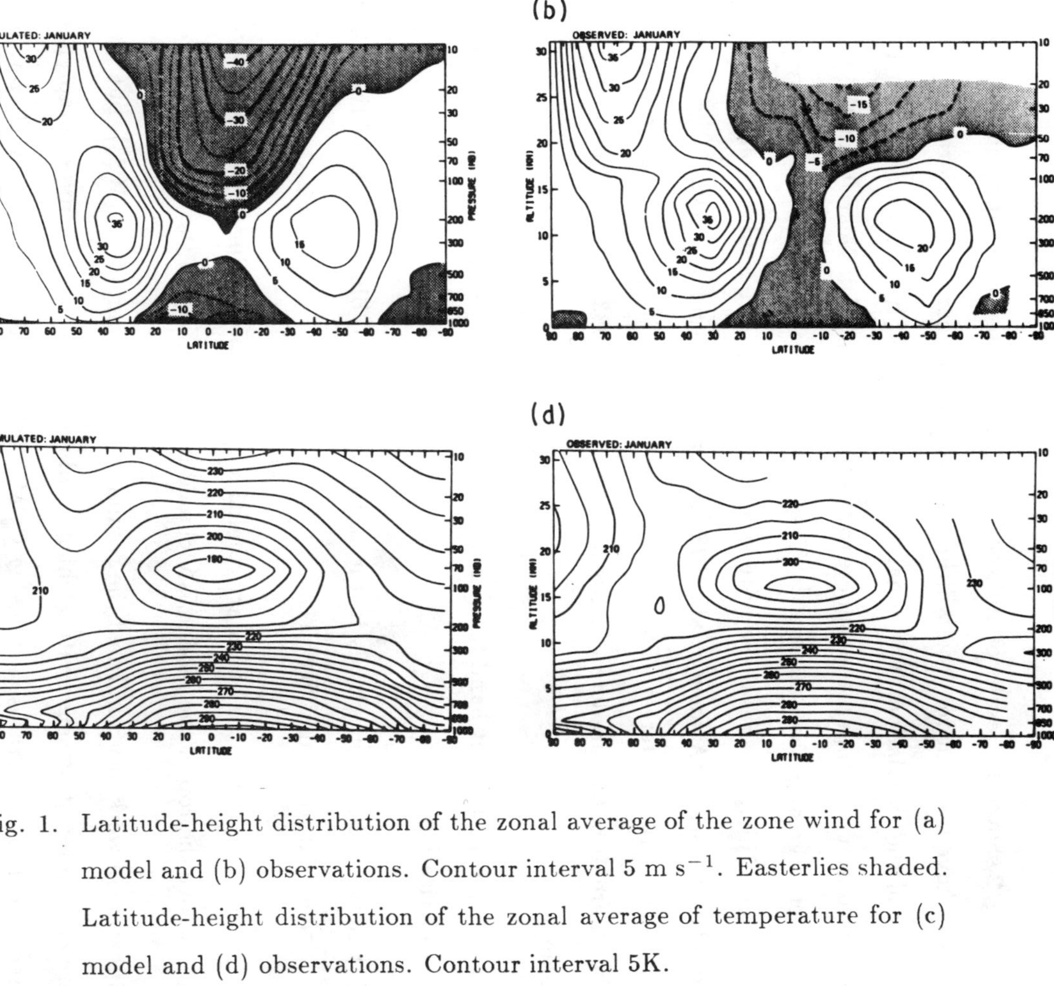

Fig. 1. Latitude-height distribution of the zonal average of the zone wind for (a) model and (b) observations. Contour interval 5 m s^{-1}. Easterlies shaded. Latitude-height distribution of the zonal average of temperature for (c) model and (d) observations. Contour interval 5K.

looks good, but closer comparison shows the model is too cold by 2–8K, depending on the latitude and height.

Fig. 2 shows the results of an experiment in which several changes in the model physics have been made. You will immediately see a marked deterioration in the quality of the simulation. The zonal wind in the winter hemisphere is larger than observed from the upper troposphere to the top of the model. Near 10 mb the polar night jet is greater than 80 m s^{-1}. The polar stratosphere temperatures are now less than 170K. This result is quite similar to the typical results of earlier GCM simulations (see McAvaney et al., 1978) so our model results, shown in Fig. 1, met with considerable interest. The changes that were made in the model to produce Fig. 2 are discussed in detail in Ramanathan et al. (1983). Briefly, these changes include:

(1) Removing the hot band temperature dependence in the CO_2 radiation parameterization.

(2) Making the cirrus clouds in the model radiate like black bodies of the same temperature (emissivity=1).

(3) Changing the upper boundary condition so that the heating in the ozone level, above the top of the model, is included in the heating of the top model layer.

Subsequent work has confirmed the results in Ramanathan et al. (1983), but there are other interesting sensitivities which I will not discuss here. A more comprehensive model of the stratosphere using CCM-∅ is under development by B. Boville. Further developments will be published by him.

The properties given clouds have interesting and important effects in the model simulation. Typically, clouds have been assumed to radiate like black bodies (emissivity=1). However, for high clouds, where the water content is small, the emissivity should be less than one. Again, see Ramanathan et al. (1983) for details. Fig. 3 shows the outgoing IR flux for (a) observations, (b) the control run of CCM-∅, (c)

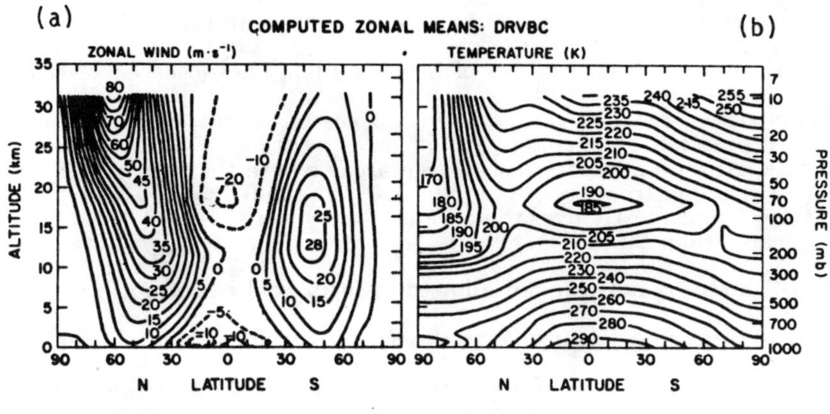

Fig. 2. Latitude-height distribution of the (a) zonal wind and (b) temperature for the degraded radiation, variable black cirrus experiment. Same contour interval as in Fig. 1.

an experiment with fixed, black cirrus and (d) variable cirrus with emissivity scaled by liquid water content of the cloud. Focus on the tropics and note that high values of IR flux correspond to clear, or nearly clear, sky conditions while low values correspond to radiation from the top of high clouds. The observed distribution, Fig. 3a, shows strong regional contrasts with areas of less than 230 W m^{-2} over South America, Africa and Indonesia. Large areas where the flux is greater than 280 W m^{-2} are found mostly over the oceans.

In CCM-Ø, cirrus clouds are not allowed to form above 300 mb. This was our first attempt to eliminate the problems caused by the formation of black cirrus. However, this introduces biases of a different sort. Fig. 3b shows that the control simulation does fairly well in producing the areas of high IR flux (the clear sky part). However, the regions of low flux (and high clouds) are not simulated well at all.

Many GCMs have used prescribed, zonally averaged cloud distributions. Fig. 3c shows the outgoing IR flux for such a parameterization. Much of the longitudinal contrast has been wiped out. However, there is significant latitudinal contrast since a high fraction of deep clouds are prescribed above the ITCZ and much less cloudiness is prescribed over the descending branches of the Hadley circulation.

Fig. 3d shows the results of allowing clouds to form at all levels in the model, but with the emissivity varying with the liquid water content. Clearly, the comparison with observations is best for this experiment. In spite of the obvious improvement, large uncertainties still remain. The result is strongly dependent on the vertical resolution near the tropopause and on the details of the function chosen to parameterize emissivity as a function of water content.

It should be obvious to the reader that there are serious problems with the radiation/cloud parameterizations in the model. Even with fixed climatological sea surface temperatures, which greatly decrease the freedom of the model to wander too far from the observed climatology, the model atmosphere is too cold. One way

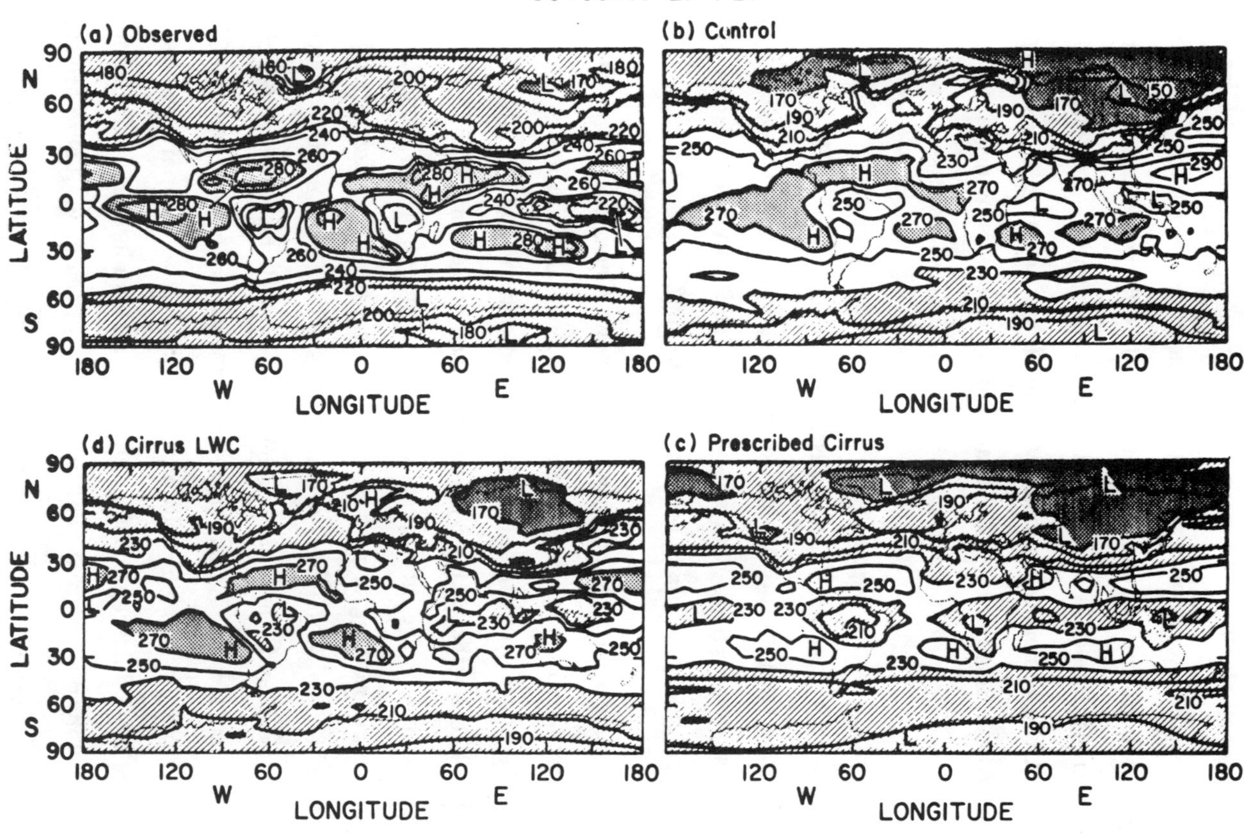

Fig. 3. Outgoing IR flux at the top of the atmosphere for (a) observations, (b) the control run of CCM-Ø, (c) CCM-Ø with zonally averaged fixed cirrus and (d) CCM-Ø with variable calculated cirrus clouds, cirrus emissivity scaled according to the liquid water content. Units watts m^{-2}.

of warming the troposphere is by increasing the fraction of cirrus clouds. This, however, can have negative consequences for the stratosphere. Over the last year or so, several parts of the solution to the cold temperature bias have been found. A better numerical scheme for calculating vertical finite differences and finer vertical resolution near the tropopause are helpful. The water vapor component of the radiation code in CCM-∅ is in need of improvement. In particular, *e*-type absorption has been included. Absorption of solar radiation by clouds is being investigated. It remains to be seen how close to observations these improvements will bring the model.

We now turn to a comparison of statistics of the standard CCM-∅ with observations. The questions we are trying to answer are: (1) what aspects of the similation are most in agreement with observations? and (2) what are some of the phenomena that can likely be simulated with this model?

I will tell the reader part of the answer right now. Generally, the model simulation of the Northern Hemisphere winter circulation is most in agreement with observations. The Southern Hemisphere circulation, summer and winter, is simulated less well. The Northern Hemisphere summer circulation is simulated least well. The reasons for this are incompletely known, although we do have hints at answers. This general ranking of results is not constant from model to model. I know of one model for which the users are most satisfied with the model's Southern Hemisphere circulation and where the model's Northern Hemisphere winter circulation is the least satisfactory. Again, there is only a partial understanding of why this is so. This situation has generally made it difficult to compare results and parameterizations from different GCM's and gain much understanding. The number of variables involved is so great that progress has been slow. There are signs, however, that this "noncomparability" of different models is lessening.

Because of limited space, I will only show results for perpetual January simulations and compare with observations typically averaged over the winter season. An annual cycle version of the model has been run and the results including the annual cycle are quite similar to those for fixed January conditions. For more details of the annual cycle simulation, see Chervin (1985). I again refer the reader to Pitcher et al. (1983) and Malone et al. (1984) for much more detailed discussion of the simulation characteristics of CCM-∅, including those for perpetual July conditions.

In Fig. 4 the mean sea level pressure distributions are shown for (a) the model and (b) observations. The Icelandic and Aleutian lows are simulated quite well, as is the Tibetan high. The most obvious failure is the poor simulation of the belt of low pressure along 70°S. The model produces low pressure there, but the values are 15 mb too high. Recent experiments have shown that this is in part a result insufficient horizontal resolution.

Fig. 5 shows the mean surface air temperature for (a) model and (b) observations. Since the ocean temperatures are fixed, the surface air temperatures over oceans cannot drift too far off. Over land, however, especially at high latitudes in the Northern Hemisphere, we see that model temperatures are too cold. For July comparisons, not shown, similar results are obtained. Part of the discrepancy is due to our use of a constant wetness factor over all land points. Over what should be dry land surfaces, such as the Sahara, the southwestern part of the U.S., Australia, etc., CCM-∅ is producing too much latent heat flux. Surface temperatures cannot increase closer to their observed values. Other parts of the temperature discrepancy were discussed earlier.

Fig. 6 compares the mean precipitation for (a) model and (b) observations. The general patterns of the distributions look similar with heavy precipitation over the tropics and in the Northern Hemisphere, from the eastern part of the continents across the oceans. It is clear, however, that the model precipitation is heavier than the observed, especially in the tropics. The reader should recognize that observations of precipitation are quite spotty and inaccurate, and that makes detailed comparison of dubious value. However, there is another bias due to our use of convective adjustment. Krishnamurti and Moxim (1971) have shown that this cumulus

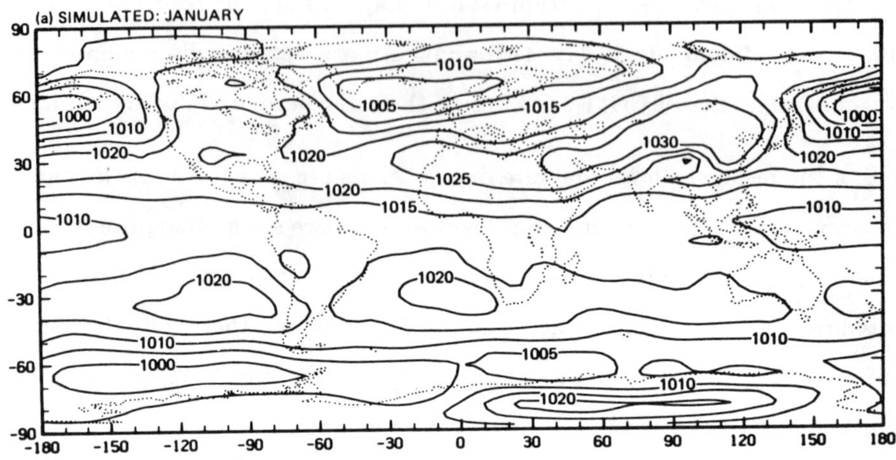

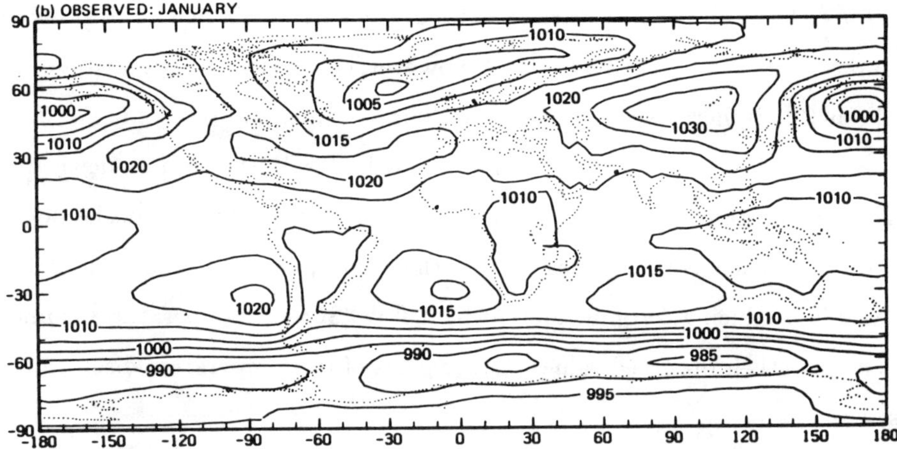

Fig. 4. Mean sea level pressure for (a) model and (b) observations. Contour interval 5 mb.

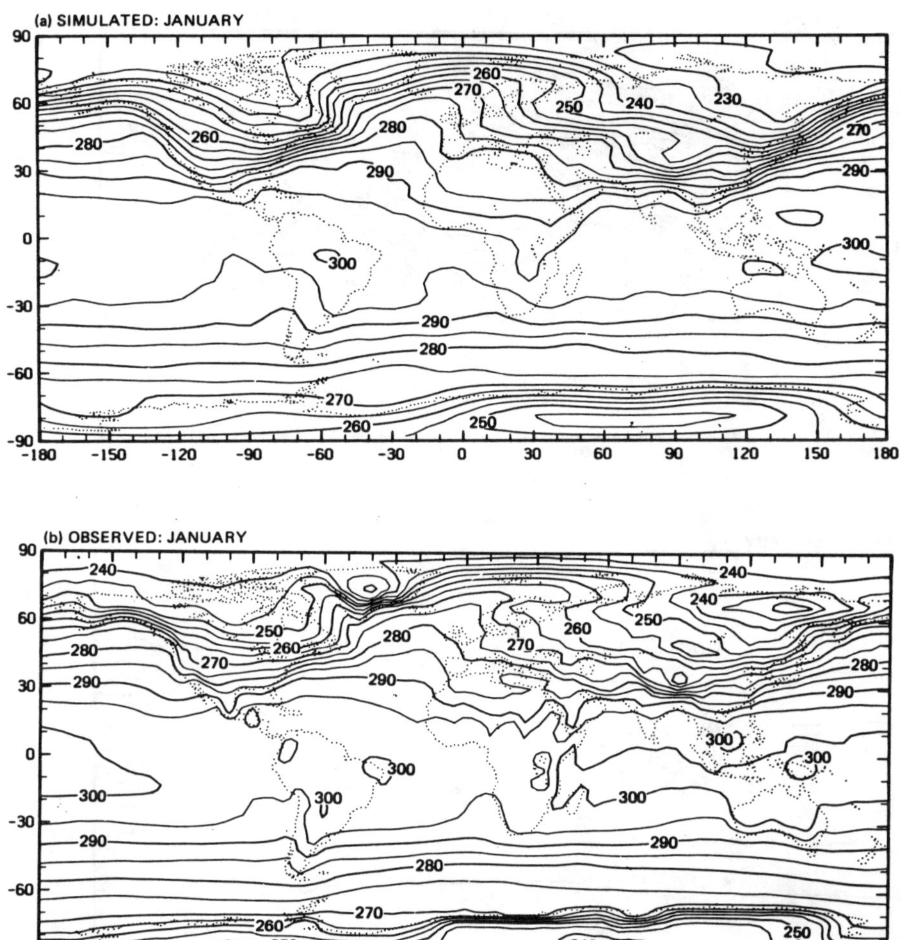

Fig. 5. Mean surface air temperature for (a) model and (b) observations. Contour interval 5K.

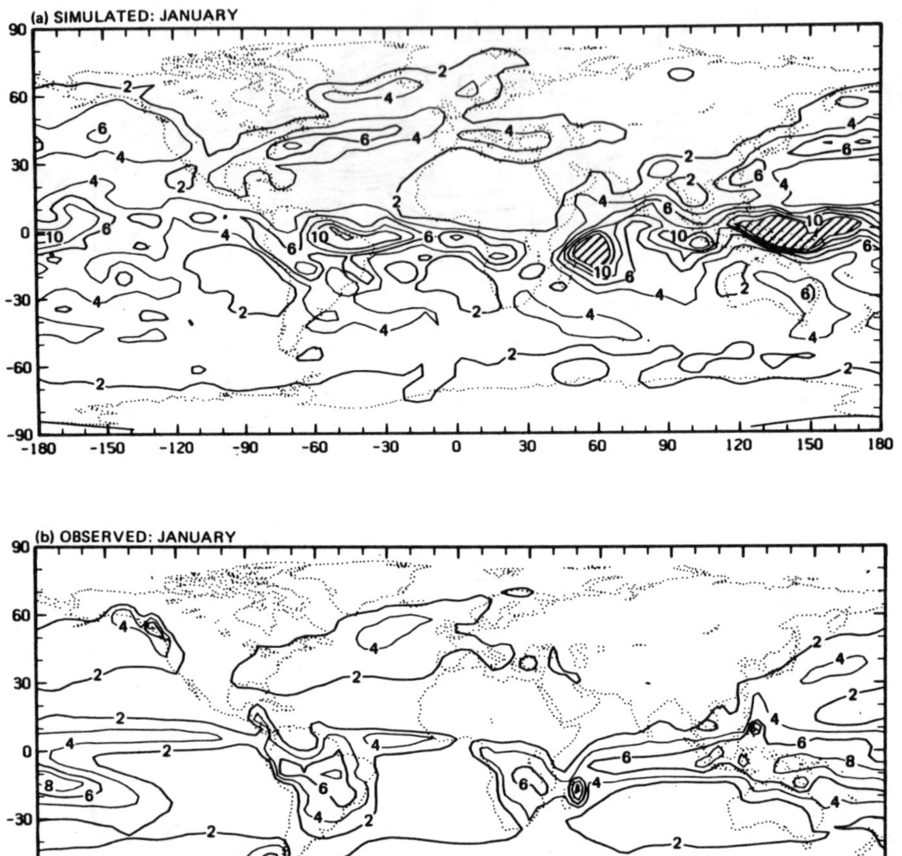

Fig. 6. Mean precipitation for (a) model and (b) observations. Contour interval 2 mm day^{-1}.

parameterization scheme typically produces excessive precipitation amounts from the lower troposphere.

Fig. 7 compares the mean 500 mb height fields of (a) model and (b) observations. The model produces the typical three-trough pattern with troughs over the east coasts of Asia and North America and another over Europe. The mean height field is about 100 mb too low due to the colder-than-observed temperatures simulated by the model.

Fig. 8 compares the mean 300 mb zonal wind for (a) model and (b) observations. The model winds have maxima near the east coasts of the continents with maximum values close to those observed. Calculation of the mean meridional ageostrophic wind (not shown) shows that the dominant terms in the maintenance of the mean zonal wind is the same as that found in observations.

$$f\overline{v}_a \cong \overline{u}\,\partial_x\,\overline{u}.$$

See Blackmon et al. (1977) for details.

We turn now to transient eddy statistics. Because heights and wind fields are dominated by low-frequency fluctuations, it is necessary to filter time series in order to examine important, high-frequency fluctuations. The filtering is accomplished by taking a weighted, running sum over the elements of the time series. Let $x(t_i)$ be the value of a field at some grid point at time t_i. A filtered time series containing only a band of frequencies desired is defined by

$$\tilde{x}(t_i) = \sum_{p=1}^{P} a_p[x(t_{i+p}) + x(t_{i-p})]$$

Blackmon (1976) gives the coefficients a_p for two filters, one which passes fluctuations with periods greater than 10 days (low frequencies) and one which passes fluctuations with periods between 2.5 and 6 days (high frequencies). The band-pass filter has been chosen to enable us to examine fluctuations with time scales of typical baroclinic eddies. Using either unfiltered or filtered data, variances or standard

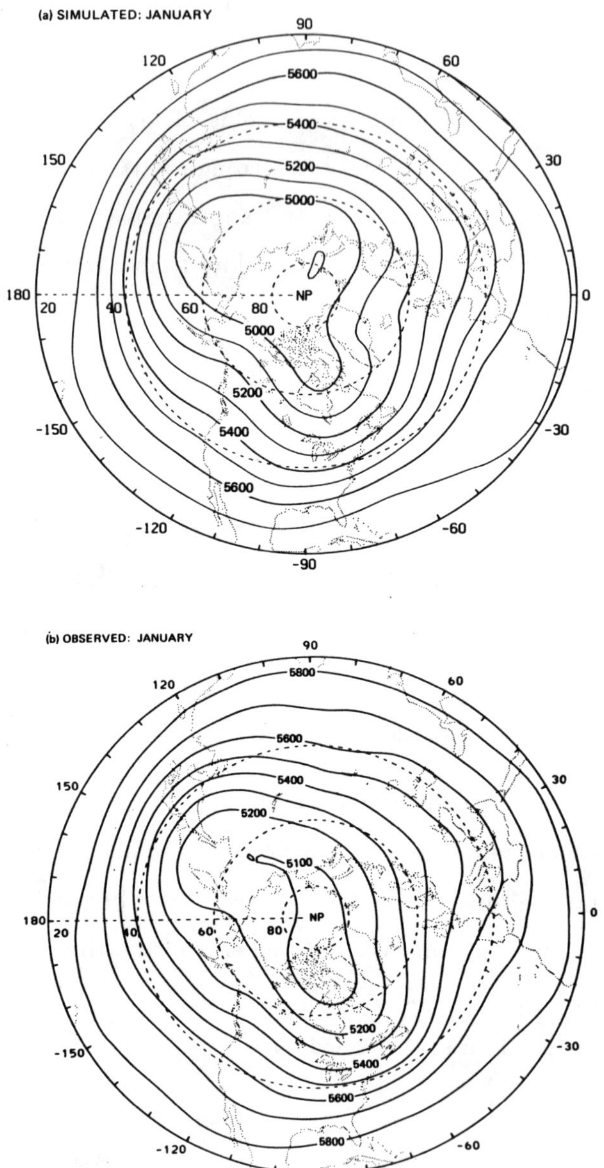

Fig. 7. Mean 500 mb geopotential height for (a) model and (b) observations. Contour interval 100 m.

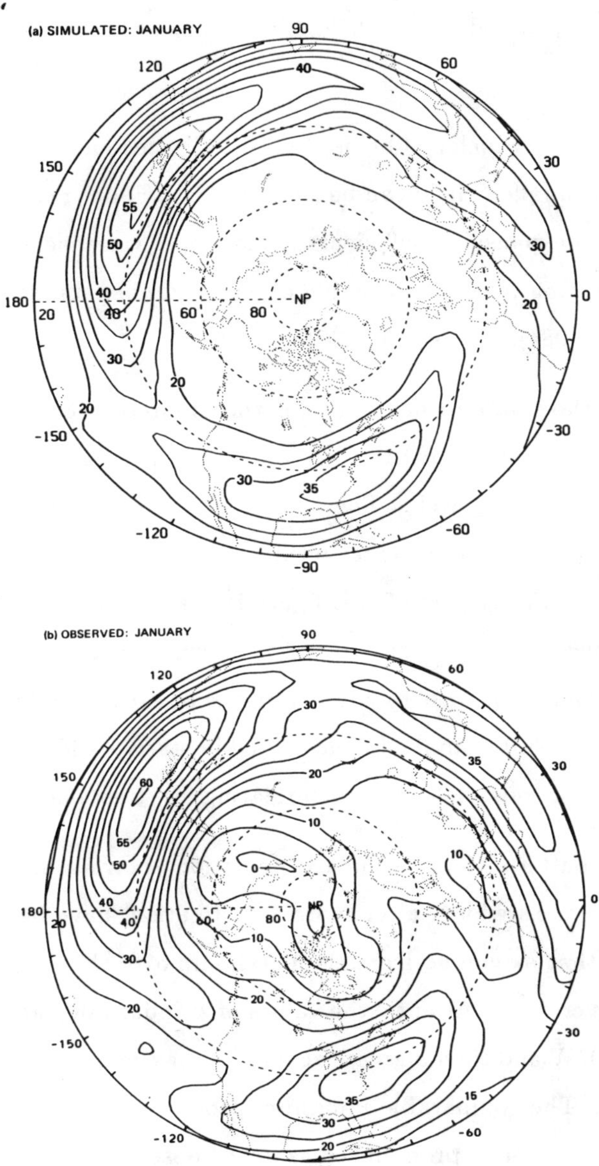

Fig. 8. Mean 300 mb zonal wind for (a) model and (b) observations. Contour interval 5 m s^{-1}.

deviations can be calculated for any field or covariances can be calculated for pairs of fields.

Another technique which can be used to investigate model performance is spatial filtering. For example, a field can be projected in o Fourier components and the statistics of various expansion coefficients (time-mean, variance, etc.) can be compared with observations. For a discussion of wave number statistics in the model, see Malone et al. (1984).

Fig. 9 shows the standard deviation for the unfiltered 500 mb heights for (a) model and (b) observations. Of the three observed maxima, the model simulates two, over the oceans, fairly well. The third maximum over the Soviet Union is not simulated at all. Why the model produces this deficiency has not yet been discovered. The interesting part of this figure is that the variability, and the low-frequency component of the variability (not shown) are close to their observed values. This is in sharp contrast to characteristics of earlier GCM's. For example, Blackmon and Lau (1980) examined transient eddy statistics for two earlier GCM's and found them to be most deficient in simulating low-frequency fluctuations.

The regional pattern of the standard deviation shows some subtle disagreements with observations which, after more investigation, have become important. Notice in Fig. 9b that as you go from the maximum over the Pacific across North America and on over to the Atlantic the values of standard deviation decrease to a minimum near 100° W and then increase again to a maximum over Hudson Bay and then the Atlantic. The comparable minimum produced by the model occurs further to the east and is not as pronounced as in the observations. Experiments with varied orography have shown that as the orography is smoothed, the distribution of standard deviation becomes less realistic. In particular, the location of this minimum and the value of the minimum depend on strong orographic forcing. Other features such as this can be mentioned in addition. This problem is accentuated for models with higher horizontal resolution. Other statistics which have strong low fequency components show similar problems. This suggests that something about

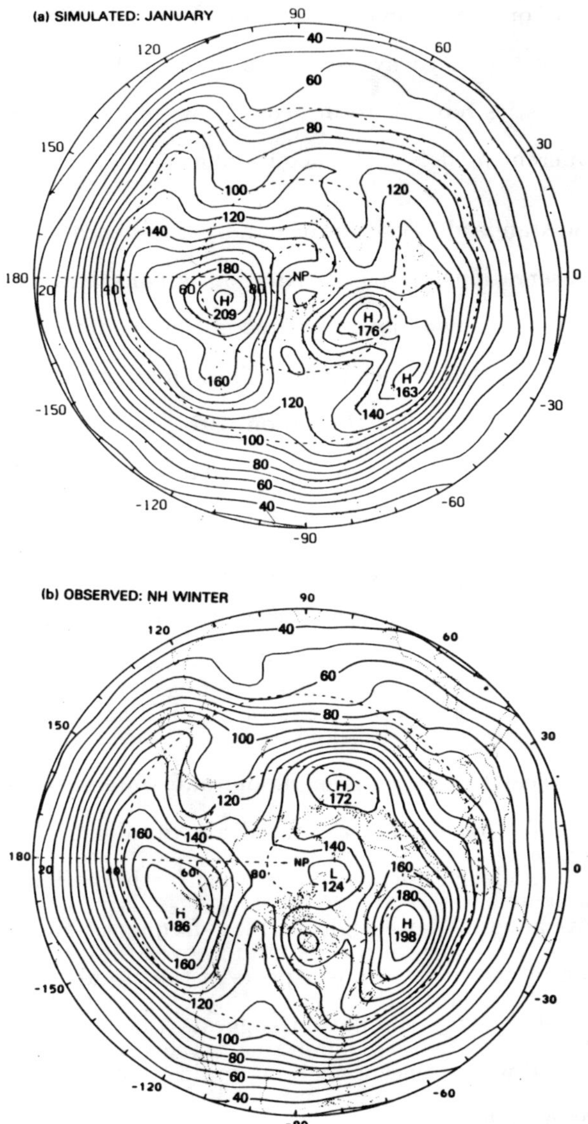

Fig. 9. Standard deviation of the unfiltered 500 mb height for (a) model and (b) observations. Contour interval 10 m.

the present treatment of orography in this and similar models is incorrect. Whether "envelope orography" or "gravity wave drag," turn out to be necessary to improve the simulated eddy statistics remains to be seen. However, both parameterizations have been tried in different models with encouraging results.

Fig. 10 shows the standard deviation of band-pass filtered 500 mb heights for (a) model and (b) observations. This distribution is considerably different from that for unfiltered data with two elongated maxima extending across the oceans and a weaker maximum over the Soviet Union. These maxima are statistical representations of the "storm tracks." (See Blackmon, 1976, and Blackmon et al., 1977, for more detailed discussions.) We see in Fig. 10 that the model and observations are in quite good agreement. The model variability is somewhat larger than the observed over the Pacific, but this might be due to inadequacies of the data in regions of sparse observations. The fact that the model simulates the high frequency eddies well is not surprising, since earlier GCM's seemed to have some success in simulating these features of the general circulation (see Blackmon and Lau, 1980).

Many other statistical comparisons of model and observations are shown in Malone et al. (1984). The results are as stated above: For both stationary and transient features, the model performs best in the Northern Hemisphere winter. The simulation is quite good for both high- and low-frequency fluctations, and this encourages one to look in detail at various aspects of the circulation. In particular, the quality of the low-frequency simulation suggests that various low-frequency phenomena such as blocking and teleconnections could be studied using the model. This we will do in the next lectures.

Although I have concentrated on results from CCM-Ø, the reader should not infer that the quality of its simulation is unique. Manabe and Hahn (1981) and Lau (1981) have discussed various aspects of the simulation of a GFDL spectral GCM which has the same resolution as CCM-Ø and has quite similar physics. It is not surprising that many aspects of the simulations are similar. Both models have been run without any variation in boundary conditions, such as sea surface temperature,

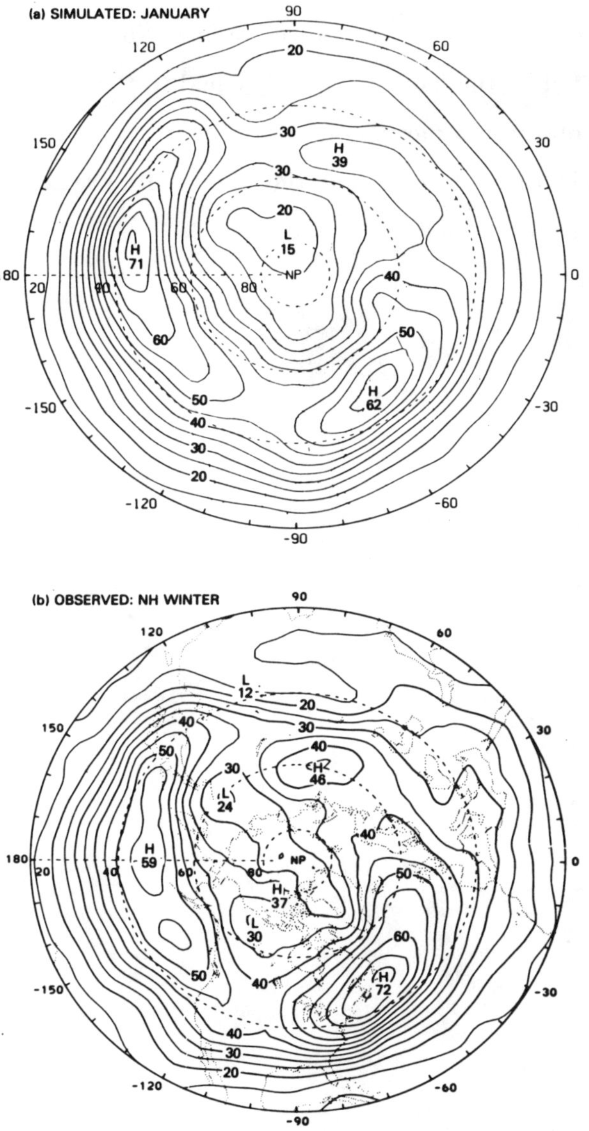

Fig. 10. Standard deviation of band-pass filtered 500 mb height for (a) model and (b) obervations. Contour interval 5 m.

that would represent interannual variation of forcing. Nonetheless, the extratropical variability simulated by the models is nearly equal to that observed. This implies that most of the observed atmospheric variability in the extratropics is internally generated. This in turn suggests that the potentially predictable component of climate variation is likely small.

III. BLOCKING IN A GCM

Blocking anticyclones (or blocking highs or blocks) have been of interest to meteorologists for decades. These events are quasistationary, long-lived highs and are related to marked changes in the weather. The highs are said to "block" the normal easterly progression of synoptic-scale storms. Since blocks are long-lived with respect to typical synoptic time scales (a few days) and since there is also a possibility for blocks to form and decay repeatedly over an entire season, blocking seems to form a substantial component of interannual variability. Some other features of interest of this phenomenon are its regionality and seasonality. In his classic study, for example, Rex (1950) found blocking to occur most frequently over the eastern Atlantic and Pacific oceans. He also found a maximum in blocking frequency in spring.

Defining blocking events is not unambiguous. There is a multitude of studies, each using slightly different definitions for blocking highs. Not surprisingly the characteristics of blocks found in the different studies are sometimes in some conflict with each other. Let me give an example: One method used to define a block is to look for positive 500 mb geopotential height anomalies at some location. If an anomaly has a value greater than some prescribed value, say 150 m, and lasts longer than some prescribed time, say 10 days, then the event is said to be a block. By searching through a big data set, a catalog of blocks can be compiled. However, in this data set, one will inevitably find events which barely fail to meet one of the defining criteria and thus be rejected as blocks. Nonetheless, this event's synoptic effects (blocking the normal easterly progression of synoptic-scale storms) are quite similar to those of events which pass the criteria. There is no crisp distinction in these events. Therefore, going back to the definition mentioned above, if we had specified criteria of 200 m and 7 days to define blocks, we would find a somewhat different catalog of events from that found previously. Some new events would be included; some old events would be excluded. However, the long-time average of both sets of events would have similar structure. So far, no compelling reasons have been given for preferring one set of criteria to another. One eventual goal for studies of blocking should be to define blocking more precisely.

The study of blocking has been hindered until recently by the fact that most blocks form over the oceans. Data quality has been questionable and dynamical studies, such as budget calculations, have been difficult and disappointing. General circulation models do not suffer from poor data quality, of course, but until the 1980s, no general circulation model had been shown to produce realistic looking blocking events—certainly not in large enough numbers that meaningful statistical composites could be formed. Why this is so is not at all clear, but early GCMs seemed to be most deficient in simulating low-frequency atmospheric variability. Blackmon and Lau (1980) examined the behavior of a gridpoint GFDL general circulation model and found that though the model produced moderately good simulation of synoptic-scale eddies (periods 2.5–6 days), the low-frequency eddies (periods greater than 10 days) produced by the model were too weak and not in the right geographical areas. Clearly this model did not produce good blocking events. These simulation characteristics were not untypical for GCMs of the 1970s.

More recently, both GFDL and NCAR have developed coarse-resolution spectral GCMs which have produced greatly improved low-frequency eddy statistics (Lau, 1981; Malone et al., 1984). I will discuss results from the NCAR model only, although I am confident that other models could produce comparable statistics. For example, Lau (1983) has studied the low-frequency fluctuations occurring in a 15-year GCM experiment, and he documents one very long-lived case of blocking in the Pacific.

Fig. 1 shows the 500 mb height field, averaged over 30 days from a perpetual January simulation with climatological, fixed sea surface temperatures. Around

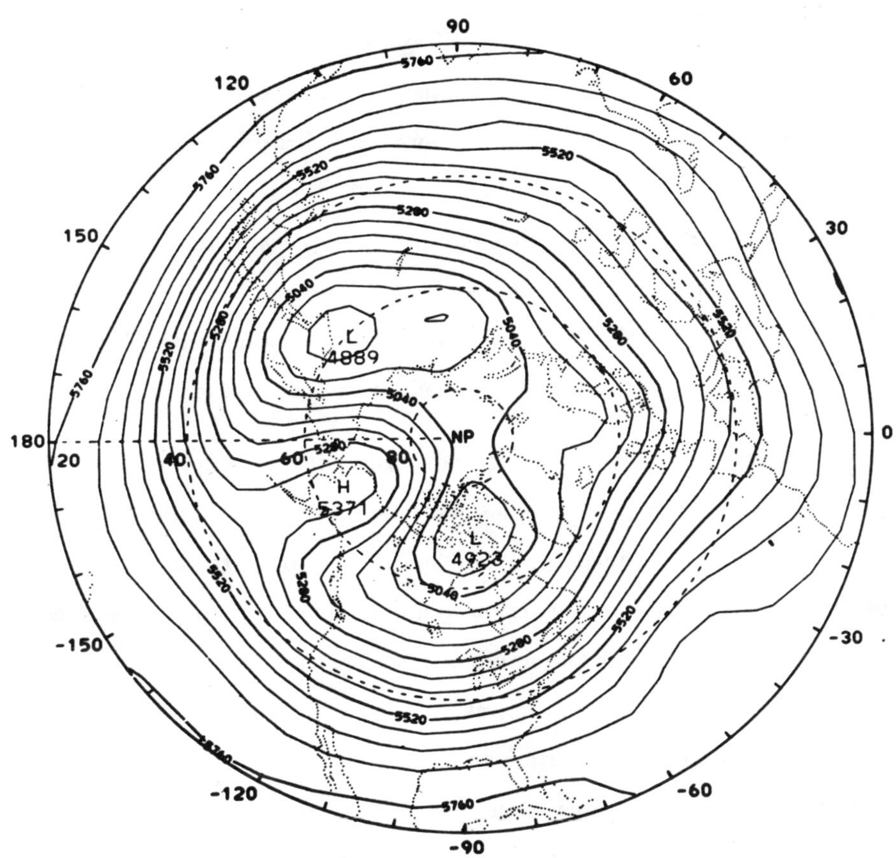

Fig. 1. 500 mb height field averaged over a 30-day period during which a blocking high was developed.

the point (60°N, 165°W) we see a strong ridge. The height anomaly is greater than 150 m and the lifetime for this block was 33 days. The block satisfies all the characteristics attributed to blocks by Rex (1950). The westerlies are split with substantial mass flow both to the north and the south of the block. The longitudinal extent of the block covers 45°. There is an abrupt transition from zonal to meridional flow. There is temporal continuity over the lifetime of the block—it does not go away and reappear. Based on this example and many other model events which looked very much like observations, Blackmon et al. (1985) decided to compare the climatology of blocking produced by the model with observations. I will show here only a small subset of the results from that paper.

Recent observational studies of the characteristics of blocking are those of Dole (1982) and Dole and Gordon (1983). These authors considered temporal anomalies in the 500 mb height and defined persistent anomalies to be those which exceed, in absolute value, some threshold criterion and endure longer than some specified time. Dole considered persistent anomalies of both positive and negative signs and found that the geographical distribution of these events were quite similar. Fig. 2 shows separately the number of positive and negative persistent anomalies at each gridpoint for the model and observations per 1000 winter days. The magnitude and duration criteria are 100 m, 10 days, respectively. It is clear that the model is quite close to observations in both location and frequency of occurrence of these anomalies over the Pacific and Atlantic oceans. The region of observed blocking over the Soviet Union is not simulated. Why the model misses this last region so badly is not understood at present. Dole and Gordon (1983) also compared the lifetime of anomalies in regions of frequent blocking versus regions of infrequent blocking. For lifetimes of less than about 5 days, the number of anomalies of both signs decreases roughly exponentially as a function of lifetime. Beyond 5 days, the number continues to decrease roughly at the same exponential rate in the nonblocking region, while in blocking regions the number of events decreases exponentially, but with a much smaller rate. Fig. 3 shows model data for two such

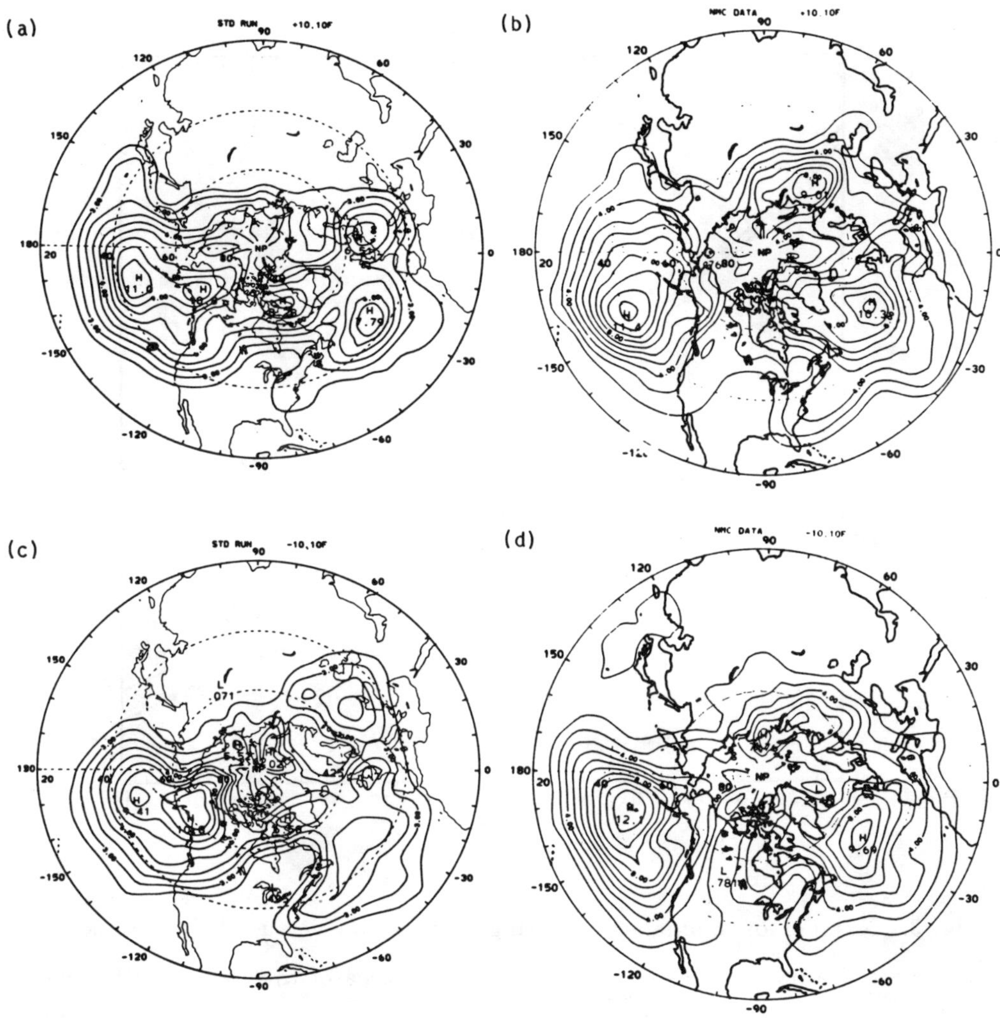

Fig. 2. Comparison of regional distributions of persistent anomalies defined by criteria (100 m, 10 days): (a) model, positive anomlies, (b) observations, positive anomalies, (c) model, negative anomalies, (d) observations, negative anomalies.

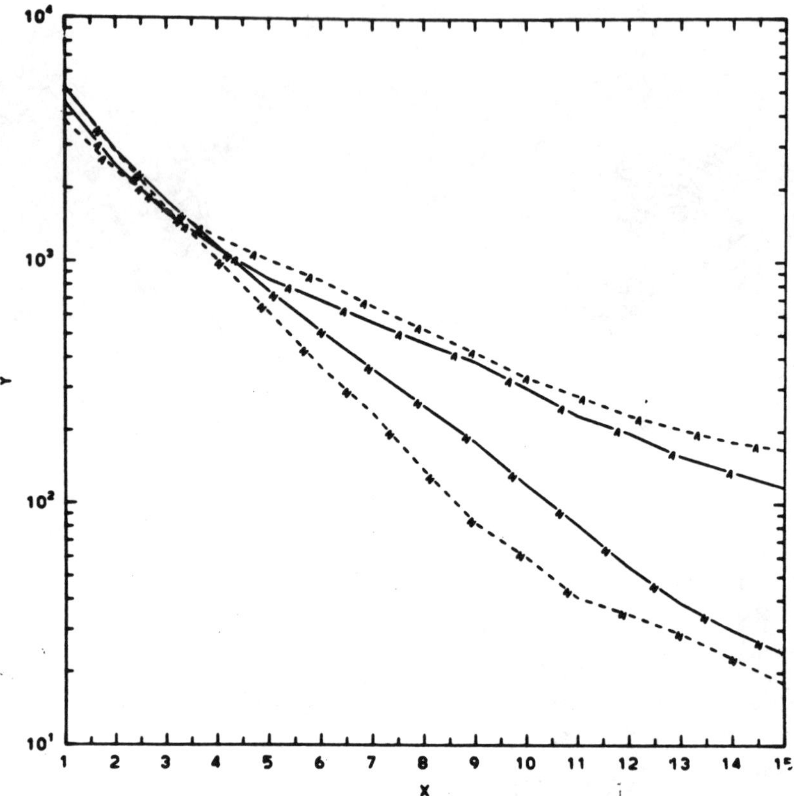

Fig. 3. Number of anomalies occurring in a region versus their lifetime. The two curves marked with A's are for an area of the Pacific with frequent occurrence of persistent anomalies while curves marked with N's are from an area over North America. Solid lines denote positive anomalies and dashed curves denote negative anomalies.

regions. The curves marked A show the number of positive anomalies (dashed) and negative anomalies (solid) in the area of frequent blocking in the Pacific while the curves marked N are for anomalies over North America. Note the change in slope of the A curves at about four days. The mean lifetime of model persistent anomalies is quite close to those observed.

One of the problems with the criteria used in Dole and Gordon (1983) is that many of the persistent anomalies would not be judged by synoptic meteorologists to be blocks. In Fig. 4 we show model events which are defined by the criteria (100 m, 10 days) at the point (38°N, 160°W). In Fig. 4a the event looks like a classic block, while in Fig. 4b we show another event which is more like a broad ridge. Most synopticians would not call such an event a block. On the other hand, it is not apparent at this stage of our understanding whether the dynamics of the two events are similar or different. These different types of events also occur in observations. By experimenting with various threshhold criteria, we have noticed that setting higher amplitude criteria, say 150 m or 200 m, identifies more events such as those shown in Fig. 4a and Fig. 1 and eliminates those shown in Fig. 4b. We have also noticed that as the amplitude criterion is increased, there is less similarity between the distribution of positive and negative events such as shown in Fig. 2.

Fig. 5 shows various aspects of blocks occurring at (150°W, 55°N) and satisfying the criteria (200 m, 7 days). In Fig. 5a we show the 500 mb height (solid) and 500 mb zonal wind. The pronounced ridge and the corresponding decrease in the zonal wind in the block are apparent. Fig. 5b shows the mean sea level pressure (solid) and 1000–500 mb thickness (dashed) for the same events. Inspection shows that the surface high and low are somewhat east of 500 mb ridge and trough positions, respectively, i.e., the ridge and trough both tilt westward with height. The tilt is about one or two gridpoints (7.5°–15° longitude) through the depth of the troposphere. The amount of tilt might be resolution dependent. The ridge in the thickness field is slightly west of the 500 mb height ridge; a positive temperature anomaly is located at the upstream part of the block. Fig. 5c shows the standard

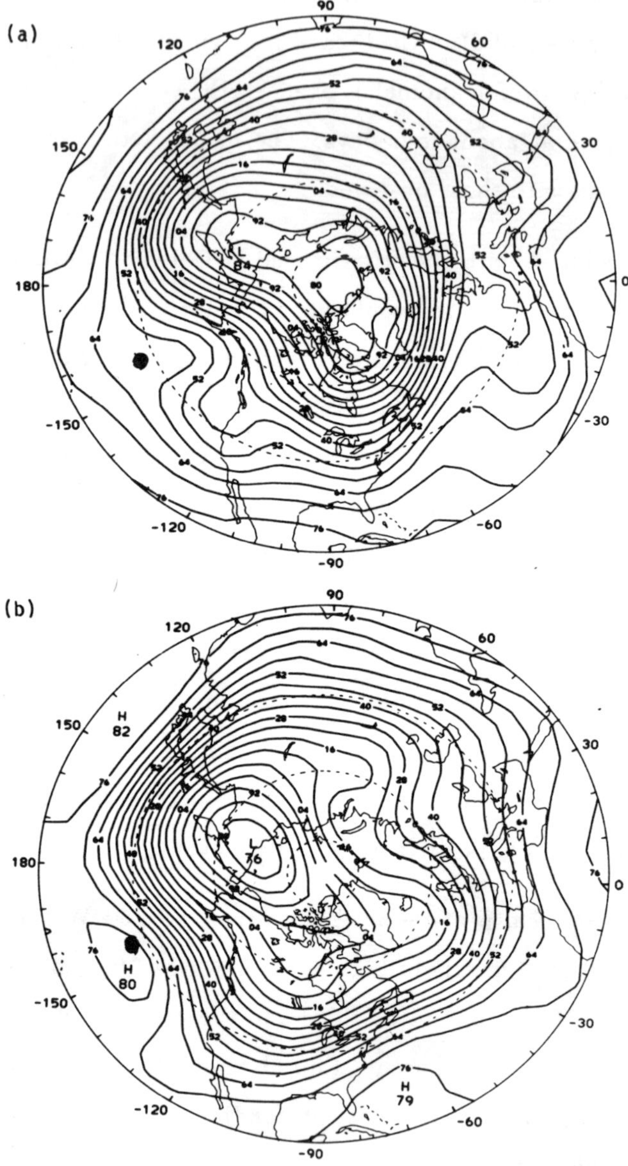

Fig. 4. Examples of two positive persistent ancmalies. Both satisfy the threshold criteria (100 m, 10 days). Panel (a) shows a well-defined block while the event in panel (b) is a broad ridge and not likely to be identified as a block by synopticians. The dots show the gridpoint used to define the anomalies.

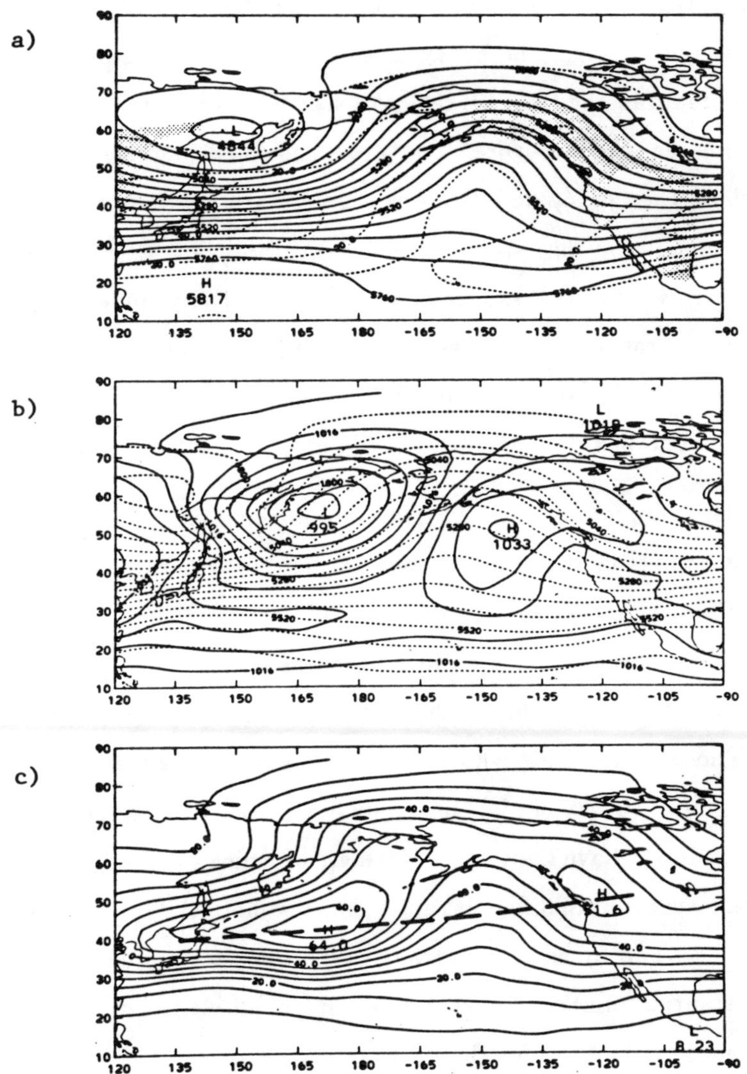

Fig. 5. Composite of blocks occurring around (55°N, 150°W): (a) mean 500 mb height (solid) and mean 300 mb zonal wind (dashed), (b) mean sea level pressure (solid) and mean 1000–500 mb thickness (dashed), (c) standard deviation of bandpass filtered 500 mb height during periods of blocking events shown in (a). The solid line denotes the climatological position of the baroclinic storm track.

deviation of the 500 mb band-pass filtered height,calculated for those days when blocking was occurring. The heavy dashed line shows the climatological position of the maximum in the storm track. We see that during the episodes of blocking shown in Fig. 5a, the storms moving across the Pacific are steered northward into Alaska where they dissipate. There is a secondary maximum downstream of the block also, indicating that storms develop or form again after passing around the block. There is a marked suppression in storm activity under the block. Based on these considerations and others discussed in Blackmon et al. (1985), we believe that the model blocks are very similar to those observed, and this has encouraged us to begin a detailed dynamical study of model blocks

In order to understand the dynamics of blocking highs, Mullen (1985a, 1985b) has used an objective definition of blocking and searched through 12,000 days of perpetual January model runs to find blocking events. These runs were performed using a variety of sea surface temperature anomalies. Some of the experiments produced statistically significant changes in the mean model circulation and some did not. Those experiments which showed changes in the mean also showed some changes in the frequency of blocking in some regions. However, individual blocking events seemed to behave the same, regardless of which experiment they came from. Compositing events from different runs seems justifiable. By searching through such a large data set and by using a restrictive clustering technique, Mullen was able to find a substantial number of events which were quite similar. He then calculated vorticity and heat budgets for these composited events.

The vorticity and temperature tendency equations were given in the first lecture. If one averages over a large enough time, and if one breaks up the statistics into mean quantities, denoted by an overbar, and transient quantities, denoted by primes, then the vorticity equation becomes

$$0 \simeq \overline{\frac{\partial \varsigma}{\partial t}} = -\overline{\vec{V}} \cdot \vec{\nabla}(\overline{\varsigma} + f) - (\overline{\varsigma} + f)\overline{\delta}$$
$$- \vec{\nabla} \cdot \overline{\vec{V}'\varsigma'} - \cdots \quad .$$

and the temperature tendency equation becomes

$$0 \simeq \overline{\frac{\partial T}{\partial t}} = -\overline{\vec{V}} \cdot \vec{\nabla} \overline{T} - \overline{\omega}\left(\frac{\partial \overline{T}}{\partial p} - \kappa \frac{\overline{T}}{p}\right)$$
$$- \vec{\nabla} \cdot \overline{\vec{V}'T'} + \overline{Q} + \cdots$$

Mullen (1985a, 1985b) has examined composites of blocks in three geographical regions: over the Pacific (150°W, 55°N), over the west coast of North America (120°W, 55°N) and over the Atlantic (15°W, 55°N). He has calculated vorticity and heat budgets at several levels as well as vertically integrated budgets for both quantities. There are some slight differences in the budgets for the different regions, mostly due to effects such as flow over high terrain. Since the blocks studied are either, say, downstream of Greenland or upstream of the Rockies, one would expect different patterns for orographic forcing in the divergence field. In spite of this and other differences, the main features of the budgets appear to be the same from region to region. I will only show results for Pacific blocks. Also, I will only show terms for the vorticity budget at 300 mb and the heat budget at 700 mb. For a discussion of budgets at other levels, see Mullen (1985a, 1985b).

Fig. 6 shows the individual terms in the vorticity budget at 300 mb for Pacific blocks (150°W, 55°N). Fig. 6a shows the divergence term and Fig. 6b the mean vorticity advection term. The shaded area covers the region with relative vorticity less than $-2 \times 10^{-5}\ s^{-1}$. Fig. 6c shows the sum of divergence and mean advection terms. It is immediately seen that the divergence and mean advection terms are largely canceling each other, with the advection being slightly larger in absolute value. The sum of the two terms has extrema smaller in absolute value than either individual term and, more interesting, nearly in quadrature with the mean vorticity pattern. The sum of divergence and mean advection would move the mean vorticity pattern downstream by about 1/4 a wavelength in about three days if unopposed by other terms. Fig. 6d shows the eddy vorticity convergence which is nearly opposite the sum of divergence and mean advection terms. The balance among these three

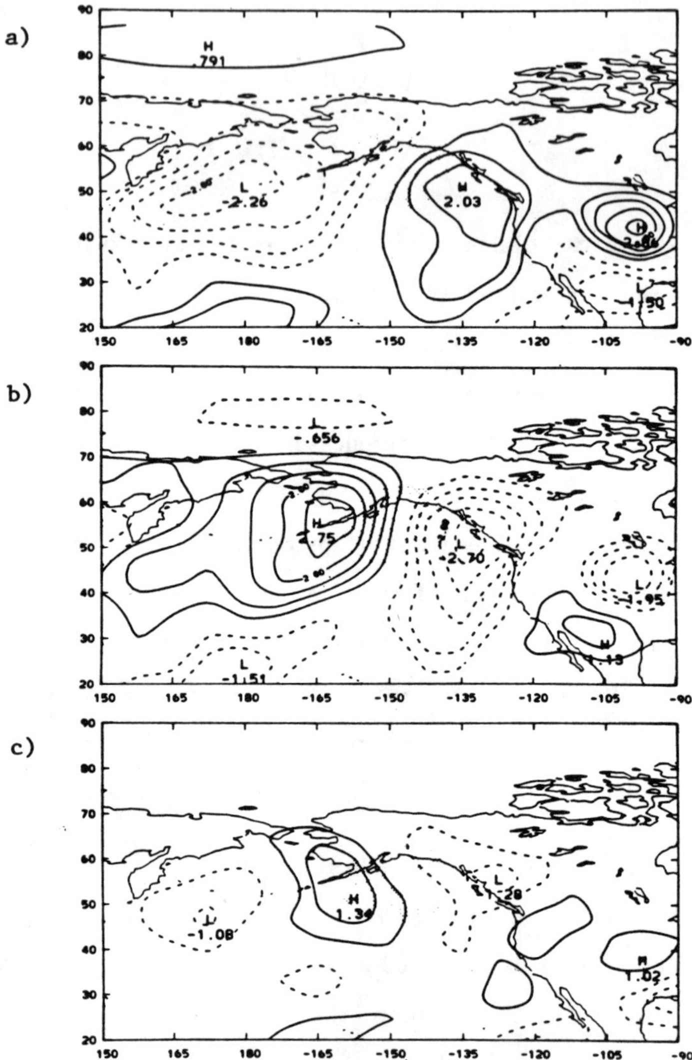

Fig. 6. Terms in the 300 mb vorticity budget for the composited events shown in Fig. 5: (a) divergence term, contour interval 0.5×10^{-10} s^{-2}, (b) mean vorticity advection term, contour interval 0.5×10^{-10} s^{-2}, (c) sum of divergence and mean vorticity advection terms, contour interval 0.5×10^{-10} s^{-1}, (d) eddy vorticity flux convergence term, contour interval 0.5×10^{-10} s^{-1}, (e) vorticity tendency term, contour interval 0.25×10^{-10} s^{-1}, (f) vorticity residual term, contour interval 0.5×10^{-10} s^{-1}.

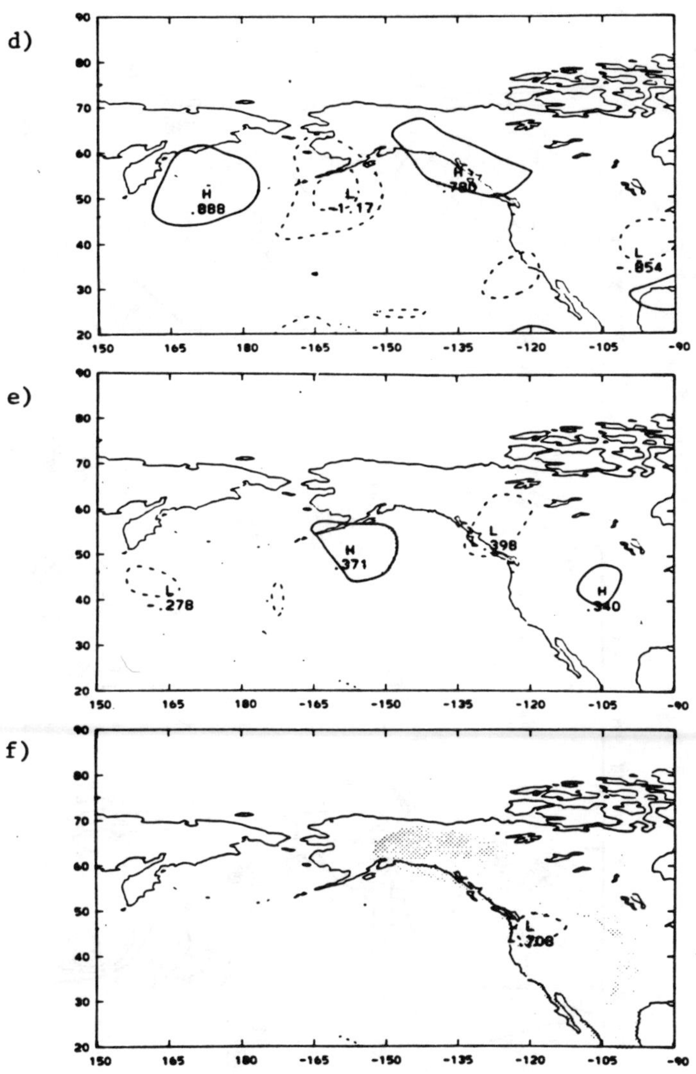
d)
e)
f)
90
80
70
60
50
40
30
20
150
165
180
-165
-150
-135
-120
-105
-90
H
.888
L
-1.17
H
.780
L
-.854
H
.371
L
.398
L
-.278
H
.340
L
.708

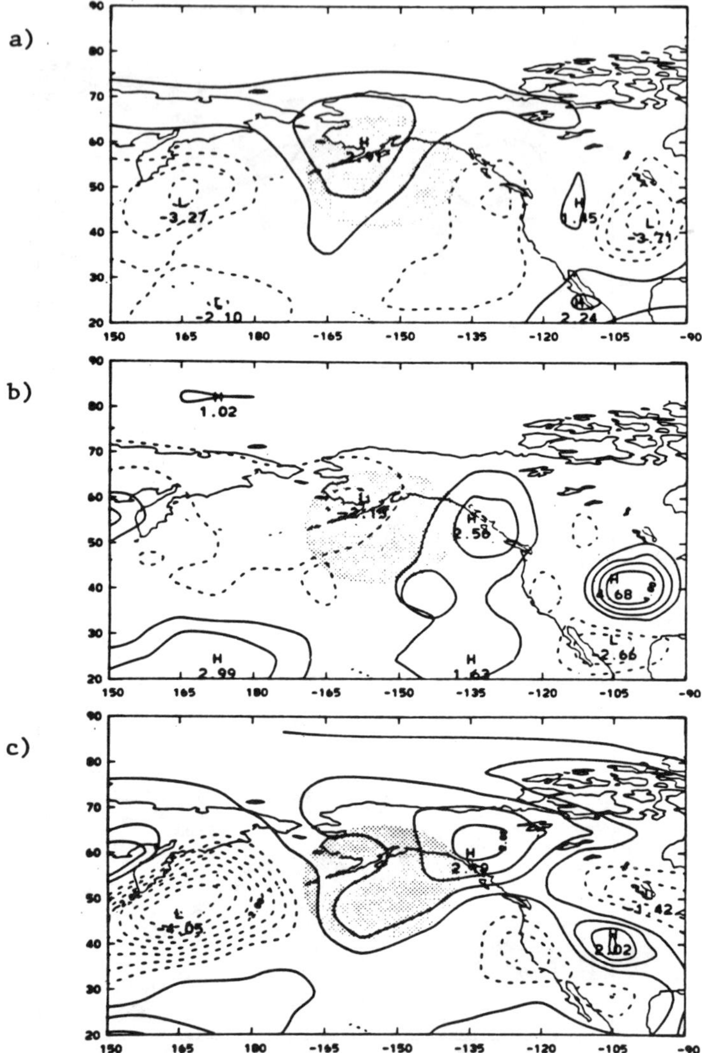

Fig. 7. Terms in the 700 mb sensible heat budget for the events shown in Fig. 5: (a) mean temperature advection term, contour interval 1K day^{-1}, (b) adiabatic heating term, contour interval 1K day^{-1}, (c) the sum of the mean temperature advection and adiabatic heating terms, contour interval 0.5K day^{-1}, (d) eddy heat flux convergence term, contour interval 0.5K day^{-1}, (e) diabatic heating term, contour interval 0.25K day^{-1}, (f) temperature tendency term, contour interval 0.25K day^{-1}, (g) residual term, contour interval 0.5K day^{-1}.

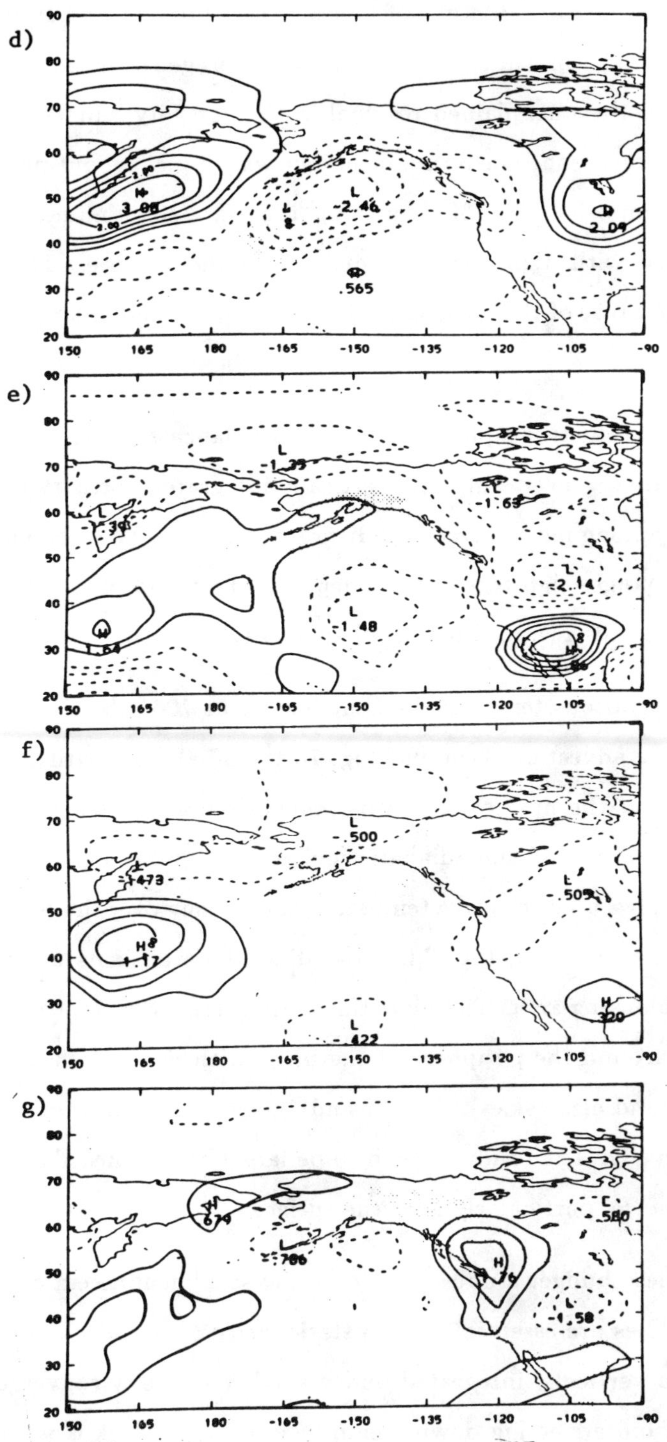
d)
e)
f)
g)
90
80
70
60
50
40
30
20
150
165
180
-165
-150
-135
-120
-105
-90
H
3.08
L
-2.46
2.09
.565
L
-1.35
-1.65
-2.14
-1.48
1.64
-.500
-.473
-.505
1.17
H
320
-.422
.679
-.706
-.580
-1.58

terms is such that the mean lifetime of blocks is extended beyond 10 days. Fig. 6e shows the difference in final and initial vorticities, averaged over the blocking events and Fig. 6f shows the unexplained residual in the vorticity budget. Fig. 6e gives an idea of how effective the clustering technique was in picking out nearly identical blocks for the composite. Fig. 6f shows that there are some numerical problems, usually associated with high terrain, remaining in the balance. The most likely candidate for the cause of this problem is interpolation through fairly thick vertical levels.

At low levels in the troposphere, the vorticity balance is mostly between surface friction and the divergence term. The vertically integrated vorticity balance again shows a balance among mean advection, divergence and eddy convergence. Because the divergence changes sign through the depth of the troposphere, its importance is less in the vertically averaged budget than at, say, 300 mb.

Fig. 7 shows various terms in the heat budget at 700 mb. Fig. 7a shows the mean temperature advection term and Fig. 7b the adiabatic heating term. Warm (cold) advection on the upstream (downstream) side of the ridge is balanced by upward (downward) motion and adiabatic cooling (warming). The sum of these two terms, Fig. 7c, shows a net positive temperature tendency over the blocking region. The eddy convergence term, Fig. 7d, shows that the transients are transporting heat out of the blocking area. The diabatic heating term, Fig. 7e, shows maximum absolute values around the periphery of the block with negative values in areas of sinking motion (and clear skies). Fig. 7f and 7g show the tendency term and the residual, respectively. The residuals tend to be larger for the heat balance than for the vorticity balance, particularly near the surface.

Based on these budget studies, the following statements seem clear. First, in the CCM the eddies are essential for the stationarity of the block. In both upper tropospheric and vertically integrated budgets, eddy vorticity convergence acts to keep the block from advecting downstream. Second, the block is warm primarily because of warm advection. The transient eddy heat convergence tends to dissipate the warmth of the anticyclone.

The net effect of the eddies is not yet clear, as the eddy flux convergence will generate compensating secondary circulations. Further work is underway. Mullen (1985b) has solved the ω-equation in two approximations and compared with the model generated ω-fields. The ω-equation has two "forcing terms," the vertical derivative of the absolute vorticity advection and the Laplacian of the thermal advection. Mullen finds that the vorticity forcing dominates the thermal forcing and agrees well with the model-generated ω field.

Another area not yet understood is the development of blocks. In the CCM it is common for blocks to form downstream of rapidly deepening cyclones. As the cyclone deepens, strong warm advection occurs downstream. A ridge begins to develop and the cyclone starts tracking to the north or northeast. Within a few days, the block is established and easily identified. Why this happens for some cases of strong cyclonic development and not others is unknown at present. Perhaps the potential vorticity view of blocks discussed in Hoskins lectures will provide additional understanding.

IV. MODELING THE SOUTHERN OSCILLATION

The Southern Oscillation was originally discovered by Sir Gilbert Walker (see Walker and Bliss, 1932) and was described as a very low frequency oscillation or fluctuation in mass between large areas in the eastern equatorial Pacific and the western equatorial Pacific and Indian Oceans. Typical stations which are correlated are Darwin, Australia and Tahiti. The monthly averaged sea level pressure difference between Tahiti and Darwin can be used to form a Southern Oscillation Index (SOI) which can be correlated with other atmospheric variables. The relationship between the Southern Oscillation and El Niño, the warming of eastern equatorial Pacific water which occurs every 2–8 years, was first suggested by Bjerknes (1969). Recently, there have been many observational studies which have examined the nature and extent of the atmospheric fluctuations during periods of El Niño. Rasmusson and Carpenter (1982) have documented the evolution of a composite El Niño from the time it appears as a warming along the coast of South America, usually about Christmas, throughout the following year, as the warm water becomes more pronounced along the equator in the eastern Pacific. About one year after El Niño has begun, this warm event has reached what Rasmusson and Carpenter call the mature phase. It is at this stage that the atmospheric effects of El Niño have been most studied—during Northern winter, about one year after the first sign of the event. After the mature stage, the warm water usually dissipates quickly.

Horel and Wallace (1981) have studied the atmospheric response to El Niño and summarized their results by the schematic diagram shown here as Fig. 1. In normal winters, precipitation is usually heaviest along the equator near Indonesia and Australia. During the mature warm events, however, there is anomalous precipitation along the equator, east of the date line, which is denoted by the shaded

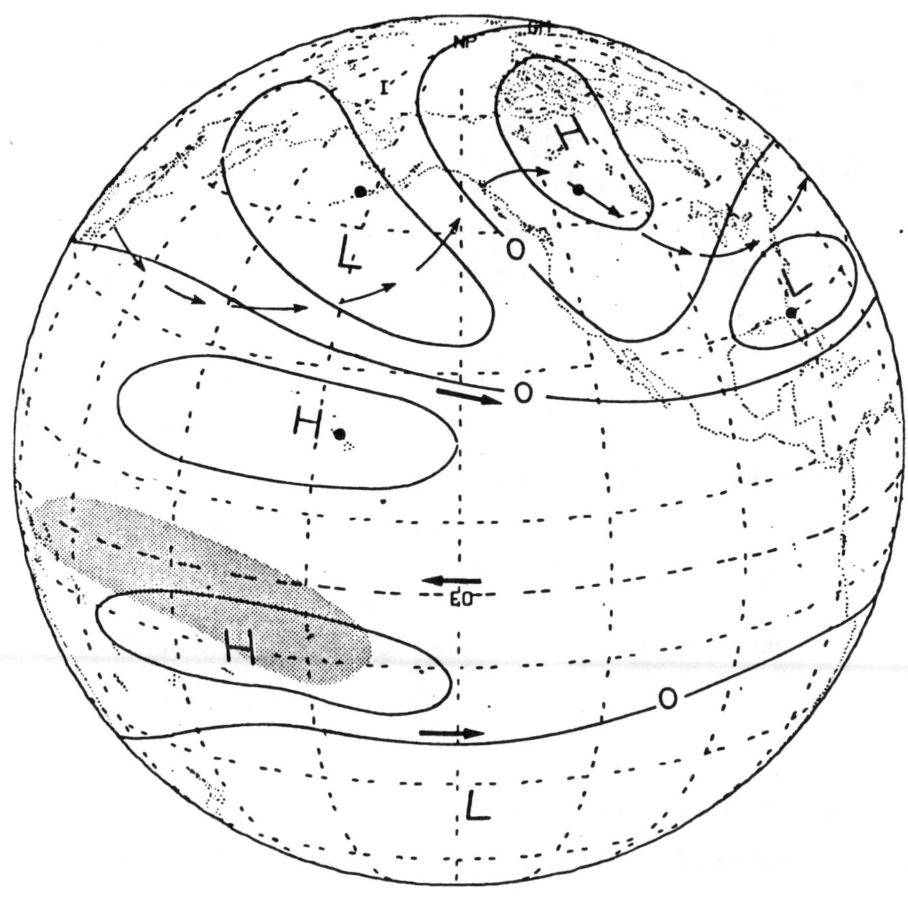

Fig. 1. Schematic illustration of the response to a mature El Niño. The shaded areas shows region of anomalous precipitation. Upper tropospheric height anomalies are labelled by H's and L's. The heavy arrows show strengthened upper tropospheric jets. The light arrows show the midlatitude streamline pattern.

area of Fig. 1. The other features in Fig. 1 show anomalies in the upper troposphere, say at 200 mb. Straddling the equator on either side of the precipitation anomaly, there is the pair of tropical anticyclones, and arcing across North Ameria, we see a wavetrain of alternating height anomalies. This wavetrain is related to the Pacific/North American (PNA) teleconnection pattern discussed by Wallace and Gutzler (1981). The anomalous high on the west coast of North America is associated with increased frequency of blocking. Storms moving across the Pacific are typically steered to the north into Alaska. There also occurs frequent development of baroclinic eddies in the lee of the Rockies which move down into the eastern U.S. The western U.S. is, therefore, high (in pressure), dry and warm. The eastern U.S. is, by contrast, bitter cold. The classic recent example of this pattern is the winter of 1976–77. In Fig. 1, arrows also denote the existence of stronger than normal eastern Pacific equatorial easterlies and stronger than normal subtropical jets in both the Northern and Southern Hemispheres.

The questions to be addressed here are: "How many of these effects can be reproduced by a GCM?" and "Can we identify mechanisms by which these fluctuations take place?"

To answer these questions we have run several experiments with sea surface temperatures (SST) anomalies put in the model. As explained in the first lecture, the sea surface temperature is a given, fixed quantity in this model. Questions of feedback of the atmosphere onto the ocean have to be ignored. We hope, however, that this type of experiment with fixed SST's will be a helpful forerunner to the ultimate, coupled-and-interactive ocean-atmosphere model of El Niño/Southern Oscillation. Given the distribution of SST, the model calculates winds, air temperatures and mixing ratios and, therefore, also calculates latent and sensible heat fluxes from the ocean using the formulas discussed in the first lecture. Our procedure, therefore, is to run the model with normal SST's to get an estimate of normal average atmospheric conditions. We also run the model with anomalous SST's to see what, if any, circulation changes result in response to El Niño. By taking the difference of various fields, one produces anomaly maps which we compare with the results of Horel and Wallace (1981) and shown in Fig. 1.

The experiments I will summarize are discussed in detail in Blackmon et al. (1983) and Geisler et al. (1985). I will only discuss the anomalous response here, but in the papers mentioned above, we also discuss the statistical significance of the response. The question of statistical significance is important to help distinguish between real effects and statistical accidents. Suffice it to say that the difference maps I will show are between 720–day averages (eight 90–day averages) for perturbed and normal conditions. We believe the responses discussed are statistically significant.

In Fig. 2 is shown the SST anomaly (Fig. 2a) used in the twice-as-warm (2W) experiment. This anamoly is the Rasmusson and Carpenter mature phase anomaly multiplied by 2. The justification for the factor of 2 is that we are trying to simulate a strong El Niño. Rassmusson and Carpenter's composite is as average over six El Niños; some strong, some weak, and also occurring in different positions. We also want to get a strong, statistically significant response. The sensitivity of the response to the magnitude of the SST anomaly is discussed in Blackmon et al. (1983) and Geisler et al. (1985). In Fig. 2b we show the total SST distribution obtained by adding to the anomaly (Fig. 2a) to the climatological SST. The resulting precipitation field is shown in Fig. 2c and the precipitation anomaly (experiment minus control) is shown in Fig. 2d. Near the position of the maximum SST anomaly is a precipitation anomaly of 9 mm/day. The vertical distribution of anomalous heating (not shown) is maximum around 500 mb and has a value of about 5K/day. This anomalous heating produces an anomalous divergence near the tropopause, which through the vorticity equation

$$\partial_t \varsigma = \cdots - \delta(\varsigma + f)$$

sets in motion the other responses.

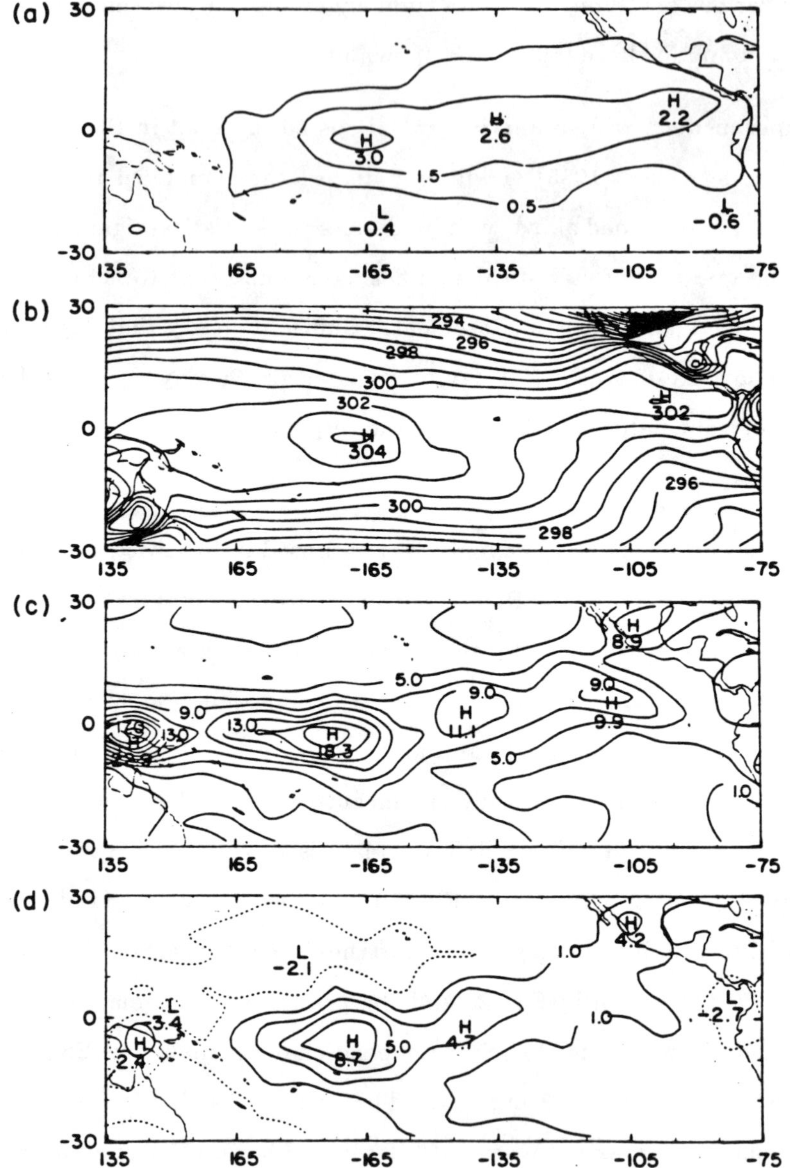

Fig. 2. Equatorial Pacific fields for the 2W experiment: (a) the SST anomaly, contour interval 1K, (b) total SST distribution (anomaly plus climatology), contour interval 1K, (c) total precipitation, for 2W experiment, contour interval 2 mm day^{-1}, (d) precipitation anomaly, contour interval 2 mm day^{-1}.

Fig. 3a shows the difference in sea level pressure for experiment minus control. A well-defined Southern Oscillation signal is seen, with negative pressure anomalies across the eastern tropical Pacific and positive anomalies across the western Pacific and Indian Oceans. If one formed an SOI such as discussed earlier, you would clearly get negative values, as are observed in typical warm events.

Fig. 3b shows the difference map for the 200 mb geopotential height. In the tropics, the most obvious feature are the anticyclones straddling the equator in the vicinity of the precipitation anomaly. Notice, however, that the height anomaly throughout the tropics is positive, in agreement with observational results of Pan and Oort (1983). In the Northern Hemisphere we see a strong wavetrain response which bears a strong resemblance to the PNA teleconnection pattern. The model response over Eurasia is not statistically significant.

Fig. 3c shows the 300 mb zonal wind anomaly. The equatorial easterlies are increased over the Pacific and the subtropical jets in both hemispheres are strengthened over the Pacific. These are just the features found by Horel and Wallace (1981) and shown in Fig. 1. In addition, we see a stronger jet across the eastern United States and a stronger jet from South America across the Atlantic and on south of the Himalayas.

In Blackmon et al. (1983) other difference maps are examined and, where possible, compared with observations. The effects of the Southern Oscillation are shown to extend even to the stratosphere, where warmer polar temperatures and slower polar-night jets are observed in warm event winters. In general, the model response was found to be in good, but not perfect, agreement with observations.

In 1982 another El Niño event developed and by the winter of 1982–83, a very strong SST anomaly was heated in the eastern tropical Pacific. However, the observed midlatitude winter anomaly was not of the type discussed above. This suggested a set of experiments, reported in Geisler et al. (1985), which examined the sensitivity of the response to the position of the SST anomaly. Figs. 4a and

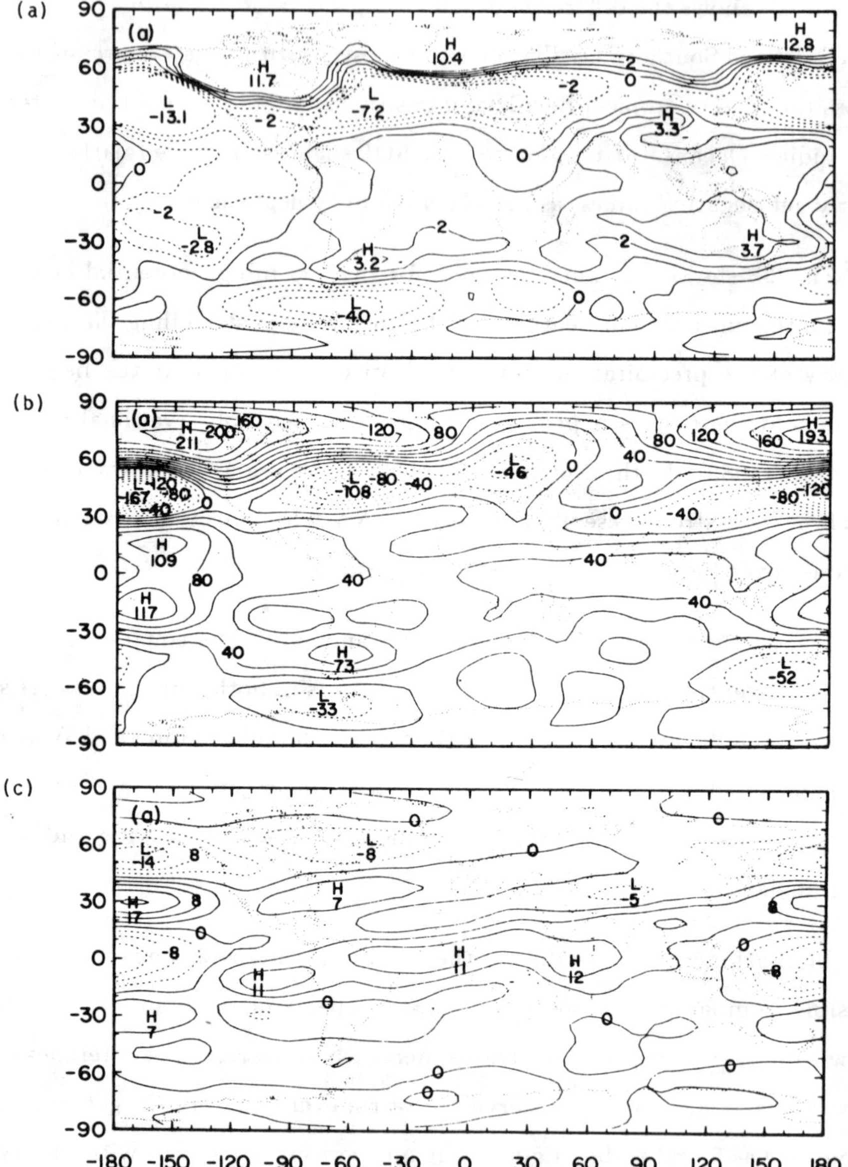

Fig. 3. Long-time average climate anomalies (experiment minus control) produced in the 2W experiment: (a) sea level pressure, contour interval, (b) 300 mb height, contour interval, (c) 300 mb zonal wind, contour interval.

4b show the SST anomalies used in these next experiments, call the East Pacific (EPAC) and Far East Pacific (FarEPAC) experiments, respectively. The EPAC anomaly, Fig. 4a, is patterned after the observed December 1982 SST anomaly, while the FarEPAC anomaly is purely hypothetical. Both these anomalies produce precipitation anomalies of about 9 mm/day. The 200 mb height responses are shown in Figs. 5a and 5b, respectively. The tropical anticyclone pair moves to the east with the SST anomaly, but the midlatitude response is geographically fixed. The midlatitude response is observed to weaken as the SST anomaly is moved east, even though the precipitation anomaly in both experiments is approximately constant.

The fact that the midlatitude response is fixed geographically suggests that the midlatitude wavetrain depends on something in addition to radiation from a vorticity source. The fact that the response weakens as the anomaly moves eastward suggests that barotropic energy conversions might play a significant role in the midlatitude response. The model's jet is a maximum near the coast of Asia. Horizontal gradients of the jet, which are components of barotropic conversion terms, are strongest over the western Pacific. Therefore, one would expect barotropic energy conversions to be weak over the eastern Pacific, as observed in our experiments.

Simmons et al. (1983) studied the role of barotropic energy conversion in generating teleconnection patterns and found that eddies could take energy from, and essentially deplete, the Pacific jet in a few days through barotropic conversion. Branstator (1985) studied the role of barotropic energy conversion in our GCM experiments by using a barotropic vorticity equation model, linearized about various parts of the model's mean state. He inserted an anomalous forcing in the vorticity equation which was designed to approximate the anomalous divergence term mentioned earlier, and he then looked at the steady state perturbation vorticity response. In Fig. 6 we show the resulting height anomaly field where the mean state used is the zonal mean solid body flow (Fig. 6a), the zonal mean flow with latitudinal variation included (Fig. 6b) and the zonal mean flow with latitudinal and longitudinal variation included. As more details are included in the basic state,

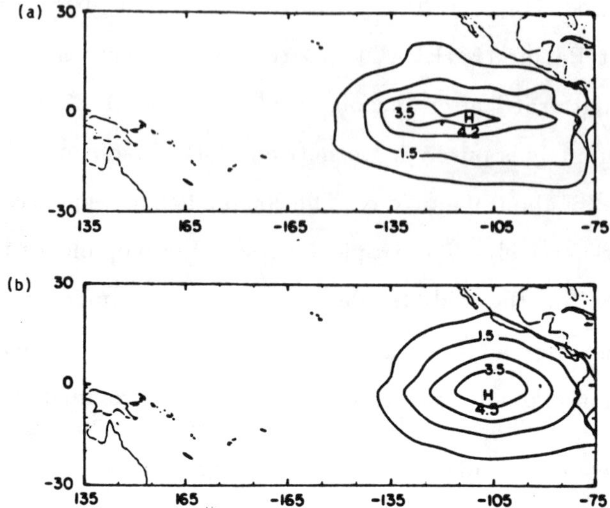

Fig. 4. ST anomalies for (a) the East Pacific (EPAC) experiment and (b) the Far East Pacific (Far EPAC) experiments. Contour interval 1K.

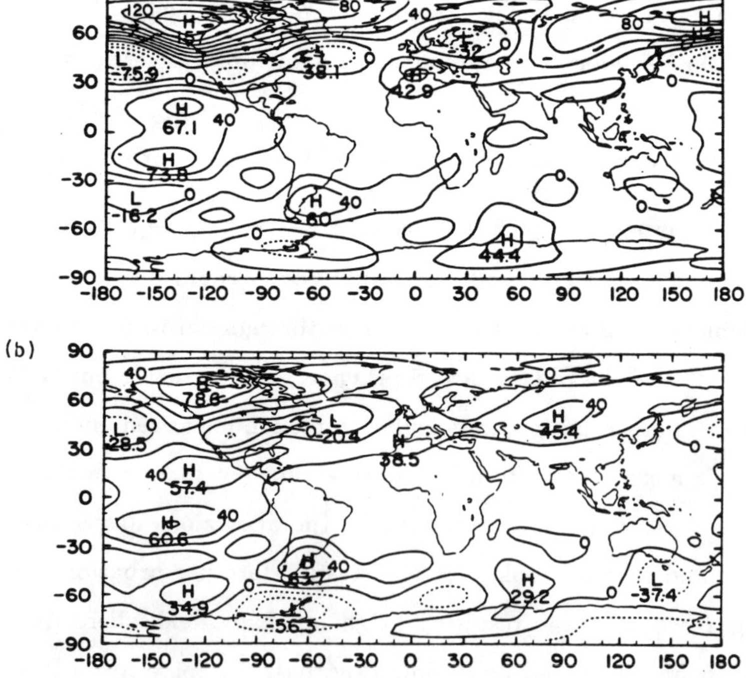

Fig. 5. 200 mb height anomalies for (a) the EPAC and (b) the Far EPAC experiments. Contour interval 20 m.

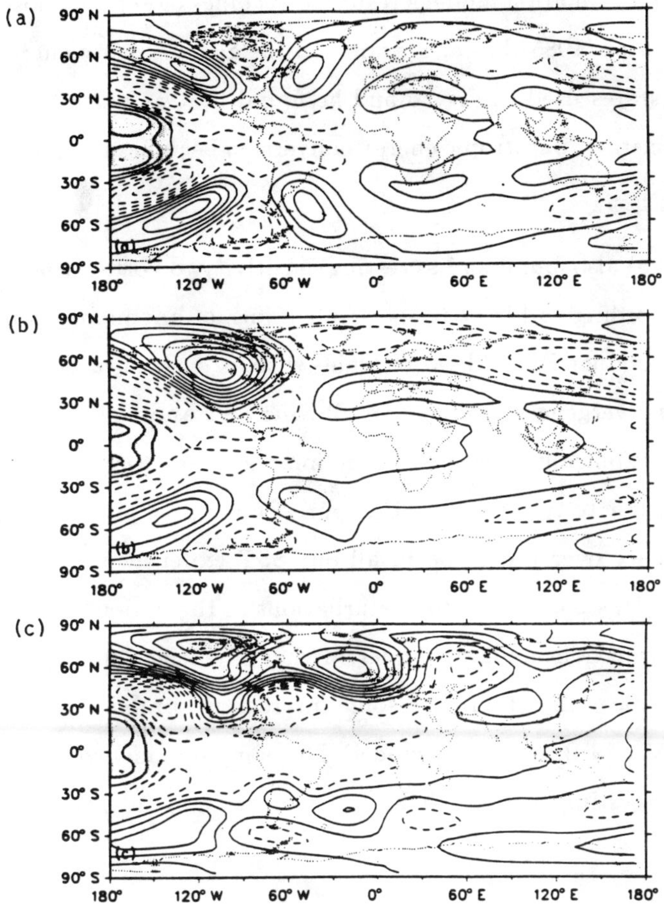

Fig. 6. Perturbation height solutions of linear barotropic vorticity equation forced by a vorticity source in the stippled region. In each panel the basic state is derived from the CCM climatology (a) solid body rotation component of the 300 mb streamfunction, (b) zonal mean component of the 300 mb streamfunction, (c) the complete 300 mb streamfunction. Contour interval 7 m (from Branstator, 1985).

the barotropic vorticity equation model's response becomes closer and closer to that of the GCM. Branstator (1985) has also examined the global kinetic energy budget for the three basic states used in Fig. 6 and he found that as more "waviness" is included in the basic state, barotropic energy conversion becomes a more important energy source.

Branstator (1985) also found an interesting sensitivity to vorticity perturbation near Indonesia. According to his analysis, small values of anomalous convergence near Indonesia can produce PNA-like wavetrains with as large or larger amplitudes than the anomalous divergences produced by El Niño SST anomalies. According to Branstator, the geographically fixed teleconnection response observed in our model response is probably due to two factors: (1) suppression of precipitation, and hence, anomalous convergence over Indonesia in all our SST experiments; (2) the large sensitivity of the model's basic state to perturbations in the region near Indonesia. Whether this interpretation holds up under further investigations remains to be seen. However, it is clear that there are some interesting similarities between the response in the GCM and Branstator's model. This simple model will undoubtedly prove helpful in suggesting further GCM experiments and in interpreting their results.

References

Bjerknes, J., 1969: Atmospheric teleconnections from the equatorial Pacific. *Mon. Wea. Rev.*, **97**, 163–172.

Blackmon, M.L., 1976: A climatological spectral study of the 500 mb geopotential height of the Northern Hemisphere. *J. Atmos. Sci.*,**33**, 1607–1623.

Blackmon, M.L., and N.-C. Lau, 1980: Regional characteristics of the Northern Hemisphere wintertime circulation: A comparison of the simulation of a GFDL general circulation model with observations. *J. Atmos. Sci.*,**37**, 497–514.

Blackmon, M.L., G.T. Bates and S.L. Mullen, 1985: The climatology of blocking events in a perpetual January simulation of a spectral general circulation model. *J. Atmos. Sci.*, submitted for publication.

Blackmon, M.L., J.E. Geisler and E.J. Pitcher, 1983: A general circulation model study of January climate anomaly patterns associated with interannual variation of equatorial Pacific sea surface temperature anomalies. *J. Atmos. Sci.*, **40**, 1410–1425.

Blackmon, M.L., J.M. Wallace, N.-C. Lau and S.L. Mullen, 1977: An observational study of the Northern Hemisphere wintertime circulation. *J. Atmos. Sci.*,**34**, 1040–1053.

Bourke, W., B. McAvaney, K. Puri and R. Thurling, 1977: Global modeling of atmospheric flow by spectral methods. *Methods in Computational Physics, v. 17*, J. Chang (Ed.), Academic Press, New York.

Branstator, G., 1985: Analysis of general circulation model sea surface temperature anomaly simulations using a linear model: I. Forced solutions. *J. Atmos. Sci.*, submitted.

Chang, J. (Ed.), 1977: *Methods in Computational Physics,* Vol. 17, Academic Press, New York.

Chervin, R.M., 1985: Interannual variability and seasonal climate predictability. *J. Atmos. Sci.*, to be published.

Dole, R.M., 1982: Persistent anomalies of the extratropical Northern Hemisphere wintertime circulation. Ph.D. Thesis, Massachusetts Institute of Technology, 226 pp.

Dole, R.M., and N.D. Gordon, 1983: Persistent anomalies of the extra tropical Northern Hemisphere wintertime circulation: Geographical distribution and regional persistence characteristics. *Mon. Wea. Rev.,* **111,** 1567–1586.

Geisler, J.E., M.L. Blackmon, G.T. Bates and S. Muñoz, 1985: Sensitivity of January climate response to the magnitude and position of equatorial Pacific sea surface temperature anomalies. *J. Atmos. Sci.* **42**, in press.

Gill, A.E., 1982: *Atmosphere-Ocean Dynamics,* Academic Press, New York.

Haltiner, G.J. and R.T. Williams, 1980: *Numerical Prediction and Dynamic Meteorology,* Second Edition, J. Wiley, New York.

Horel, J.D., and J.M. Wallace, 1981: Planetary-scale atmospheric phonomena associated with the Southern Oscillation. *Mon. Wea. Rev.,* **109,** 813–829.

Krishnamurti, T.N., and W.J. Moxim, 1971: On parameterization of convective and nonconvective latent heat release. *J. Appl. Meteor.,* **10,** 3–13.

Lau, N.-C., 1981: A diagnostic study of recurrent meteorological anomalies appearing in a 15-year simulation with a GFDL general circulation model. *Mon. Wea. Rev.,***109,** 2287–2311.

Lau, N.-C., 1983: Mid-latitude wintertime circulation anomalies appearing in a 15-year GCM experiment. *Large-Scale Dynamical Processes in the Atmosphere.* B.J. Hoskins and R.P. Pearce (Eds.), Academic Press, London.

Lorenz, E.N., 1967: *The Nature and Theory of the General Circulation of the Atmosphere,* World Meteorological Organization, Geneva.

Malone, R.C., E.J. Pitcher, M.L. Blackmon, K. Puri and W. Bourke, 1984: The simulation of stationary and transient geopotential-height eddies in January and July with a spectral general circulation model. *J. Atmos. Sci.,* **41,** 1394–1419.

Manabe, S., and D.G. Hahn, 1981: Simulation of atmospheric variability. *Mon. Wea. Rev.,* **109,** 2260–2286.

Manabe, S., J. Smagorinsky and R.F. Strickler, 1965: Simulated climatology of a general circulation model with a hydrologial cycle. *Mon. Wea. Rev.,* **93,** 769–798.

McAvaney, B.J., W. Bourke and K. Puri, 1978: A global spectral model for simulation of the general circulation. *J. Atmos. Sci.,* **35,** 1557–1583.

Mullen, S.L., 1985: On the maintenance of blocking anticyclones in a general circulation model. Ph.d. Thesis, University of Washington and NCAR Cooperative Thesis No. 86, 261 pp.

Mullen, S.L., 1985b: The local balances of vorticity and heat for blocking anticyclones in a spectral general circulation model. *J. Atmos. Sci.,* submitted for publication.

Pan, Y.H., and A.H. Oort, 1983: Global climate variations connected with sea surface temperature anomalies in the eastern equatorial Pacific Ocean for the 1958–73 period. *Mon. Wea. Rev.,* **111,** 1244–1258.

Pitcher, E.J., R.C. Malone, V. Ramanathan, M.L. Blackmon, K. Puri and W. Bourke, 1983: January and July simulations with a spectral general circulation model. *J. Atmos. Sci.*,**40,** 580–604.

Ramanathan, V., E.J. Pitcher, R.C. Malone and M.L. Blackmon, 1983: The response of a spectral general circulation model to refinements in radiative processes. *J. Atmos. Sci.*,**40,** 605–630.

Rasmusson, E.M., and T.H. Carpenter, 1982: Variations in tropical sea surface temperature and surface wind fields associated with the Southern Oscillation/El Niño. *Mon. Wea. Rev.*, **110,** 354–384.

Rex, D.F., 1950: Blocking action in the middle troposphere and its effect upon regional climate. II. The climatology of blocking action. *Tellus,* **2,** 275–301.

Simmons, A.J., J.M. Wallace and G. Branstator, 1983: Barotropic wave propagation and instability, and atmospheric teleconnection patterns. *J. Atmos. Sci.*, **40,** 1363–1392.

Walker, G.T., and E.W. Bliss, 1932: World Weather V., *Mem. Roy. Meteor. Soc.*, **4,** 53–84.

Wallace, J.M., and D.S. Gutzler, 1981: Teleconnections in the geopotential height field during the Northern Hemisphere winter. *Mon. Wea. Rev.*, **109,** 784–812.

SOME TOPICS IN THE GENERAL CIRCULATION OF THE ATMOSPHERE

B.J. HOSKINS

Department of Meteorology
University of Reading
Whiteknights,
Reading, Berkshire RG6 9AU, U.K.

ABSTRACT

The characteristics of the observed seasonal mean flow of the atmosphere for December-February and June-August are discussed. Various models and sets of equations are used to investigate the importance of orographic and thermal forcing and the role played by transients on the synoptic and longer time-scales.

The usefulness of considering maps of potential vorticity on isentropic surfaces for the diagnosis of the behaviour of synoptic systems in the atmosphere is illustrated.

1. THE OBSERVED SEASONAL MEAN ATMOSPHERIC FLOW

1.1 Introduction

The following discussion is based on the data routinely analysed and initialised at the European Centre for Medium Range Weather Forecasts (ECMWF) for the period 1979-84. Seasonal means have been constructed for this period and those for December-February (DJF) and June-August (JJA) will be illustrated here, concentrating on the important longitudinal asymmetries.

J. Willebrand and D. L. T. Anderson (eds.), Large-Scale Transport Processes in Oceans and Atmosphere, 71–104.

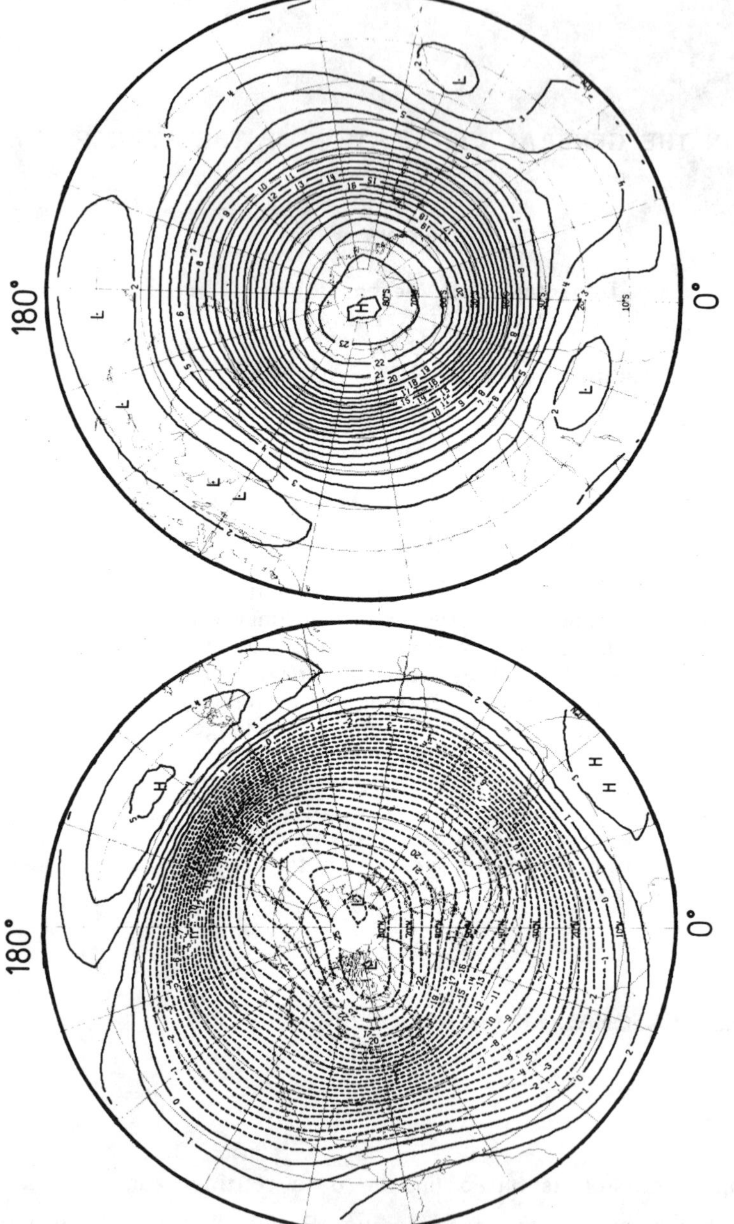

Fig. 1.1 : Streamfunction at 250 mb for DJF 1979-84. Contour interval 5.10^6 m^2s^{-1}.

1.2 December - February

The flow of the atmosphere at a certain pressure level is often illustrated by a height field chart. The geostrophic wind then blows in the direction of the contours, with a magnitude that is inversely proportional to their spacing and to the sine of the latitude. Consequently, the representation of the tropical and sub-tropical flow is poor. Since the wind on a pressure surface is approximately non-divergent, a better view is provided by streamfunction charts such as those for 250 mb illustrated in Fig. 1.1 The westerly flow in the Northern Hemisphere middle latitudes shows major regions of increased flow near 150°E (more than 65 ms^{-1}) and 80°W (more than 40 ms^{-1}) and a less distinct wind maximum near 40°E (more than 40 ms^{-1}). The circulation in the Southern Hemisphere is more zonally symmetric. Anticyclonic cells with easterlies in equatorial regions are noticeable, particularly the pair on either side of the equator in the Indonesian region. The global streamfunction field at 150 mb shown in Fig. 1.2 gives a better representation of this mean tropical circulation. Also shown in this figure is the velocity potential, drawn with the same contour interval. The dominance of the rotational portion of the flow is apparent.

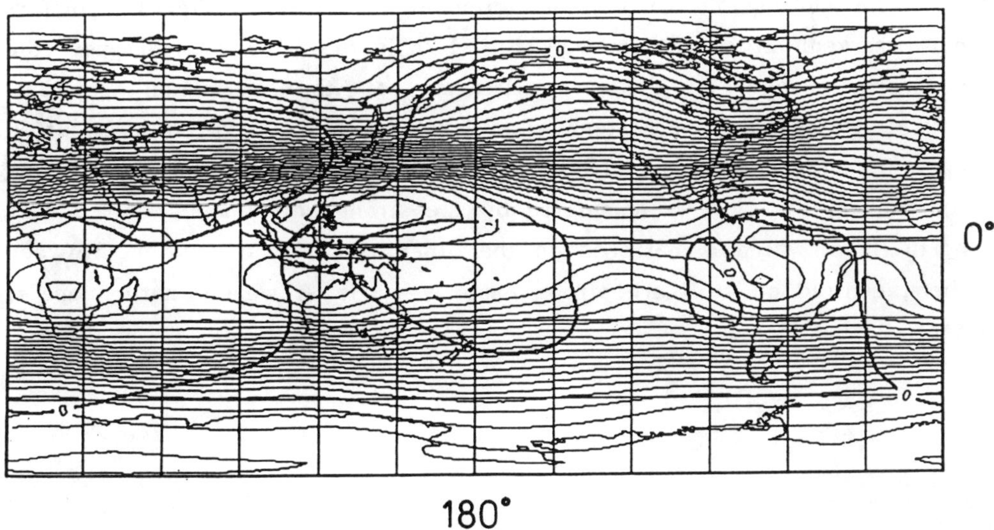

Fig. 1.2 : Streamfunction at 150 mb for DJF 1979-84. Contour interval $5.10^6\ m^2s^{-1}$. Also shown, with the same contour interval, is the velocity potential

Because of the complication introduced by boundary layer processes, it is convenient to illustrate the low level atmospheric state using 1000 mb height. Fig. 1.3 shows that in the middle latitudes of the Northern Hemisphere there is a tendency to have lows over the warm oceans and highs over the cold continents. In the Southern Hemisphere there is again more evidence of symmetry around the pole, but the high pressure systems over the sub-tropical oceans are prominent.

The data is such that all the terms in the thermodynamic equation, apart from the diabatic heating Q, may be evaluated, allowing a realistic looking determination of Q to be produced as a residual. The vertically integrated Q is shown in Fig. 1.4. The major sources may be associated with latent heat release both in the western oceanic regions of the Northern Hemisphere and in the tropical continental regions of the Southern Hemisphere. A heat sink extends from the North Pole southward over the North American and Asian continents.

1.3 June - August

The global 150 mb streamfunction field, shown in Fig. 1.5, differs quite markedly from that for DJF (Fig. 1.2). In the Southern Hemisphere the dominant change is the intensification of the meridional gradient, indicating the increased westerly wind which reaches a maximum of about 50 ms^{-1} in the sub-tropical jet in the Australian region. In the tropics and sub-tropics, the flow is dominated by the strong anticyclone centred over north India, which is mirrored on the other side of the equator. Between them, centred on 10°N are strong easterly winds. A weaker anticyclone is centred near Mexico.

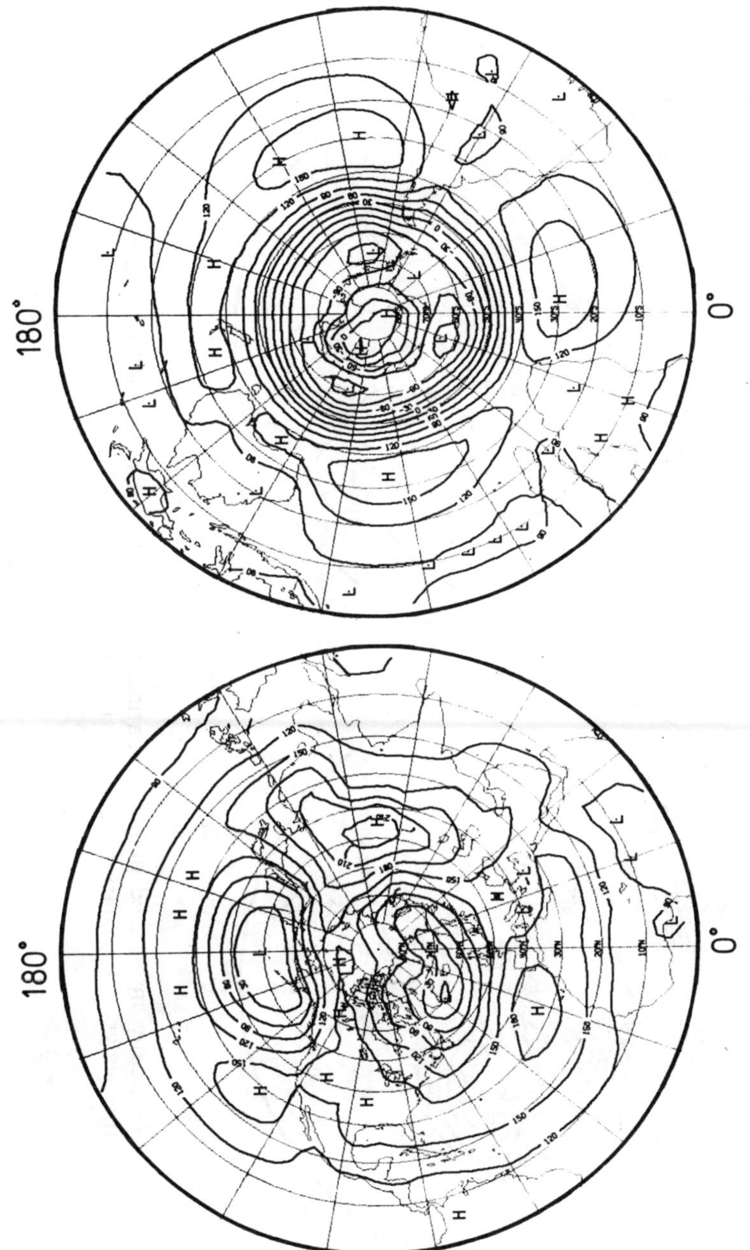

Fig. 1.3 : 1000 mb geopotential height field for DJF 1979-84. Contour interval 30 m.

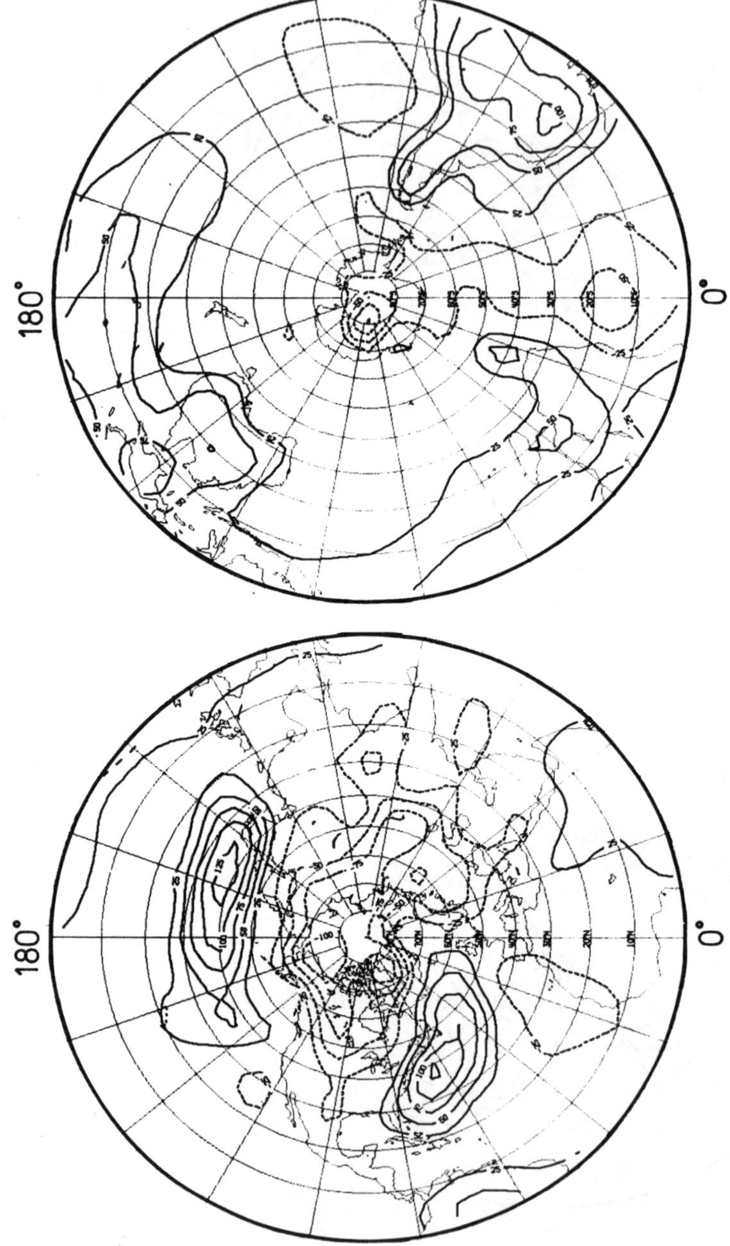

Fig. 1.4 : Vertically integrated diabatic heating for DJF 1979-84 calculated as a residual in the thermodynamic equation. Contour interval 25 Wm^{-2}.

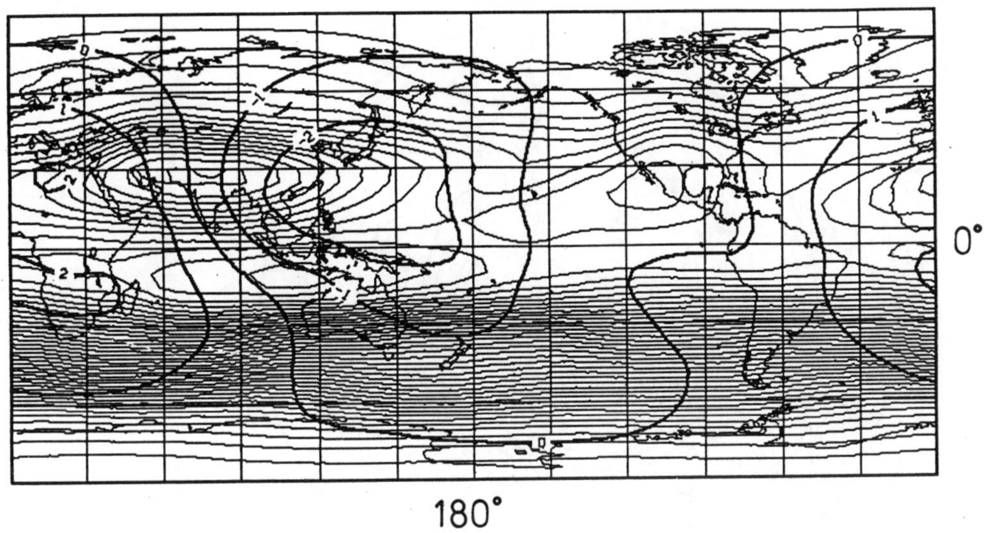

Fig. 1.5 : Streamfunction and velocity potential at 150 mb for JJA 1979-84. Contour interval $5.10^6 m^2 s^{-1}$.

The Northern Hemisphere 1000 mb height field (Fig. 1.6) shows approximately the opposite of that in the winter season with low pressure over the land masses and high pressure over the oceans. The changes in the Southern Hemisphere are less marked.

The diabatic sources maxima in the Northern Hemisphere (Fig. 1.7) are now predominantly in the subtropics, in particular over S.E. Asia and to a lesser extent Mexico and Africa. The sinks over the ocean off West Africa and California are presumably connected with radiative cooling from the tops of stratus cloud over the cold water upwelling regions. The expected sink near the Southern Hemisphere pole is not apparent which is presumably due to the inadequacies of the data in this region.

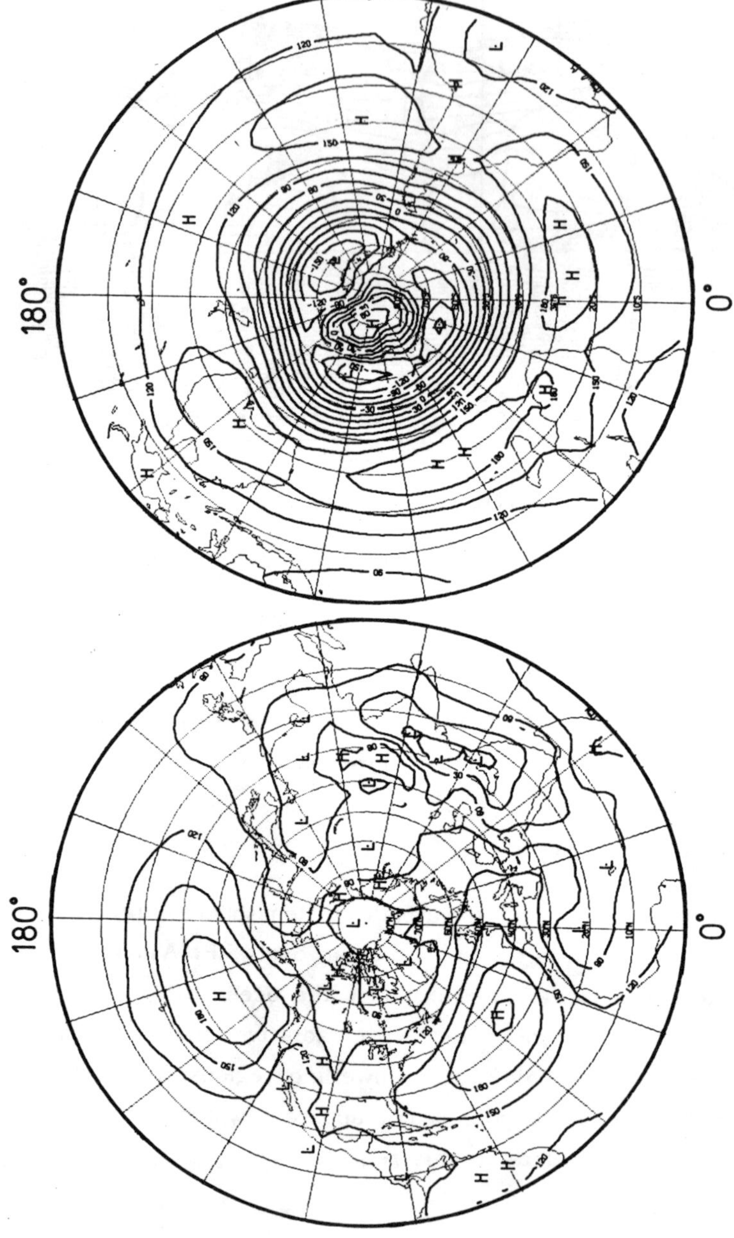

Fig. 1.6 : 1000 mb geopotential height field for JJA 1979-84. Contour interval 30 m.

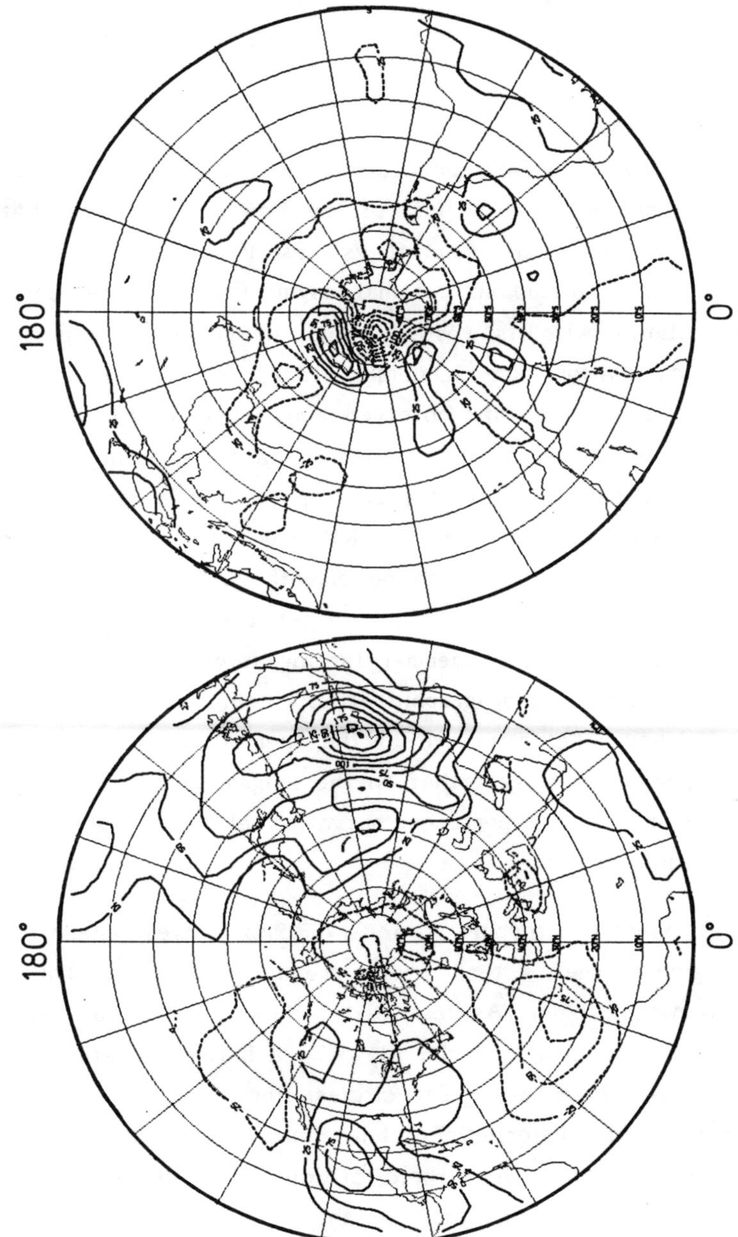

Fig. 1.7 : Vertically integrated diabatic heating for JJA 1979-84 calculated as a residual in the thermodynamic equation. Contour interval 25 Wm-2.

2. OROGRAPHIC AND THERMAL FORCING

2.1 Introduction

One topic which has frequently been discussed is the relative roles of the mountains with their orographic uplift of the flow and of the diabatic heating in forcing the seasonal mean flows described in the previous section. It is clear that latitudinal thermal asymmetries are responsible for the Hadley circulation with its westerly winds in middle and high latitudes (for detailed calculations see Schneider, 1977; Held and Hou, 1980). However, it is less clear whether the observed longitudinal asymmetries are associated with the longitudinal structure in the diabatic heating or with the interaction of the zonal wind with the orography. In order to provide a partial answer to this question, Held (1983) has run a full general circulation model and then rerun it with no mountains. The latter run he equates with thermal forcing, and the difference between the two runs with orographic forcing. Of course the thermal forcing is somewhat different with and without orography, and orographic effects could well be different in the absence of longitudinal structure in the thermal sources. However those interesting experiments suggest that on the longest zonal wavelengths the two forcings are of similar importance, but at zonal wavenumbers 4 and 5 the orographic forcing dominates.

Since the atmosphere can support stationary Rossby waves, whenever it is forced in one region there is a tendency to produce a stationary train of waves emanating from that region. As described in Hoskins and Karoly (1981), for very simple constant angular velocity flows in a barotropic atmosphere the ray paths for those waves are great circles. For more realistic flows the ray paths are somewhat distorted. For baroclinic flows, Held (1983) has shown the so-called equivalent barotropic structure of the downstream Rossby wave-train, having no phase tilt in the vertical and a maximum amplitude in the region of the tropopause.

2.2 Orographic forcing

A useful starting point for the consideration of the impact of orographic uplift on the large-scale atmospheric flow is the ß-plane barotropic vorticity equation linearized about a uniform zonal flow $\bar{u}$ and with the divergence and convergence above orography envisaged to occur uniformly over a depth H:

$$\bar{u} \frac{\partial}{\partial x} \nabla^2 \psi + \beta \frac{\partial \psi}{\partial x} = - \bar{u} \frac{f}{H} \frac{\partial h}{\partial x} , \qquad (2.1)$$

where ψ is the perturbation stream function and h is the orographic height. For orography of the form $h_{k,l} e^{ikx} \sin ly$, the corresponding streamfunction amplitude $\psi_{k,l}$ is given by

$$\psi_{k,l} = \frac{f\, h_{k,l}/H}{K^2 - K_s^2} \qquad (2.2)$$

where K is the total wavenumber $(k^2+l^2)^{1/2}$ and K_s is the stationary Rossby wavemunber $(\beta/\bar{u})^{1/2}$. For $K < K_s$ (short waves) the first term in (2.1) dominates the left hand side, the anticyclone is over the orographic crest and there is poleward flow on the upslope.

For $K < K_s$ (long waves) the second term in (2.1) dominates the left hand side, there is equatorward flow on the orographic upslope and cyclonic circulation over the mountain.

There is resonance at $K = K_s$ which necessitates the inclusion of a frictional term in any application of (2.1) as was originally done by Charney and Eliassen (1949). Solutions tend to be dominated by the presence of this resonance.

The same model may be applied in spherical geometry and for an observed zonally averaged $\bar{u}$ at some level in the atmosphere. Figure 2.1 shows the vorticity perturbation and total streamfunction generated using the 300 mb DJF $\bar{u}$ and a circular mountain at 30°N. There is anticyclonic vorticity over the mountain, downstream cyclonic vorticity and a wavetrain which tends to split at the latitude of the northern flank of the jet. The wavetrain propagating towards the equator decreases in amplitude as it approaches the easterly wind region ($K_s \rightarrow \infty$), and there is negligible response in the southern hemisphere. The total streamfunction indicates the strong flow downstream from the mountain and a diffluent region about 90° downstream. Experimentation shows that the results are not sensitive to the choice of or the frictional damping and that resonance is not a feature of this model.

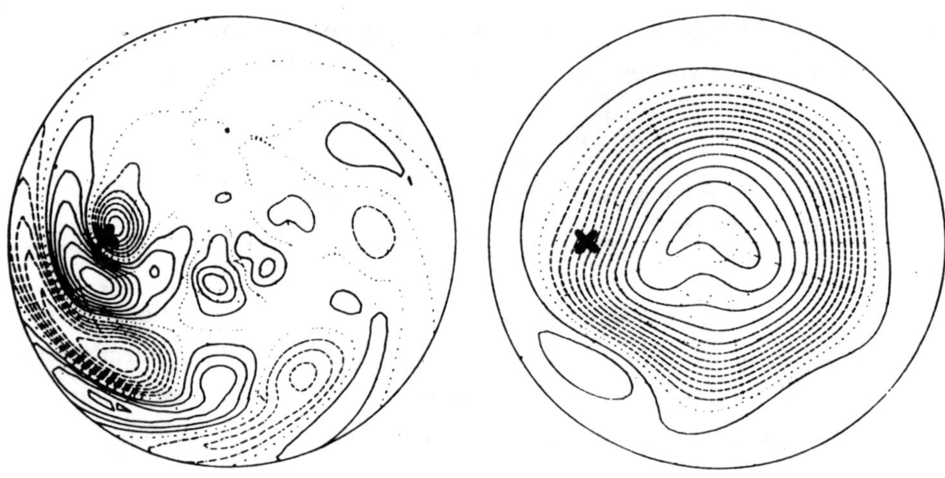

Fig. 2.1 : The northern hemispheric vorticity perturbation and total streamfunction obtained from a barotropic model when a 300 mb DJF zonal flow is forced with a circular mountain centred at the point indicated at latitude 30°N. (Grose and Hoskins, 1979). Contours at negative values are dashed.

The NH streamfunction obtained by using the full earth orography in this model is shown in Fig. 2.2. Comparing with Fig. 1.1, it is seen that many features are reproduced, though their amplitudes are too large which is to be expected since the 300 mb flow is much stronger than that which interacts with the orography. Surprisingly, even the anticyclone near Indonesia is reproduced to some extent as a remote response to the flow over the Tibetan plateau. It is clear from other considerations, however, that the convective heating in this region is very important.

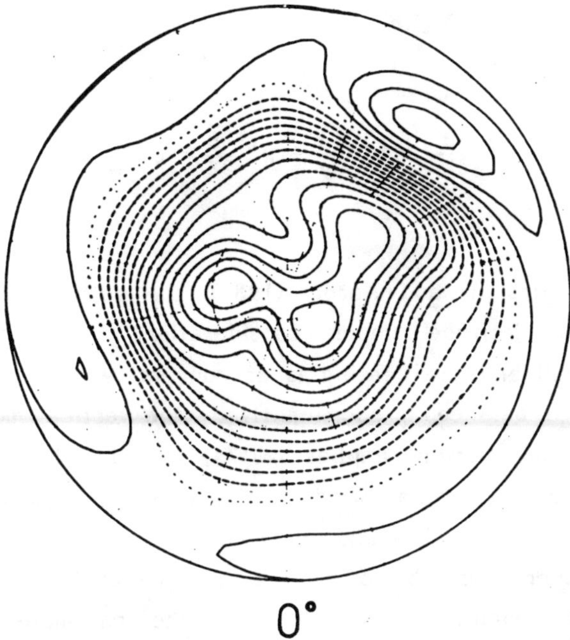

Fig. 2.2 : The northern hemispheric total streamfunction obtained from a barotropic model when a 300 mb DJF zonal flow is forced with the Earth' orography. (Grose and Hoskins, 1979).

The primitive equations may also be used for such calculations. Hoskins and Karoly (1981) describe the results of a linearized five level model which differ rather little from the barotropic results.

2.3 Thermal forcing

Consider the vorticity and thermodynamic equations for a steady thermally forced perturbation to a zonal flow:

$$\bar{u}\,\frac{\partial \xi'}{\partial x} + v'\,(\beta - \frac{\partial^2 \bar{u}}{\partial y^2}) = f\,\frac{\partial w'}{\partial z} \tag{2.3}$$

$$\bar{u}\,\frac{\partial \theta'}{\partial x} + v'\,\frac{\partial \bar{\theta}}{\partial y} + w'\,\frac{\partial \bar{\theta}}{\partial z} = Q \tag{2.4}$$

As discussed by Hoskins and Karoly (1981), the dominant balance in (2.4) depends on the latitude under consideration. In the tropics where $f^2\bar{u}_z/\beta N^2 H$ is small and the heating is of convective origin with a maximum near 400 mb, Q is balanced predominantly by the vertical motion term. In a heating region, the ascent implies $\partial w'/\partial z$ positive below and negative above. Given the small zonal wind u, the vorticity equation then implies that v' must be positive below and negative above. Consequently to the west of the source there must be a cyclone at low levels and anticyclone at upper levels. In middle latitudes, where the parameter is large, the horizontal advection terms dominate. Using the thermal wind relations, these two terms have a sum proportional to $\bar{u}^2 \frac{\partial}{\partial z}(v'/\bar{u})$. Thus in a heating region the meridional wind must change in the sense of v' negative at low levels to v' positive at upper levels. This is the opposite of that for the tropics and is consistent with a low level cyclone and upper level anticyclone to the east. At lower levels if zonal advection is negligible in (2.3), $\partial w'/\partial z$ must be negative so that there is even a tendency for descent in the region of the heating. Of course, Ekman pumping may be expected to modify this result.

Fig. 2.3 shows the results of a linearized five level baroclinic model on the sphere for heating centred on 500 mb, 45°N. The section at the latitude of the heating supports the ideas described above. The 300 mb vorticity field shows the generation of wavetrains and the geopotential height figure accentuates the importance of the northern train. The surface characteristics of the response in the region of the heating are very dependent on the depth of the heating. For a column integrated heating of 145 Wm^{-2} (equivalent to 5 mm of rainfall per day, or 1.25 K/day average heating) the source with a maximum at 500 mb gives a surface pressure depression of 3.3 mb and a 900 mb temperature drop of 1.4 K, whereas a heating maximum at 700 mb gives a pressure fall of 6.2 mb and a temperature rise of 2.6 K.

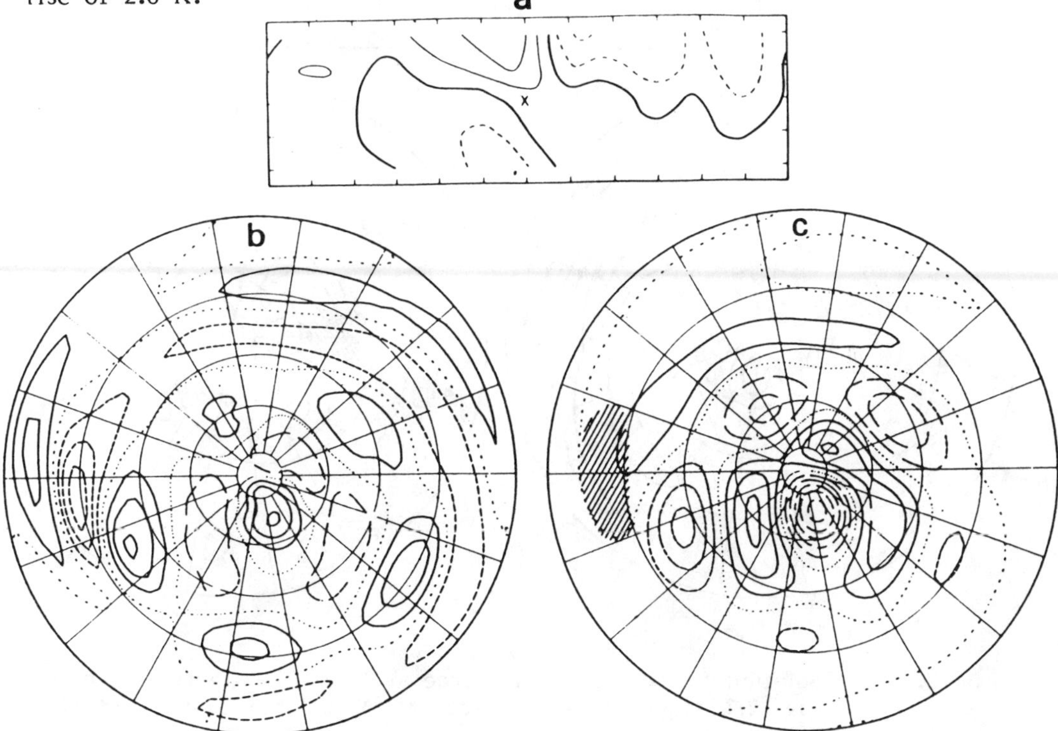

Fig. 2.3: Steady state linear solution of a five-layer baroclinic model for a deep circular heat-source at 45°N perturbing the DJF zonal flow. Shown are (a) height field in a longitude-pressure section at 16.0°N (contour interval 1 dam), (b) 300 mb vorticity perturbation (contour interval 0.025 Ω) and (c) 300 mb height field perturbation (contour 1 dam). The center of the source is indicated by a cross in (a) and the region of heating greater than 0.25 K day-1 by hatching in (c). Tick marks in (a) are at 100, 300, 500, 700 and 900 mb.

Results from the same model but with an elliptical source centred at 500 mb and 15°N are shown in Fig. 2.4. The local response is again as suggested by the discussion above. The 300 mb flow shows the propagation of a wavetrain into middle and high latitudes with a 300 mb height perturbation of almost 6 dam near 70°N. This wavetrain is equivalent barotropic in structure and is quite well represented in a barotropic model using a 300 mb flow and a divergence source at 15°N.

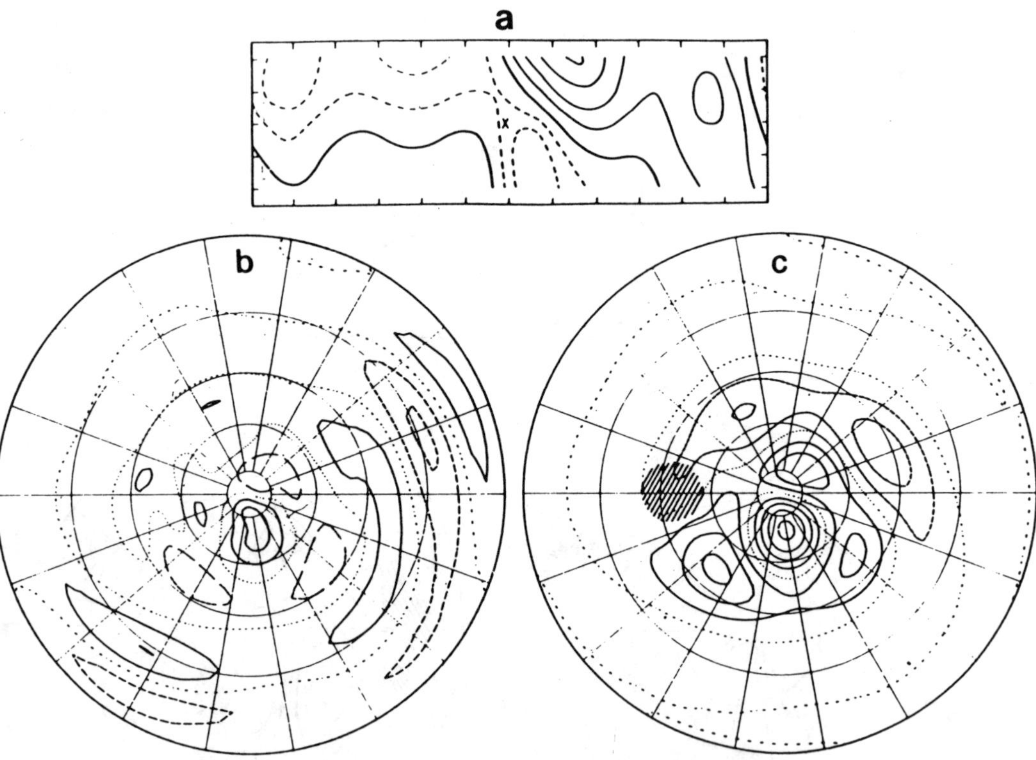

Fig. 2.4 : Solution for a deep heat source at 15°N. The figures are as in Fig. 2.3 except that (a) is at 18.1°N and the contour interval in (a) is 0.5 dam.

Unpublished studies with a time-dependent nonlinear baroclinic model have also been performed in which at an initial time a tropical heat source has been switched on to perturb a baroclinically unstable flow. An equivalent barotropic wavetrain emanates from the tropics much as in the linearized steady-state model. In middle latitudes, after about two weeks, the response is dominated by the development of baroclinic instability. The main difference from the linear model is that even when the heating is placed in the equatorial easterlies approximately the same response is obtained. Simple linear theory would suggest no Rossby wavetrain in this case.

2.4 Other comments

The recent interest in the El Niño-Southern Oscillation phenomenon has focussed attention on the response of the atmosphere to the modification of the tropical diabatic heating associated with the movement towards the central Pacific of the Indonesian convective maximum in sympathy with that of the warmest waters. Simple models (e.g. Gill and Rasmusson, 1983) and general circulation models have been used to model the phenomenon. The response shown in Fig. 2-4 is generally consistent with that which is observed, but it is clear from Simmons (1982) and Simmons et al. (1983) that more realism is introduced by an inclusion of the longitudinal variations in the ambient flow. Sardeshmukh and Hoskins (1985) have also stressed the nonlinearity of the observed upper troposphere vorticity budget. With the observed heating and thus upper level divergence field imposed it is possible to use the nonlinear vorticity equation to deduce the rotational flow.

A more complete account of the basic theory of stationary waves is given in Held (1983) and, as is suggested above, there is much current research in this area.

3. TRANSIENTS AND THE SEASONAL MEAN FLOW

3.1 Introduction

We now consider the distribution and nature of the deviations from the seasonal mean flow, and their possible contribution to the maintenance of that flow. Denoting a time mean by a bar and a transient motion by a prime, the mean potential temperature equation may be written

$$\bar{D}\,\bar{\theta} = \bar{Q} - \nabla\cdot\overline{\underset{\sim}{v}'\theta'} - \partial/\partial z\,(\overline{w'\theta'}) \tag{3.1}$$

Here $\bar{D}$ is the time derivative following the mean flow, ∇ and v are the horizontal gradient operator and velocity, respectively. This equation is in a convenient form for discussion. However, similar equations for the mean horizontal velocity $\bar{v}$ are less convenient because it is observed that this velocity is approximately non-divergent. Consequently the equation for the mean vertical component of relative vorticity $\bar{\xi} = \partial\bar{v}/\partial x - \partial\bar{u}/\partial y$ is preferable. To a good approximation the contribution of the transients to this equation may be written as $-\nabla\cdot\,\overline{v'\xi'}$. For horizontally non-divergent transients, as shown by Holopainen (1978) and Hoskins et al. (1983), this vorticity flux is related to properties of the velocity correlation tensor:

$$\underset{=}{C} = \begin{pmatrix} \overline{u'^2} & \overline{u'v'} \\ \overline{u'v'} & \overline{v'^2} \end{pmatrix} = \begin{pmatrix} k & o \\ o & k \end{pmatrix} + \underset{=}{A} \tag{3.2}$$

where

$$K = \frac{1}{2}\,\overline{u'^2+v'^2}\ , \quad M = \frac{1}{2}\,\overline{u'^2-v'^2}\ , \quad N = \overline{u'v'} \tag{3.3}$$

and

$$\underset{=}{A} = \begin{pmatrix} M & N \\ N & -M \end{pmatrix}$$

It can be shown that in this case

$$- \nabla \cdot \overline{\mathbf{v}'\xi'} = 2 \frac{\partial^2 M}{\partial x \partial y} - \frac{\partial^2 N}{\partial x^2} + \frac{\partial^2 N}{\partial y^2} = \frac{\partial}{\partial y} \nabla \cdot \underset{\sim}{E} - \frac{\partial^2 N}{\partial x^2} \tag{3.4}$$

where $$\underset{\sim}{E} = (-2M, -N) = (\overline{v'^2 - u'^2}, -\overline{u'v'}) \tag{3.5}$$

Thus it is the anisotropic portion $\underset{=}{A}$ of the velocity correlation tensor that is important in forcing the mean flow: the kinetic energy is not relevant. Further in cases where the $\partial^2 N/\partial x^2$ term is negligible in (3.4), i.e. when $|M| \gg |N|$ or the length scales are such as $L_x \gg L_y$, noting that these scales apply to N and not to the individual eddies, then the forcing of $\bar{\xi}$ is well approximated by $-\partial/\partial y \, \nabla \cdot \underset{\sim}{E}$. The approximation is mostly very good for the seasonal mean transients and so their forcing of the mean rotational flow is the same as if there was a forcing $\nabla \cdot \underset{\sim}{E}$ in the mean zonal momentum equation. The advantage of this analysis is that $\underset{\sim}{E}$ is deduced simply from data and that it is interpreted simply in terms of the shapes of the transients. Hoskins et al. (1983) showed further that when group velocity ideas are applicable, $\underset{\sim}{E}$ indicates the propagation of the transient wave activity relative to the mean flow.

As reviewed in Holopainen (1984) there are other ways of considering transient forcing of the mean flow, mostly based on the equation for the mean quasi-geostrophic potential vorticity. In order to include the crucial boundary temperature fluxes in such discussions, it is necessary, as has been done by Lau and Holopainen (1984), to invert the elliptic operator relating potential vorticity and streamfunction: interior potential vorticity fluxes are not sufficient.

3.2 Observed transients

Fig. 3.1 shows the DJF 1979-80 Northern Hemisphere 250 mb kinetic energy K and the axis of the tensor $\underline{\underline{A}}$ for the transients in two frequency bands separated at approximately 10 days. K for the high pass synoptic transients tends to be concentrated in two so-called storm-tracks, with the tensor axis indicating eddies elongated in the meridional direction ($v'^2 > u'^2$). The lower frequency K is also concentrated in the mid-oceanic regions but here $u'^2 >> v'^2$. In both cases $|\overline{u'v'}|$ is generally smaller. The level chosen is close to a maximum in the quantities considered but there is little change in characteristics with height. However, the horizontal heat flux tends to be a maximum nearer to the 700 mb level. Its convergence also tends to be dominated, particularly for the high pass transients, by that of its meridional component. Consequently a convenient summary of the transient fluxes is provided by diagrams with arrows indicating the 250 mb $\underset{\sim}{E}$ and contours indicating the 200 mb $\overline{v'T'}$. Such diagrams for the two frequencies for the season concerned are presented in Fig. 3.2. The high pass transient heat flux indicates a tendency to decrease the baroclinity in the mean flow on the upstream and of the storm-track. The divergence of the $\underset{\sim}{E}$-vector from the same region shows a tendency to accelerate the barotropic westerly flow there. The net effect is to accelerate the mean surface westerlies in the storm-track. For a non-divergent mean flow this implies a tendency to force low-level horizontal confluence at the beginning of the storm-track thereby helping to maintain the mean baroclinity in this region. This raises the intriguing possibility that there may be a positive feedback onto the existence of storm-tracks. The dominant signature from the lower frequency transients is a tendency to decelerate the mean jets and accelerate the mean flow in the eastern oceans where it is weaker.

The mean diabatic heating in the western North Atlantic and Pacific Oceans shown in Fig 1.4 may also be considered to be a forcing associated with the synoptic transients. It is determined by the latent heat release in the rising warm moist air and sensible heat flux from the ocean into the cold air ahead of and behind the developing depressions.

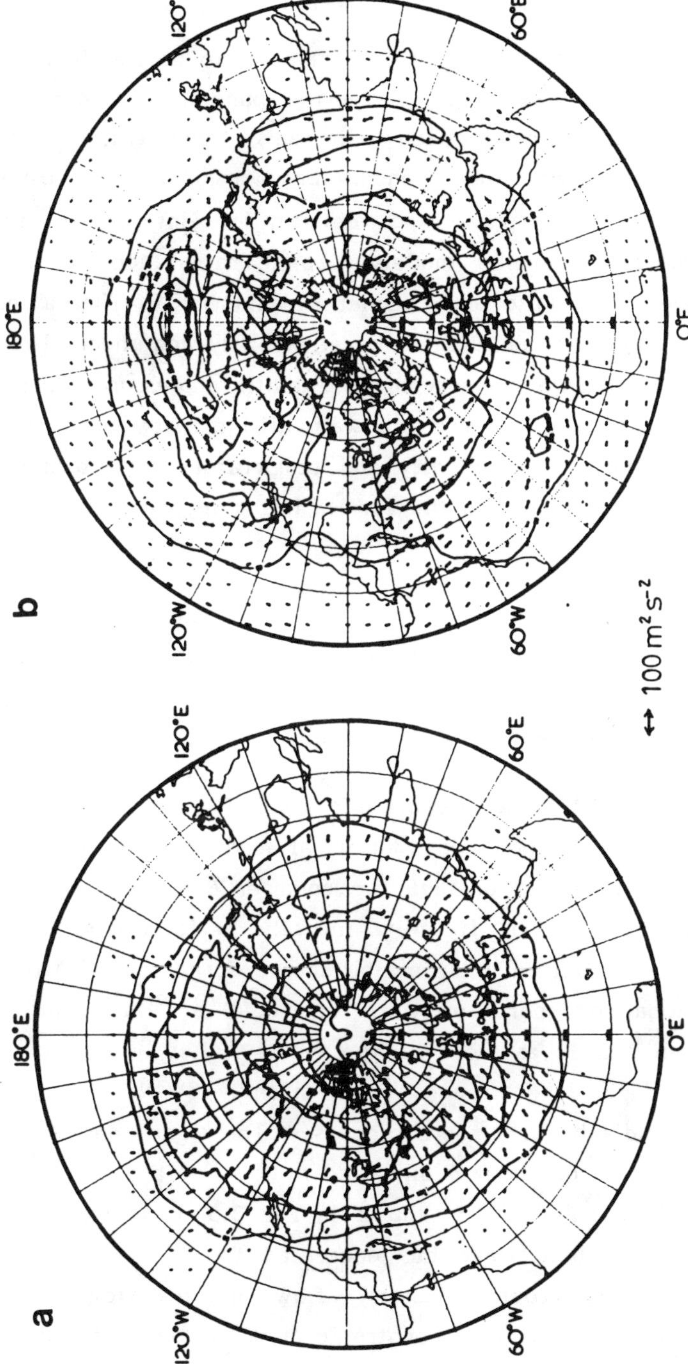

Fig. 3.1 : Synoptic time-scale (a) and low frequency (b) 250 mb eddy kinetic energy, K (contour interval $50m^2s^{-2}$) and major axis of the tensor $\underline{\underline{A}}$ for DJF 1979-80, Northern Hemisphere.

3.3 Modelling of the transients

The synoptic time-scale transients are thought to be manifestations of baroclinic instability. Following the pioneering papers of Charney (1947) and Eady (1949) there have been numerous investigations of this phenomenon. Linear studies (e.g. Simmons and Hoskins, 1977) using the primitive equations on the sphere and realistic jet flows have generally confirmed the results of simpler models. Growth rates increase from the largest zonal wavelengths upto about zonal wavenumber 8. Thereafter, the behaviour is very dependent on the low level stability and shear. The zonal wavenumber 6 most unstable mode typically exhibits a pressure wave tilting westwards with height and a temperature wave tilting eastwards with height in the lower troposphere. The velocity and temperature perturbations have maxima at the lower boundary and the former has a second maximum at the tropopause.

Baroclinic instability calculations for perturbations to realistic mean flows including the stationary waves by Frederiksen (1983 and refs.) have generally indicated maximum growth to occur on the upstream sides of the storm-tracks, in agreement with observation.

Nonlinear studies have also been performed with zonal flows such as that shown in Fig. 3.2. A nonlinear numerical model was initialised using this zonal flow plus the most unstable wavenumber 6 mode with a surface pressure amplitude of 1 mb. It grew initially as described by linear theory. The surface structure then exhibited the formation of fronts and significant meridional transports of warm and cold air. Baroclinic growth at low levels then ceased as most of the wave activity was apparent at upper levels. The upper level troughs tilted strongly SW-NE and the wave decayed equally rapidly by barotropic processes. The significant growth and decay occupied a period of about 5 days and at the end the zonal flow was only weakly unstable. This flow is shown in Fig 3.3. It is clear that the life cycle of the baroclinic wave has decreased the low-level baroclinity and increased the barotropic westerly flow in the region of the instability. On either side, easterly barotropic flow has been created.

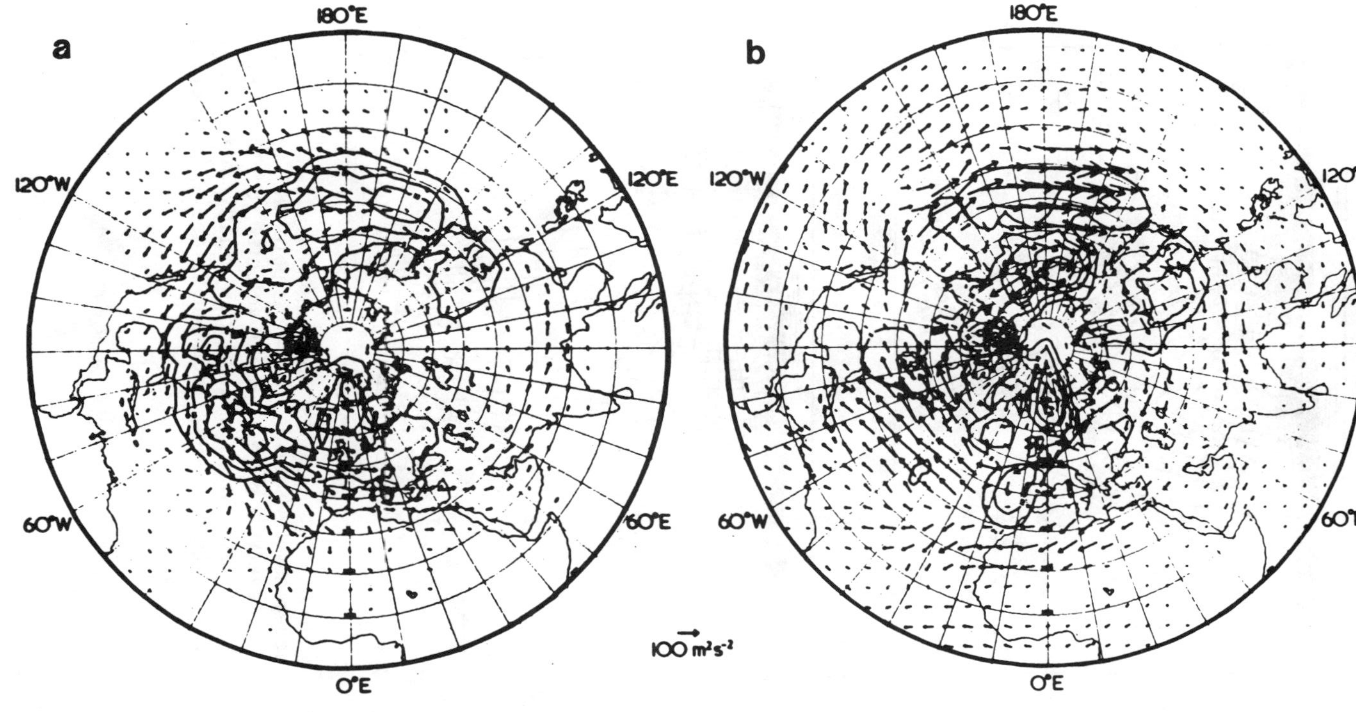

Fig. 3.2 : Transient eddy flux summary pictures for Northern Hemisphere DJF 1979-80, for (a) synoptic time-scale and (b) low frequency. The contours are those of 700 mb poleward temperature flux $\overline{v'T'}$ (contour interval 5 K ms-1) and the vectors represent $\underset{\sim}{E} = (\overline{v'^2-u'^2}, \overline{-u'v'})$.

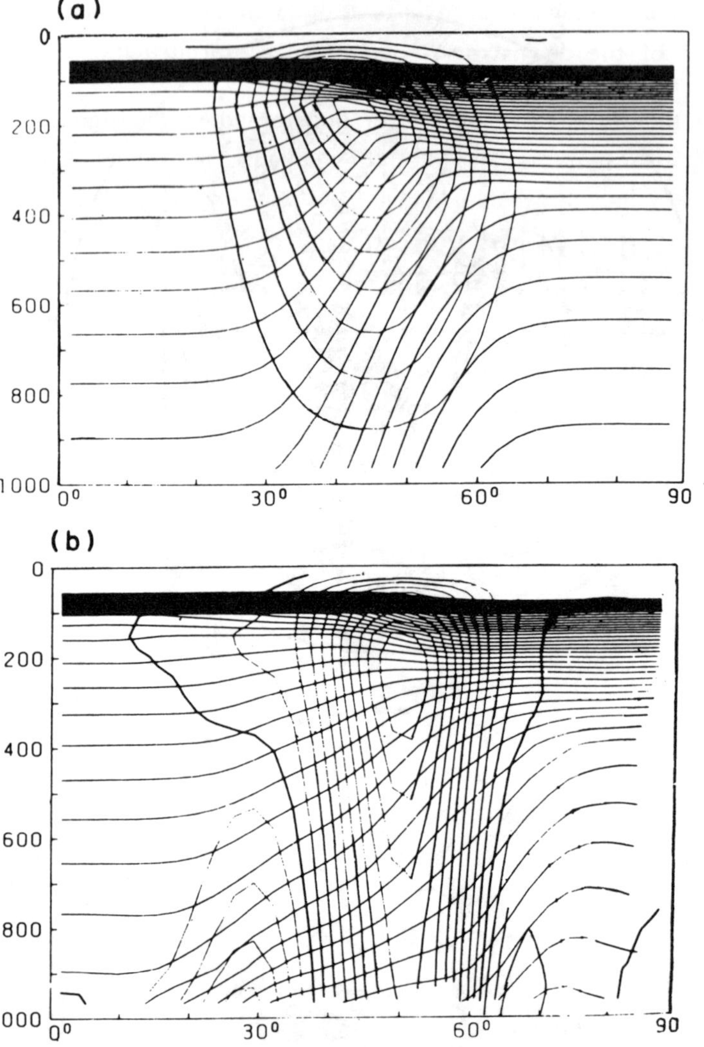

Fig. 3.3 : Latitude-pressure sections showing potential temperature (contours every 5 K) and zonal velocity (contours every 5ms^{-1}) for (a) the basic state and (b) the zonally averaged flow at day 15. The zero velocity contour is heavy (Simmons and Hoskins, 1980).

The time sequence of events in this life-cycle appears to give a good description of the downstram sequence in a stormtrack.

The theoretical framework for understanding the signature of the lower frequency transients is less well developed. The tendency of the observed transients to decelerate the mean flow where it is strong and accelerate it where it is weak suggests a barotropic conversion of energy from the mean flow to the transients. Simmons et al. (1983) have performed barotropic instability calculations for longitudinally varying flows that confirm this view. However, the lack of separation in scale between the wave in the imposed basic flows and in the linear model structures poses conceptual difficulties that have not yet been resolved.

4. ISENTROPIC POTENTIAL VORTICITY IN THE ATMOSPHERE

4.1 Introduction

In the absence of diabatic heating, potential temperature (θ) is conserved following the motion of the atmosphere, so that air moves along surfaces of constant θ, i.e. isentropic surfaces. In 1930 Sir Napier Shaw propounded the benefits of using θ as a vertical coordinate. Of interest for comparison with recent oceanographic discussion of isopycnals that outcrop and those that do not, he considered the atmosphere to be divided into "the overworld" composed of isentropes not intersecting the Earth's surface, and the "underworld" in which we live. Fig. 4.1 shows the isentropes for the zonally averaged flow in DJF. As expected, they generally slope upward from equator to pole in the troposphere. The tropopause usually marks an abrupt change in the vertical gradient of θ. In the 1940's pressure generally replaced θ as the vertical coordinate, though the latter was used in the research of Reed, Danielsen, Eliassen, Johnson, Shapiro and others (for full references, see Hoskins et al., 1985).

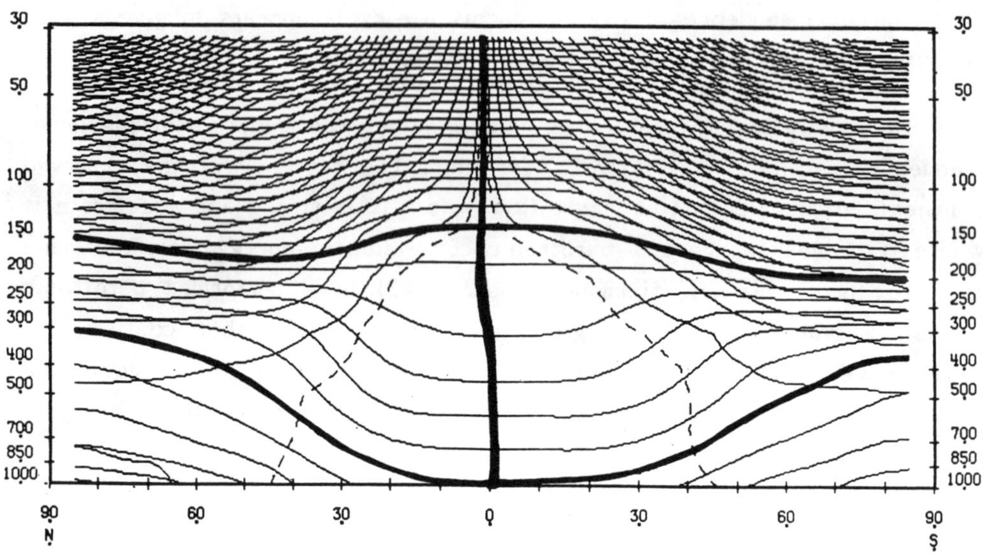

Fig. 4.1 : The potential temperature and potential vorticity of the zonally averaged DJF flow. The contour for potential temperature is 10 K, with the 300 K and 360 K contours thickened, the former grazing the surface in equatorial regions. The potential vorticity contour is 1 unit exept that the ±0.5 unit contours are shown by dashed lines. The zero PV contour is thickened.

Rossby (1939) introduced as an approximation to atmospheric motion the conservation of the vertical component of absolute vorticity, $f+\xi$, in 2-D horizontal flow. In Rossby (1940) he indicated how some stretching effects could be included by showing that the quantity $(f+\xi)$/fluid depth is conserved in a shallow fluid. Rossby and his co-workers routinely used isentropic maps to represent the atmosphere and it was an obvious extension for him to put forward the conservation of $(f+\xi_\theta)/\Delta p$ for an isentropic layer. Here Δp is the pressure thickness of the layer and ξ_θ the relative vorticity evaluated on an isentropic surface. He defined the "potential vorticity" ξ_o to be the value of ξ_θ when the layer is brought to a standard latitude and made a standard pressure thickness:

$$\frac{f+\xi_\theta}{\Delta p} = \text{const.} = \frac{f_0+\xi_0}{\Delta p_0} \tag{4.1}$$

Independently, Ertel (1942) showed that quantities such as

$$P = \frac{1}{\rho}\,\underset{\sim}{\zeta}\cdot\nabla\theta \tag{4.2}$$

are conserved in adiabatic, frictionless motion where $\underset{\sim}{\zeta}$ is the absolute vorticity. If the hydrostatic approximation is made, it can be shown that

$$P = -g\ (f+\xi_\theta)/\frac{\partial p}{\partial \theta} \tag{4.3}$$

so that the conservation of P is essentially the same as Rossby's result (4.1). The term potential vorticity (PV) is now used for quantities like P. rather than ξ_0.

A detailed historical account of the development and use of PV is given in Hoskins et al. (1985). Here we note only that (i) contours of P on isentropic surfaces, the so-called isentropic potential vorticity (IPV) maps, provide a tracer of the fluid motion; (ii) under suitable balance conditions the IPV distribution along with the surface temperature distribution may be inverted to give the entire fluid motion; (iii) Charney and Stern (1962) showed that a small amplitude approximation to IPV conservation gives the conservation of quasi-geostropic potential vorticity along "horizontal" trajectories.

4.2 Examples from atmospheric flow

As well as showing isentropes, Fig. 4.1 also has contours indicating the PV of the zonally averaged flow in DJF. In the troposphere there is a general increase in magnitude towards the poles and upwards consistent with the variation of f and ρ^{-1} respectively. At the tropopause, individual cases show a jump in PV values and there is a dramatic increase in the stratosphere. Generally, values below 1.5 units indicate trophospheric air and those above 4 units stratospheric air. (The unit is $10^{-6}m^2s^{-1}K\ kg^{-1}$).

Fig 4.2 shows the 500 mb geopotential charts every other day for the sudden development of a so-called cut-off low. Surface pressure maps (not

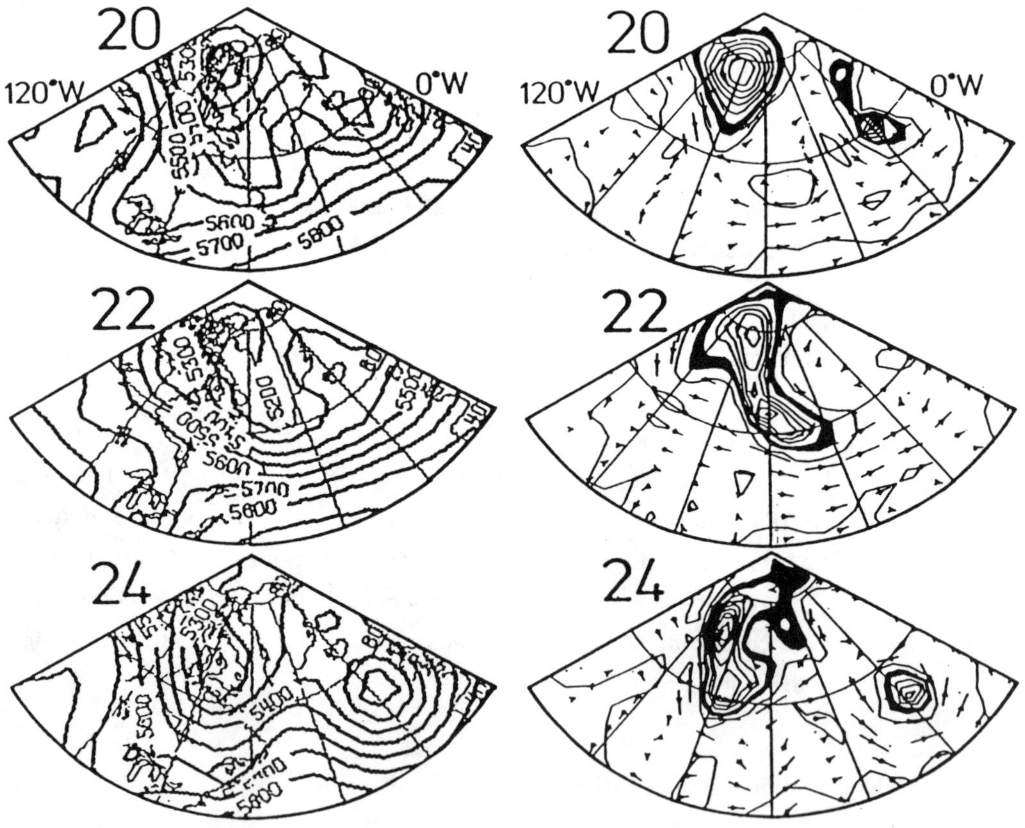

Fig. 4.2 : 500 mb geopotential height (left) and 300 K IPV and wind vectors for a section from 120°W - 0°W and 40°N - 90°N on 20, 22, and 24 September 1982. The contour intervals are 10 dam and 0.5 units, respectively. The region 1.5 - 2 units is blackened to indicate the probable tropopause position.

shown) indicate the simultaneous development of a low pressure centre, there being little tilt with height and no sign of growth by baroclinic processes. The 300 K IPV maps in Fig 4.2 indicate how stratospheric air in the vortex in the Davis Straits region is pulled out. Since it is a positive IPV anomaly, it develops its own cyclonic circulation and cuts itself off. This corresponds to the development of the cut-off low. Having been created, this IPV anomaly can only be destroyed by advection back into a stratospheric source region, or by diabatic frictional processes.

Fig. 4.3 gives an example of a blocking high development as seen in the geopotential height and 330 K IPV maps. The poleward intrusion of subtropical tropospheric air ahead of a large amplitude trough is apparent. This anomalously low IPV develops its own anticyclonic circulation and cuts off to form the blocking high. Again decay is possible only by advection or diabatic and frictional processes. The duality of the two events in Figs. 4.2 and 4.3 is striking.

4.3 Some theoretical considerations

Simple models of the isolated cyclonic and anticyclonic anomalies may be obtained by specifying states with uniform surface and upper level temperatures, uniform tropospheric PV, uniform larger stratospheric PV and a circularly symmetric temperature distribution on the tropopause. Fig. 4.4 shows solutions for such basic states using an extension of the semi-geostropic equations (Thorpe, 1985). The cut-off low (a) and blocking high (b) show many of the observed characteristics including small and large static stabilities, respectively. In the former moist convection is encouraged and in the latter it is suppressed. The time-scale for decay of the former is therefore much shorter, in agreement with observation. Bretherton (1966) showed that surface temperature anomalies may be considered as PV anomalies confined to the boundary. Solutions for such surface temperature anomalies are shown in Figs. 4.4 (c) and (d).

These four solutions may be thought of as the building bricks for atmospheric synoptic systems, though linear superposition is not strictly applicable in large amplitude cases. In Hoskins et al. (1985) it is shown

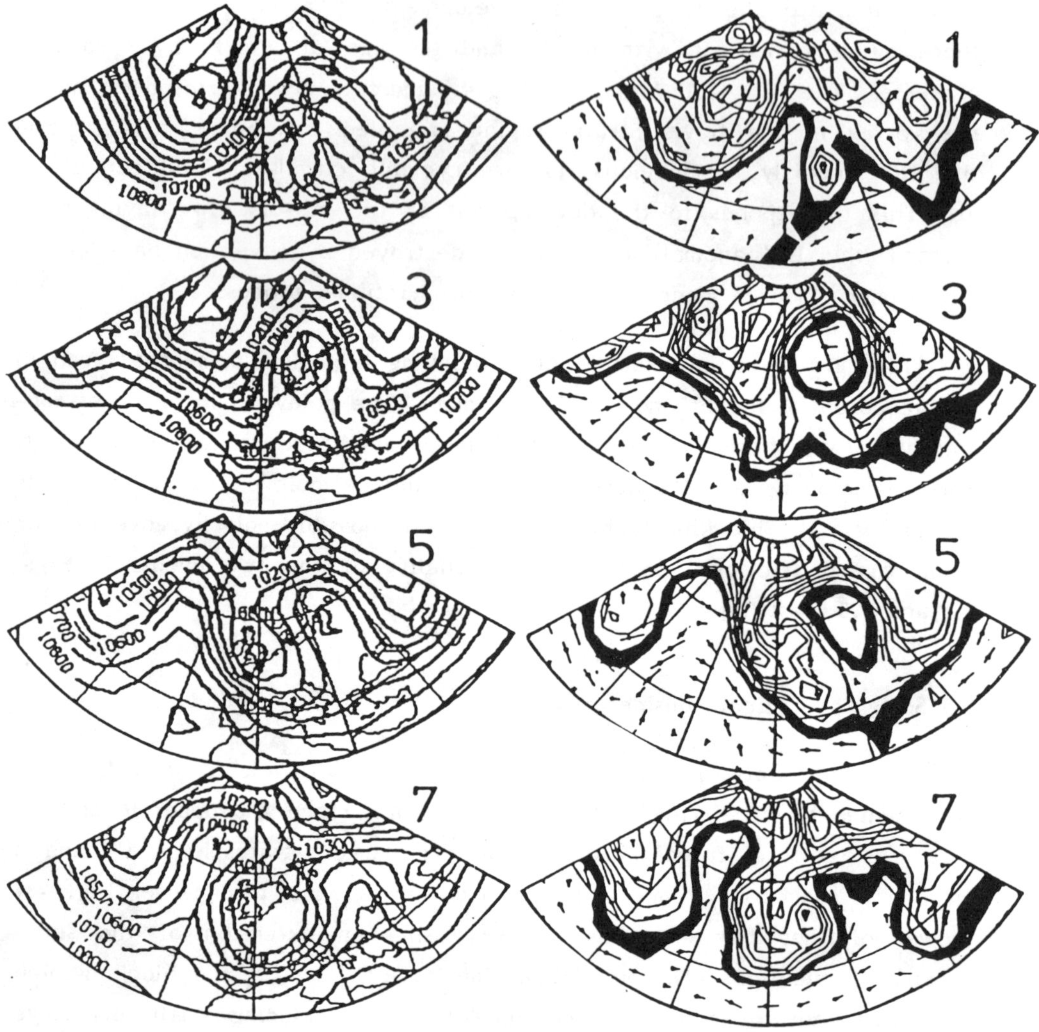

Fig. 4.3 : 250 mb geopotential heigth (left) and 330 K IPV and wind vectors (right) for a sector from 60°W - 60°E and 30°N - 80°N on 1, 3, 5 and 7 October 1982. The contour intervals are 10 dam and 1 unit, respectively. The region 1 - 2 units is blackened.

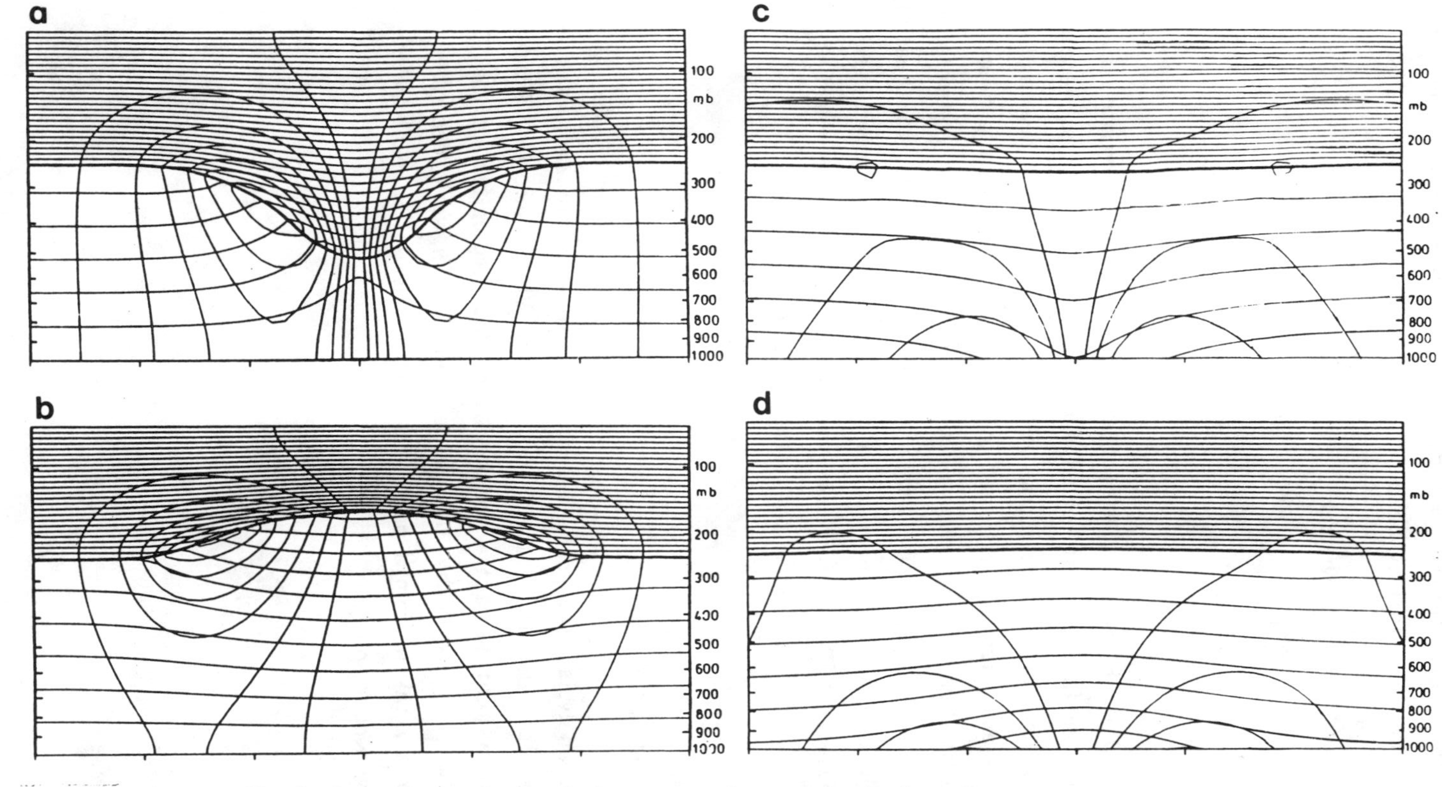

Fig. 4.4 : Vertical sections of circularly symmetric model solutions for
(a) cyclonic upper tropospheric IPV anomaly.
(b) anticyclonic upper tropospheric IPV anomaly
(c) warm surface anomaly
(d) cold surface anomaly.
Shown are isentropes every 5 K, transverse velocity every $3 ms^{-1}$ and the tropopause (thick line). The region shown is $r < 2500$ km. More details of very similar calculations may be found in Hoskins et at. (1985) and Thorpe (1985).

how Rossby wave propagation normal mode baroclinic instability and upper tropospheric induced growth may be described in terms of these solutions.

Since $\nabla\cdot\zeta = 0$, (4.2) may be written

$$\rho P = \nabla\cdot(\underset{\sim}{\zeta}\theta) \tag{4.4}$$

Thus the integral of P over a mass of atmosphere may be written as an integral over the surface of that region. Consequently such a mass weighted PV integral is constant even in the presence of diabatic heating and frictional processes which are internal to the region. Considering boundary temperature anomalies as PV anomalies, it is only boundary frictional processes that can change the mass-weighted PV. For example, the heating associated with the moist convection in the cut-off low case will reduce the upper tropospheric PV and increase the PV near the surface, enabling an efficient spin-down of the system by surface friction.

ACKNOWLEDGEMENT

The data analysis leading to the figures presented in section 1 has been performed by Dr. Ian James, Dr. Glenn White and Dr. Prashant Sardeshmukh in a joint project with the U. K. Meteorological Office. The figures have been prepared by Dr. Sardeshmukh. The work described in section 4 is mostly taken from a paper co-authored by Dr. Michael McIntyre and Dr. Andrew Robertson.

REFERENCES

Bretherton, F.P. (1966): Critical layer instability in baroclinic flows. Quart. J. Roy. Met. Soc., 92, 325-334.

Charney, J.G. (1947): The dynamics of long waves in a baroclinic westerly current. J. Met., 4, 135-163.

Charney, J.G., and Eliassen, A. (1949): A numerical method for predicting the perturbation of the middle-latitude westerlies. Tellus 1 (2), 38-54

Charney, J.G. and Stern, M.E., 1962: On the stability of internal baroclinic jets in a rotating atmosphere. J. Atmos. Sci., 19 159-172.

Eady, E.T. (1949): Long waves and cyclone waves. Tellus 1, (3), 33-52.

Ertel, H. (1942): Ein neuer hydrodynamischer Wirbelsatz. Meteorol. Z., 59, 271-281.

Frederiksen, J.S. (1983): Disturbances and eddy fluxes in Northern Hemisphere flows: instability of three dimensional January and July flows. J. Atmos. Sci., 40, 836-855.

Gill, A.E., and Rasmusson, E.M. (1983): The 1982-83 climate anomaly in the equatorial Pacific. Nature, 306, 229-234.

Grose, W.L. and Hoskins, B.J. (1979): On the influence of orography on large-scale atmospheric flow. J.Atmos.Sci., 36, 223-234.

Held, I.M. (1983): Stationary and quasi-stationary eddies in the extratropical Troposphere: theory, In 'Large-scale dynamical processes in the atmosphere' (B.J. Hoskins and R.P. Pearce, eds.). Academic Press, 397 pp.

Held, I.M. and Hou, A.Y. (1980): Nonlinear axially symmetric circulations in a nearly invisid atmosphere. J. Atmos. Sci., 37, 515-533.

Holopainen, E.O. (1978): On the dynamic forcing of the long-term mean flow by the large-scale Reynolds stresses in the atmosphere. J. Atmos. Sci., 35, 1596-1604.

Holopainen, E.O. (1984): Statistical local effect of synoptic-scale transient eddies on the time-mean flow in the northern extratropics in winter. J. Atmos. Sci., 41, 2505-2515.

Hoskins, B.J., and Karoly, D.J. (1981): The steady linear response of a spherical atmosphere to thermal and orographic forcing. J. Atmos. Sci., 38, 1179-1196.

Hoskins, B.J, James, I.N. and White, G.H. (1983): The shape, propagation and mean flow interaction of large-scale weather systems. J. Atmos. Sci., 40, 1595-1612.

Hoskins, B.J., McIntyre, M.E. and Robertson, A.W. (1985): On the use and significance of isentropic potential vorticity maps. Quart. J. Roy. Met. Soc., 111, 877-946.

Lau, N.-C., and Holopainen, E.O (1984): Transient eddy forcing of the time-mean flow as identified by geopotential tendencies. J. Atmos. Sci., 41, 313-328.

Rossby, C.G. (1939): Relation between variations in the intensity of the zonal circulation of the atmosphere and the displacement of the semi-permanent centers of action. J. Mar. Res., 2, (1), 38-55.

Rossby, C.G. (1940): Planetary flows in the atmosphere. Quart. J. Roy. Met. Soc., 66, Suppl., 68-87.

Sardeshmukh, P.D. and Hoskins, B.J. (1985): Vorticity balances in the tropics during the 1982-83 El Nino-Southern Oscillation event. Quart. J. Roy. Met. Soc., 111, 261-278.

Schneider, E.K. (1977): Axially symmetric steady-state models of the basic state for instability and climate studies. Part II: nonlinear calculations. J. Atmos. Sci., 34, 280-297.

Shaw, Sir Napier (1930): Manual of Meteorology, Vol III: The physical processes of weather. Cambridge. 445 pp.

Simmons, A.J. (1982): The forcing of stationary wave motion by tropical diabatic heating. Quart J. Roy. Met. Sci., 108, 503-534.

Simmons, A.J. and Hoskins, B.J. (1977): Baroclinic instability on the sphere solutions with a move realistic tropopause. J. Atmos. Sci., 34, 581-588.

Simmons, A.J. and Hoskins, B.J. (1980): Barotropic influences on the growth and decay of nonlinear baroclinic waves. J. Atmos. Sci., 37, 1679-1684.

Simmons, A.J. (1982): The forcing of stationary wave motion by tropical diabatic heating. Quart J. Roy. Met. Sci., 108, 503-534.

Simmons, A.J., Wallace, J.M., and Branstator, G.W. (1983): Barotropic wave propagation and instability, and atmospheric teleconnection patterns. J. Atmos. Sci., 40, 1363-1392.

Thorpe, A.J. (1985): Diagnosis of balanced vortex structures using potential vorticity. J. Atmos. Sci., 42, 397-406.

LECTURES ON OCEAN CIRCULATION DYNAMICS

PETER B. RHINES

School of Oceanography and
Department of Atmospheric Sciences,
University of Washington

INTRODUCTION

The oceans are massive moving reservoirs of heat, moisture and energy underlying a relatively diffuse atmosphere, whose entire extent contains two or three cm of water and perhaps the heat found in 5 m of ocean surface water. The pattern of energy flux involves a multiple communication between air and sea: of the net incoming solar at the top of the atmosphere some 73% is absorbed by the sea. The partition of that heating between internal and gravitational potential energy is depth dependent, but vastly favors internal energy (perhaps by a factor of 10^3). But in a perfect gas the ratio of potential to internal energy produced by heating is far different ($\sim (1-\gamma)$, where γ is the ratio of specific heats), such that fully 40% of the heat flowing back into the atmosphere is mechanically active. By gravitational/rotational adjustment this potential energy generates kinetic energy of the symmetric circulation, which breaks down further in baroclinic instability. The zonal winds, thus intensified, set up the oceanic gyres whose potential energy exceeds kinetic by a factor $(L/L_\rho)^2 \sim 10^3$, where L is the gyre-scale and L_ρ the gravest Rossby radius of deformation. This not to belittle the direct production of potential energy in the oceans by buoyancy flux from the atmosphere. The air-sea transfer is comparable with the solar gain, and guided by the winds, it drives significant branches of circulation. The great meridional cells of buoyancy-driven circulation stand out in any measurement that weights

J. Willebrand and D. L. T. Anderson (eds.), Large-Scale Transport Processes in Oceans and Atmosphere, 105–161.

heavily long transit or deep penetration: chemical and biological tracers reveal the slow but persistent long paths of the deep waters.

What do dynamical studies contribute to this qualitative story? It almost seems as if the chemists have done better at tracking the global ocean circulation than have physical oceanographers. The classical dynamics of Sverdrup circulation, western boundary currents and thermohaline circulation provide a useful but rather schematic dynamical framework, which all of us want to improve upon.

The ideas that we shall describe are among those that add to the rather empirical studies dominating the present understanding of the general circulation. Among these ideas are: (1), the vorticity principles governing large-scale circulations, (2) eddy transport of potential vorticity and its consequences, (3) relations between Lagrangian particle diffusivity and dynamics and, (4), discussion of recent models of wind-forced circulation. Sections (1) through (3) review previously published work, while (4) is new.

1. VORTEX STRETCHING AND POTENTIAL VORTICITY

The vorticity principles governing large-scale geophysical fluid dynamics bear interesting relations with those of classical fluid dynamics. There are strong antecedants in G.I. Taylor's work. Geophysical flows are characterised by the strong and combined effects of buoyancy and planetary rotation. In presence of these non-classical elements we have to abandon the idea of conservation of the strength of vortex tubes, in favor of a conservation principle for absolute circulation on isentropic surfaces. Dominating the fluid environment are the large-scale stable stratification of density and potential vorticity, yielding wave modes that dominate the transient behaviour of the fluids, and to some extent limit the extent and effects of turbulence.

Turbulence theories may come and go, but there is little argument that the process of vortex-line extension is at the heart of many interesting flows of a classical fluid. Vortex lines and tubes, which lie along the direction of the local vorticity vector,are 'tracers' of the fluid motion. If the vorticity is initially marked by a chain of infinitesmal arrows made with dye, the arrows continue to indicate both the direction and the strength of the vorticity by their orientation and length. This gives the dynamical problem a visual and intuitive tool, so long as dissipation may be ignored.

Flows whose Reynolds number is initially large cannot dissipate their energy through direct action of viscosity, for the total dissipation,

$$\nu \iint |\boldsymbol{\omega}|^2 \, dxdy$$

is initially very small ($\boldsymbol{\omega} = \nabla \times \mathbf{u}$ is the vorticity). But dissipation is enhanced by the production of mean-square vorticity through the stretching process. In this manner flows of very large scale may quickly be damped, although any irrotational components of flow rely on communication with viscous boundary layers to dissipate.

G.I. Taylor had in 1917 already recognized that two-dimensional flows of an ideal fluid are denied this mechanism of dissipation of energy. With no velocity variation in the direction of the vorticity, vortex lines are not stretched. When a mean rotation, $\mathbf{\Omega}$, is added to the system and dominates the absolute vorticity, much work is required to stretch absolute vortex lines, and their propensity to stretch like arbitrary marked fluid lines is no longer great. The atmosphere and ocean are 'thin' sheets of fluid, and it was long suspected that many ideas of two-dimensional flows should be valid there. In fact the vertical variations in velocity through the oceans or atmosphere are comparable with the vertical average, making them genuinely three-dimensional fluids. But the vertical stratification of density so inhibits vertical motion that the vertical vortex stretching is everely limited. It was not until nearly geostrophic flows were examined by Rossby, Oboukhov and successors that the combined effects of planetary rotation and stratification of the fluid density on vortex stretching have become clear.

Finally, Charney (1971) articulated these ideas by showing that the Fjortoft's (1953) and Batchelor's (1969) constraints on energy cascades in two dimensions apply in modified form to a density stratified fluid which is nearly geostrophic, but far from two-dimensional. In addition to Coriolis effects, the primary new property is the creation of vorticity by buoyancy-twisting forces. The curl of the momentum equation now has a production term

$$(\nabla p \mathrm{x} \nabla \rho)/\rho^2,$$

p being the pressure and ρ the density. One might, for example, view the velocity within internal waves in a density stratified fluid as involving the production of horizontal vorticity by horizontal density gradients of the disturbed fluid. This term has a disasterous effect on the classical intuition about vortex stretching. Only in a special direction, taken normal to the (undulating) surfaces of constant entropy or potential densitiy, is the vorticity free of this effect. We center attention on this special component of $\boldsymbol{\omega}$, and arrive at a conservation principle for the absolute circulation in such a plane. By carrying out Kelvin's circulation theorem for a small circuit in surface of constant potential density, the buoyancy twisting term is absent and there follows the potential vorticity conservation equation of Ertel (1942),

$$DQ/Dt = F - D \tag{1}$$

where $$Q = (2\boldsymbol{\Omega} + \boldsymbol{\omega})\cdot\nabla\tilde{\rho}/\tilde{\rho},$$

F is external forcing, and D is dissipation. This hinges on the existence of a property like $\tilde{\rho}$, obeying

$$D\tilde{\rho}/Dt = 0,$$

and which together with the pressure determines the density through an quation of state. Here we take $\tilde{\rho}$ to be the density which the fluid would attain if moved isentropically to a reference pressure. Variations in $\nabla\tilde{\rho}$ then correspond to variations in the area of elemental disc of fluid lying parallel with the constant-$\tilde{\rho}$ surface, making Q identical to the absolute circulation about the rim of such a disc. The principle fails, at least in detail, for fluids with more complex equations of state, for example where density is determined by two independent constituants like temperature and salinity.

With the actual slope of the gravitationally free horizons being far less than unity at large scale, ∇p and $\nabla\rho$ are each nearly vertical, and hence the buoyancy twisting produces almost exclusively horizontal vorticity. The equation of motion (1) then becomes a statement involving the **vertical** vorticity, $\boldsymbol{\omega}\cdot\hat{z}$, where $\hat{z}$ is a vertical unit vector:

$$Q \simeq (f + \boldsymbol{\omega}\cdot\hat{z})\partial \ln\tilde{\rho}/\partial z$$

where f is now the local Coriolis frequency, $2\Omega\sin\lambda$, at latitude λ.

Given that Q is a scalar field relating to one component of the vorticity $\boldsymbol{\omega}$ it is somewhat surprising how much information it contains.† A Poisson-type inversion leads from knowledge of Q to the velocity and density fields themselves, provided that the amount of mass between successive constant-$\tilde{\rho}$ horizons and certain boundary conditions are known. (For applications see the review of Hoskins **et al.** (1986)) In this way Q merits the term 'fundamental field variable' for large-scale geophysical circulations. Of course Q relates only to the rotational component of the velocity field, the irrotational part being hidden in the distribution of ρ along the boundaries. Not a minor point, since important motions like Eady baroclinic instability waves, Kelvin waves, and fast baroclinic waves trapped near topography have essentially zero potential vorticity. In Bretherton's (1966) view, these motions are best viewed as being rotational, using a potential vorticity augmented by delta-function sheets along the boundaries.

Mean Structure. In the general circulation of the oceans, the potential vorticity field evolves from occasional encounters of the fluid with the sea surface, where convective and mechanical forcing occur. Subsequently the fluid submerges to depths where dissipative

---

† When cast in terms of a stream-function, ψ (which is nearly proportional to pressure) the definition of Q, eqn. (1), is an elliptic relation between ψ and Q; for the restricted case of quasi-geostrophic theory,

$$Q = \nabla^2\psi + ((f^2/N^2)\psi_z)_z + \beta y$$

where ∇^2 is the horizontal Laplacian operator, f is the Coriolis frequency, $2\Omega\sin\lambda$, λ is latitude, $N(z) = -(g/\rho)(\partial\tilde{\rho}/\partial z)$ is the frequency of buoyancy oscillations, and β is the rate of change of f with latitude. The first term is the relative vertical vorticity, the second the term proportional to thickness between constant-ρ horizons and the third is the planetary vorticity.

processes and external forcing are both weak. We believe that while the fluid circulates in deep isolation (it may take years or millenia before coming up 'for air' once again) its potential vorticity is only slowly changing. At great depth, there are signs that vertical diffusivity rises (inverse to N, according to Gargett's (1984) extrapolations). There, significant diffusion of Q occurs as with other dynamical 'tracers' like salinity. For the top half of the ocean, however, near-conservation of Q following the flow, with integral constraints and boundary conditions applied along the top, bottom and sides of the oceans, is a plausible model. Suffice it to say that the problem has mixed hyperbolic-elliptic nature, and that the natural boundary conditions on stress and buoyancy flux do not readily translate into simple boundary conditions on Q. The circulation has both wind-driven and convective origins and hence, in addition to the dynamics contained in the governing equation (1), we have thermodynamic equations describing the evolution of ρ itself, through the constituants temperature and salinity.

Without entering into the detailed dynamics, we can nevertheless see that the 'planetary' fluid is in one sense remarkably simple: the potential vorticity of the large-scale circulation (L >> 50 km.) is dominated by the 'layer thickness' contribution, so that

$$Q \simeq 2\mathbf{\Omega}\cdot\nabla\tilde{\rho}/\rho \simeq 2\Omega\tilde{\rho}_z/\tilde{\rho}. \qquad (2)$$

This is a statement that the thickness of a layer of fluid between two adjacent constant-$\tilde{\rho}$ surfaces, **as measured parallel with the rotation vector $\mathbf{\Omega}$**, is constant following the motion. It is a generalization of the idea of 'stiffness' imparted to the fluid by planetary rotation, and as such is a fitting derivative of the idea of a Taylor column. Operationally speaking it is fortunate that Q is less differentiated in horizontal directions than is the case with $\boldsymbol{\omega}$ itself; as anyone interested in turbulence can appreciate, maps of vorticity have fine detail that would make the task of measurement at sea indeed hopeless. Instead, we have found Q to be readily 'mappable' and to be a direct reflection of the largest flow structures in the planetary domain. The relative vorticity, neglected in this approximation, is very significant at the scale of the energy containing eddies, but there is a great separation of scale between these and the major circulation gyres. Intense boundary currents are an integral part of the circulation, and these indeed have the interplay of $\boldsymbol{\omega}$ and $2\boldsymbol{\Omega}\cdot\nabla\tilde{\rho}/\tilde{\rho}$ that is implicit in baroclinic instability and geostrophic turbulence.

A new, 'non-classical' part of the problem is contributed by the large-scale gradient of Q when the fluid is at rest. This strong component of the spherical rotating environment provides a restoring effect for Rossby waves, which are essential to the process of establishing the mean circulation. This environmental component is known as the β-effect, for a locally Cartesian approximation of the problem suggests a Coriolis frequency varying like $f = f_0 + \beta y$. Q also has flow-related components, the relative vorticity $\boldsymbol{\omega}$ which we are temporarily neglecting, and the 'thickness' contribution. Coriolis forces on the mean velocity must be balanced by tilted $\tilde{\rho}$-surfaces which lead in turn to lateral variations in $\tilde{\rho}_z$.

An interesting property of Q is that it is modal; after separating linearized flows into different vertical structures we find that each such mode sees a different weighted integral of Q as the mean restoring field. In particular the depth-independent component of the velocity field sees an averaged mean Q-field,

$$Q \simeq f/H$$

where H is the depth of the ocean, measured along the local vertical. The gradient of ocean depth combines with the Earth's sphericity to define the environment, and following the above reasoning, we see that unforced, steady flows in such a mode proceed so as to preserve the projection of the local fluid depth upon the rotation axis. The set-up of such flows is determined by pseudo-westward influence propagation in low-frequency modes (i.e., along these contours with Q increasing to the right). Modes whose energy is concentrated near the surface see β as the dominant potential-vorticity gradient, and their influence propagates due westward. Conversely those concentrated near the bottom see H itself as dominating Q, and develop accordingly.

A fluid at rest has east-west contours of constant Q, and owing to blockading boundaries, these are not acceptable flow paths for a circulation in which Q is conserved following the motion. To permit flow north or south the velocity must be great enough to tilt the stable ρ-surfaces enough at least to cancel the prevailing β-effect. The great gyres of wind-driven circulation manage this by concentrating their flow in the top km. or so of the typically 5 km. deep oceans. The curves of constant Q in figure 4 below are suggestive of the flow itself, and are direct evidence of the control of the dynamics by these processes.

2. SIGNIFICANCE OF THE TRANSPORT OF POTENTIAL VORTICITY BY EDDIES

In the study of water waves one frequently encounters ideas of wave drag and momentum flux. In view of the practical issues to do with ship wakes, forces on offshore structures, and beach erosion it is natural to attend to the forces that accompany wave generation. In classical turbulence studies we find equally important questions relating to transport of momentum, yet there is less emphasis on the role of waves in carrying out the transport. In geophysical fluid dynamics, by contrast, the gradient of the large-scale potential vorticity provides wave motions that are essential to the development of the general circulation.

Fluctutuating motions affect the mean circulation in many ways. Beyond the 3-dimensional turbulence active in horizontal boundary layers, which lead to spin-up secondary flows that energize the interior, we have at larger scale geostrophic turbulence, dominating kinetic energy of the fluids.

When the question, "how do eddies and waves affect the general circulation?" was first being asked, the lesson of classical turbulent shear flow was taken, and Reynolds' stresses identified in observational data (e.g., Schmitz **et al**. 1983). A heavy density stratification will inhibit vertical velocity, and this suggests that the prime role of Reynolds stresses will be to transport momentum along mean surfaces of potential density, ρ. Yet the vertical transport of momentum must be very significant, for the ocean is driven at or near its top, and yet the momentum appears internally. Correspondingly, meteorologists have long known that a meridional circulation (that is, an Eulerian mean circulation in the y-z plane) can lead to zonal acceleration of the mean flow through the Coriolis force, $2\mathbf{\Omega}\times\mathbf{u}$. But the essential relation of that secondary flow to wave processes was unclear until recently.

The quasi-geostrophic momentum balance for a layer of fluid between two undulating constant-ρ surfaces is

$$\partial h\mathbf{u}/\partial t = -\hat{z} \times hq\mathbf{u} - H\nabla(p + 1/2\, |\mathbf{u}|^2)$$

where $q = H(f+\zeta\cdot\hat{z})/h$ is the form of Q appropriate to a stratification consisting of a parfait of layers of constant density, whose thickness is $h(\mathbf{x},t)$. Here the quasi-geostrophic approximation which singles out horizontal velocites, $\mathbf{u} = (u,v)$ as being larger than the vertical velocity w by a factor typically $(R_0D/L)^{-1}$, where D,L and U are estimates of the vertical scale, horizontal scale and horizontal velocity and $R_0=U/fL$ is the Rossby number. Now take an idealized flow confined to a zonally (i.e., parallel to the x-axis) oriented channel. Let brackets indicate an x-average, and primes the deviation from this average, and let $H = \langle h \rangle$.

The transport of potential vorticity, hq**u**, is a rewritten form of the combined Coriolis and 'vortex-' forces, and as such it represents the sole rotational force acting to change the momentum h**u**. For a flow that is homogeneous in the x-direction, an average of the above momentum equation with respect to x gives

$$\begin{aligned} \partial\langle hu\rangle/\partial t &= \langle hqv\rangle \\ &= f\langle hv\rangle + H\langle q'v'\rangle \\ &= f\langle hv\rangle + H\langle \omega'v'\rangle - f\langle h'v'\rangle \\ &= f\langle hv\rangle - H\langle u'v'\rangle_y + \langle p'h'_x\rangle/\rho \qquad (3) \end{aligned}$$

The Coriolis force on the net meridional mass flux between the two moving constant-ρ surfaces is the first term on the right-hand side. In stationary diabatic motion of fluid confined in a channel (whose vertical walls run east-west) this term will vanish. The remaining terms are the direct eddy contributions to the momentum transport, first expressed as turbulent fluxes of the relative vorticity and 'thickness' component of the potential vorticity, respectively, and in the final line expressed as the divergence of the north-south component of Reynolds stress and the inviscid pressure drag, $p\partial h/\partial x$, exerted from layers above and below. In this rotation-dominated approximation the vertical transport of momentum by Reynolds stress is negligible. We have adopted a mixed averaging, Eulerian in the horizontal (i.e., control volume fixed in space) and Lagrangian in the vertical (where we are accounting for the momentum between moving material surfaces). This gives a clear physical sense for the vertical transport of momentum, which in the purely Eulerian analog to eqn (3) appears lumped with the first of the three terms as just $f\langle v\rangle$.

The lateral Reynolds' stress, the first of these two terms, is familiar. The second term is the contribution to the x-component of the inviscid pressure force, $\int p n dS$, which communicates x-momentum vertically. At this point it is important to think of the fluid confined between moving surfaces of constant ρ, rather than that between fixed values of z. Vertical motion of these ρ-surfaces is relatively small, but the pressures exerted in a rotating fluid can be very great.

The relation between the potential vorticity flux, the momentum flux, and the flux of buoyancy was discussed by Green (1970) in a meteorological context, and by Rhines and Holland (1979) in the context of theoretical models and ocean circulations. In zonally averaged atmospheric models one defines the Eliassen-Palm flux (Andrews and McIntyre, 1976) whose divergence is the meridional eddy flux of potential vorticity. In small-amplitude waves in slowly varying mean flows the Elissen-Palm vector indicates the group velocity, and its z- and y- components represent respectively the vertical- and lateral momentum flux. This general feature of the Eliassen-Palm flux emphasizes the interaction of the mean potential voriticity profile with the forces that create it: the attitude of ray paths in the meridional plane is linked to the sense of the momentum flux that they induce.

In atmospheric modelling the implications of wave focussing to mean flow development are being examined in detail (e.g., Hoskins, 1983). In oceanic studies we are lacking detailed observations of either the energy containing 100-km scale eddies or the waxing, waning and spatial evolution of the larger scale circulation. Eddy-resolving numerical models thus play a crucial role in our field, although detailed diagnosis of eddy-mean dynamics has not yet been very extensive. Hogg (1983) has been able to identify the form drag term which leads to downward momentum flux, in observations of the intense recirculation region of the North Atlantic subtropical gyre. In just such a region Holland and Rhines (1980) found the equivalent strong vertical transport in a numerical circulation model. Further analyses of potential vorticity flux are being made, but the prospects are not bright for the kind of data sets that will allow extensive mapping of eddy-mean flow interaction from data. The best resolved experiment to date was the Local Dynamics Experiment of POLYMODE; there Brown, Owens and Bryden (1986) found the eddy potential vorticity flux to be strong and downgradient in sense in the Gulf Stream recirculation. The flux was dominated (by an order of magnitude) by the form-drag component, in agreement with the theoretical predictions. The connection between this 'thickness' flux and lateral eddy density flux provides a connection between heat flux studies (e.g., Bryden, 1983) and dynamics.

3. RELATION WITH THE DISPERSAL AND ORBITAL MOTION OF FLUID PARTICLES

Given the impossibility of making global surveys of the important eddy-mean flow interactions, we must seek in theoretical models aspects of the interaction that may be more visible.

Although the inviscid pressure force in (3) is in some way equivalent to the Coriolis force exerted upon the mean north-south circulation (i.e., the 'Coriolis torque' familar to meteorologists), it is not yet clear what phenomena are responsible for it. To clarify, we take another leaf from G.I. Taylor's book (Taylor, 1921), adopting a mixing length approximation for potential vorticity,

$$q' = (\xi_j - \xi^o{}_j)\partial Q/\partial x_j$$

where ξ_j is the horizontal displacement of a fluid parcel in the jth direction, and $\xi^o{}_j$ its 'rest' position, at which q' would vanish. Because of the quasi-conservative nature of q, we can make far more exact models of the flux than is possible for a non-conservative property like momentum. At geostrophic scales, it is the enstrophy cascade that provides the dissipation of fluctuations, q', and the rate of this process will in many circumstances limit the intensity of q'. A Rayleigh-damping model for the fluctuating potential vorticity field was described by Rhines (1977). In this approximation the potential vorticity flux becomes just

$$\langle q'u'_i\rangle = -\kappa_{ij}\partial Q/\partial \chi_j, \tag{4}$$

essentially a stirring of the large-scale profile by the Lagrangian single-particle diffusivity,

$$\kappa_{ij} = \langle u_i \xi_j\rangle.$$

The forcing of the mean zonal momentum (the righthand-side of eqn. (3)), in this approximation becomes

$$f\langle hv\rangle - H\kappa_{yy}\ \partial Q/\partial y.$$

This relation provides a great intuitive aid to a set of problems so diverse as to contain the instability of parallel flows, the transport of momentum by eddy motions, the difference between the Eulerian- and Lagrangian mean circulation and the distortion of marked lines and surfaces in a fluid. In present form this relation is

$$\kappa_{ij} = \int R_{ji}(\tau)d\tau$$

where $R_{ji}(\tau) = \langle u_j(t)u_i(t+\tau)\rangle$

is the Lagrangian correlation function of the velocity of a single fluid particle with its velocity at a time τ later. Stationarity and homogeneity of the velocity fields are required to give this simple form of the relations, independent of reference origins in space or time.

From simple integration it follows that

$$d\langle \xi_i \xi_j \rangle/dt = \kappa_{ij} + \kappa_{ji}$$

The symmetric part of κ_{ij} relates to the increase with time of the principle second moments of the particle ensemble, and it is this dispersion that carries with it such strong dynamical consequences. Taylor's initial step related the subtleties of dynamic instability and mean-flow induction to this very tangible and visible aspect of fluid motions. In its original form, it recast the Raleigh necessary condition for instability in the form

$$\iiint \kappa \; \partial Q/\partial y \; dxdydz = 0 \qquad (5)$$

where $\kappa \equiv \kappa_{22}$ is the yy component. By redefining instability to be the growth (everywhere) of the second moment of particle displacements in the y-direction (i.e., $\kappa > 0$), we recover Rayleigh's condition that $\partial Q/\partial y$ should take both positive and negative signs within the fluid. But the condition now represents the conservation of overall x-momentum, and we are invited to speculate on instabilities that may arise when conditions are such that this conservation no longer holds. Externally driven flows, flows over hilly topography, flows with viscous boundary layers and flows with momentum flux from adjoining regions of fluid are all examples in which the instability properties may greatly change. Taylor's original example of this was the large Reynolds number flow in a pipe, in which finite momentum flux from the walls can occur even as the viscosity becomes very small, thus allowing instability in a flow without an inflection point. Not the least, this formulation of momentum flux and instability carries with it more generality than the more familiar linearized eigenmode theories. Validity of transport relations like eqn.(4) requires that the displacement of fluid particles be smaller than the length-scale characteristic of the variation of mean quantities (the mean flow, the mean energy of eddies, etc.). Fully developed turbulence can occur under without violating such a separation of scales. Furthermore, processes leading to the fading memory of a fluid parcel for its potential vorticity can be built into more careful formulations of momentum flux. The enstrophy cascade, alluded to above is one such process. This leads us to the connection between the transport of potential vorticity 'across wavenumber space' toward dissipation at large k, and the its transport across physical space and attendant momentum flux.

The leftover antisymmetric part of κ_{ij} has a peculiar nature of its own. Wallace (1978), Plumb (1979), Rhines (1977), Rhines and Holland (1979), Haidvogel and Rhines (1984) discuss relations with the orbital motion of fluid particles, and Lagrangian-mean flow. We are often concerned with wave motions, and the particle motion in waves often is rotary in sense. The antisymmetric part, in two dimensions given by

$$\begin{vmatrix} 0 & -\kappa_{12}+\kappa_{21} \\ \kappa_{12}-\kappa_{21} & 0 \end{vmatrix}$$

is associated with the vector

$$\epsilon_{ijk} u_j \xi_k$$

which is the angular momentum of the ensemble of marked fluid particles, with respect to their origin, $\xi = 0$.

The Lagrangian mean velocity, given an appropriate scale separation between waves and mean quantities, is

$$\mathbf{u}^{\ell} = \mathbf{u}^{e} + \partial\kappa_{ij}/\partial x_j$$

which has contributions both from the variation of dispersion in space (the symmetric part) and an extra contribution from the variation in wave orbital amplitude with position (for example, the Stokes drift of surface gravity waves is $\partial\kappa_{xz}/\partial z$). In addition the strong contribution to tracer- or potential vorticity transport follows, with the transport directed **along** isopleths of the mean concentration: for a tracer Θ

$$F_i = -\kappa_{ij}\partial\langle\Theta\rangle\partial x_j$$

What to an Eulerian observer might be an eddy flux of Θ along mean isopleths, is to a Lagrangian observer the result of orbital drift of fluid particles **across** isopleths of the mean concentration. To see this, suppose there is a uniform mean gradient of Θ and steady wave-orbital motions of an incompressible two-dimensional fluid in x and y. Then the symmetric part of $\boldsymbol{\kappa}$ vanishes and the local accumulation of Θ becomes

$$\begin{aligned}\partial\langle\Theta\rangle/\partial t + u^e_j\partial\Theta/\partial x_j &= -\nabla\cdot\mathbf{F}\\ &= \partial/\partial x_i(\kappa_{ij}\partial\langle\Theta\rangle/\partial xj)\\ &= \partial\kappa_{ij}/\partial x_i(\partial\langle\Theta\rangle/\partial x_j)\\ &= -\partial\kappa_{ji}/\partial x_i(\partial\langle\Theta\rangle/\partial x_j)\\ &= -(u^\ell_j - u^e_j)\partial\langle\Theta\rangle/\partial x_j.\end{aligned}$$

or

$$\partial\langle\Theta\rangle/\partial t = -u^\ell_j\partial\langle\Theta\rangle/\partial x_j.$$

Thus the Eulerian observer attributes a local change in Θ to eddy flux $\langle\Theta'u'_i\rangle$ along contours of mean Θ, which seems rather fictitious in view of the conservation of total Θ following fluid particles. But the Lagrangian observer points out that nondivergent velocity forces a mean particle drift, in the presence of gradients of wave amplitude, and that the drift across mean contours of θ leads to local rate of change.

It should be apparent that we are guilty of building a topsy-turvey structure of Eulerian and Lagrangian quantities. Just above we have discussed the way in which the dispersal of marked fluid particles (i.e., a Lagrangian event) leads to an average flux of momentum or tracer past a fixed point (i.e., an Eulerian average). We have tacitly assumed that the 'arrival' diffusivity of many particles observed passing by the fixed (Eulerian) observer is the same as the 'departure' diffusivity of particles released from a single point; the relation between these complementary statistics is indeed worthy of study.

The true Lagrangian velocity is

$$<\mathbf{u}^{\ell}(t;\mathbf{x}_0,t_0)>$$

where $\mathbf{u}^{\ell}$ is the velocity of a marked fluid particle at time t, given release time and position t_0 and $\mathbf{x}_0$, respectively. In realistic systems with strong dispersal of fluid particles across the domain, there is no simple relation between the Eulerian and Lagrangian means; indeed the time-dependence of $<\mathbf{u}^{\ell}>$ is extreme; for short time it is equal to the Eulerian mean velocity, at larger time it departs, and in a bounded domain it drifts toward the centroid of the region of mixing of fluid particles (often the region coinciding with a closed gyre of circulation), eventually tending toward the centroid of the entire fluid domain in the final well-mixed state.

Example: beta-plane jets. To emphasize the value of these connections between dynamics and eddy stirring, we must recount the classic experiment of Whitehead (1974). A rotating tank with a free surface has an equivalent β-effect, in essence the center of the tank acting as the pole. Whitehead placed a horizontal disk in the fluid, supported by an arm that was connected to an electric motor. The disk was moved up and down periodically, leading to a mixture of Rossby waves and turbulence. Beside these time dependent flows, there emerged a strong mean circulation, directed westward along circles of constant radius in most of the cylinder. At the radii close to the wavemaker, however, an intense eastward jet was observed.

In this context, eqn (5) is dominated by

$$\beta \iiint \kappa \; dxdydz = 0.$$

Instead of using this net momentum integral to establish requirements on Q in order that instability may occur, we conclude that in a stable flow the volume integral of the Lagrangian diffusivity should vanish. If initially some wave energy is present in a narrow band of latitudes, κ must inescapably grow in the neighboring, initially motionless, fluid as the motions propagate or advect energy into it. Conversely, by eqn.(2d), κ must be negative in the source region where eddy energy is dying. The pattern of zonal acceleration, given by $-\beta\int \langle\kappa\rangle dz$, is just as Whitehead found it to be. Because of the unusual role of the mean potential vorticity gradient, it must be described as an unexpected application of Taylor's (1915) relation between vorticity transport, momentum transport, and dispersal of marked fluid particles in a region of turbulence. Unlike the classical problem, ours finds the Lagrangian diffusivity itself heavily constrained by the limited energy and strong restraint of the β-effect.

This experiment expresses the tendency for Rossby-wave radiation to separate mean zonal momentum, in such a way as to sharpen and intensify an eastward directed jet. It is a fundamental property of the circulation of the atmosphere, in which context it was described by Green (1970). Colin de Verdiere (1979) provided more quantitative experimental study.

Shear dispersion. As is so often the case, the seeds of new work come from difficulties with the old. The simplicity of Lagrangian diffusivity is much reduced when there is a significant mean flow. This is so, whether the problem at hand involves momentum flux or kinematic transport of passive tracers. Parallel shear-flow can be accommodated by fixing attention on dispersal of fluid particle across the mean streamlines. But in real situations one quickly finds that 'removal of the mean velocity' from the calculation of κ_{ij} is ambiguous and unsatisfying. Which mean velocity should we remove? This seeming difficulty of description leads to study of the genuine and inseparable interaction between mean shear and the random-walk nature of turbulence, which combine to give us the process known as shear dispersion. Unfortunately our thinking about Lagrangian ocean circulation has for many years been dominated by simple random walk diffusion calculations. In section 5. we will describe likely sites of enhanced mixing and air-sea exchange in wind-driven circulations, based on ideas of the enhancement of property gradients by shear. Many of the basic ideas of shear dispersion are in the literature. Illustrations of the effect in the context of oceanic flows are given by Musgrave (1985), Young **et al.** (1983), Vasholz and Crawford (1984).

The effect is particularly relevant to passive chemical tracers in the ocean, as well as the dynamically active tracers: potential vorticity, temperature and salinity. Air-sea transfer of a soluble gas, for example, depends on the air-sea difference in partial pressures. With typical piston velocities and wind speeds, the mixed layer comes to saturation in a month or so. Further gas influx requires renewal of the mixed layer either by mixing downward or advection from the sides. In the latter case, the influx becomes sensitive to advection rate: either upwelling of low-concentration waters or lateral flow from warmer latitudes where the solubility is lower. Box-models used by the chemists to describe these tracers are implicitly using shear dispersion, with flow between boxes holding down the surface concentration, and hence supporting the gas influx.

4. WIND-DRIVEN GYRES

In this section we would like to focus on a problem of great current interest, the wind-driven circulation in the upper km. or so of the oceans. Recent analytical models of the general circulation of the oceans have centered on the production and evolution of the central dynamical field variable: the potential vorticity. The models are often difficult to grasp because of their sometimes complex mathematical structure. Nevertheless one can readily sense the flavor of the competition between processes: on the one hand, the injection of potential vorticity, Q, from the upper boundary, as modelled by Luyten, Pedlosky and Stommel, 1983a) and on the other, the redistribution of Q by lateral mesoscale eddy mixing in a recirculating gyre, modelled by Rhines and Young, 1982a,b.

The major oceans have a remarkable contrast in the balance between wind- and buoyancy forcing, the North Atlantic and North Pacific being protypes of strong and weak buoyancy forcing, respectively. This distinction is important to what follows. There is still argument about the relative strengths of the lateral wind gyres and meridional thermohaline gyre in the Atlantic (e.g., Leetmaa, Niiler and Stommel (1977), Wunsch and Roemmich (1985)), while the thermohaline circulation in the upper 2 km of the Pacific is clearly weak in comparison with the wind-gyre.

In view of the proliferation of models, it seems important to give a simple argument that clarifies the relative roles of injection and recirculation, the role of mixing, and has implications to the global nature of the circulation problem. The discussion here uses little mathematics, and applies to essentially all the models and to the ocean itself. The basis of these remarks was given in OCEAN MODELLING, 49, 1983, but we here extend it in several directions.

The Recirculation Index. In mesoscale f-plane flows, the planetary vorticity provided by the Earth's rotation is convertible to relative vorticity, given some vertical vortex-stretching. At planetary-geostrophic scale, β-plane dynamics also yields intense 'swirling' motion in response to imposed vertical velocity, but for different reasons. A homogeneous sheet of fluid, when exposed to vertical velocity at its top, moves north or mouth at a rate just sufficient to avoid stretching of the planetary vortex lines lying parallel to the Earth's rotation vector, $\mathbf{\Omega}$. The rigidity of fluid columns imparted by $\mathbf{\Omega}$ means that far less work is done by the boundary conditions if the fluid moves 'sideways' in its wedge-like domain, rather than allowing itself to be compressed in directions parallel to $\mathbf{\Omega}$. This Sverdrup response,

$$\beta v = f \partial w / \partial z \qquad (6)$$

again implies a great swirling of fluid if the scale of the motion is less than the radius of the Earth ((u,v,w) and (x,y,z) are East, North and upward velocities and coordinates, respectively, f is the Coriolis frequency and β its northward gradient). Stratification, while allowing vertical shear, retains this stiffness property with respect to line segments instantaneously parallel to $\mathbf{\Omega}$. Imagine a flow driven by a source of fluid in the north and a sink somewhere to the south. We might look for solutions of (6) (together with a density equation) which would represent the connecting flow in the ocean interior. It is difficult to look at such a section without mentally picturing a two-dimensional (y-z) flow from source to sink. The typical magnitude of v from (6) is

$$v_{Sverdrup} \sim fw/\beta d$$

where d is the vertical extent of the circulation. Yet, if we sketched a meridional section we might find another estimate for v from a two-dimensional continuity equation equation,

$$\partial v/\partial y + \partial w/\partial z = 0.$$

his estimate, say v_{cont}, is

$$v_{cont} = wL_y/d$$

where L_y is the north-south scale of the gyre and d its depth scale. Now the ratio of these two estimates is

$$v_{Sverdrup}/v_{cont} \sim f/L_y\beta = a/L_y \qquad (7)$$

where $a = f/\beta =$ Earth radius x $\tan\lambda$. For reasons that will become apparent, we introduce the recirculation index, R_c, as

$$R_c = a/L_y.$$

The difference in the two estimates for v implies in general that the third dimension is important, and $\partial u/\partial x$ will be involved in the continuity relation. In order to develop the large meridional velocity required by the Sverdrup constraint, fluid must be drawn in from the 'sides', fig. 1.

This elementary argument has strong implications to planetary flow patterns: (i), **planetary source-sink flow**. Accompanying any circulation with scale significantly less than a will be intense gyres of circulation: often so intense that the net mass flow (which created them) may nearly be invisible. Mass does not simply flow from source to sink. A dramatic illustration is seen in Colin de Verdiere's (1977) laboratory β-plane experiments, some of which involved sliced cylinder source-sink flows. The effective scale of the forcing can be quite small in such a problem, and the result is that intense gyres form at the latitudes of the external forcing, with closure in western boundary currents. The lesser flow of mass from source to sink (down in magnitude by a factor R_c^{-1}) is invisible in the streak photographs, and must be seen with dye.

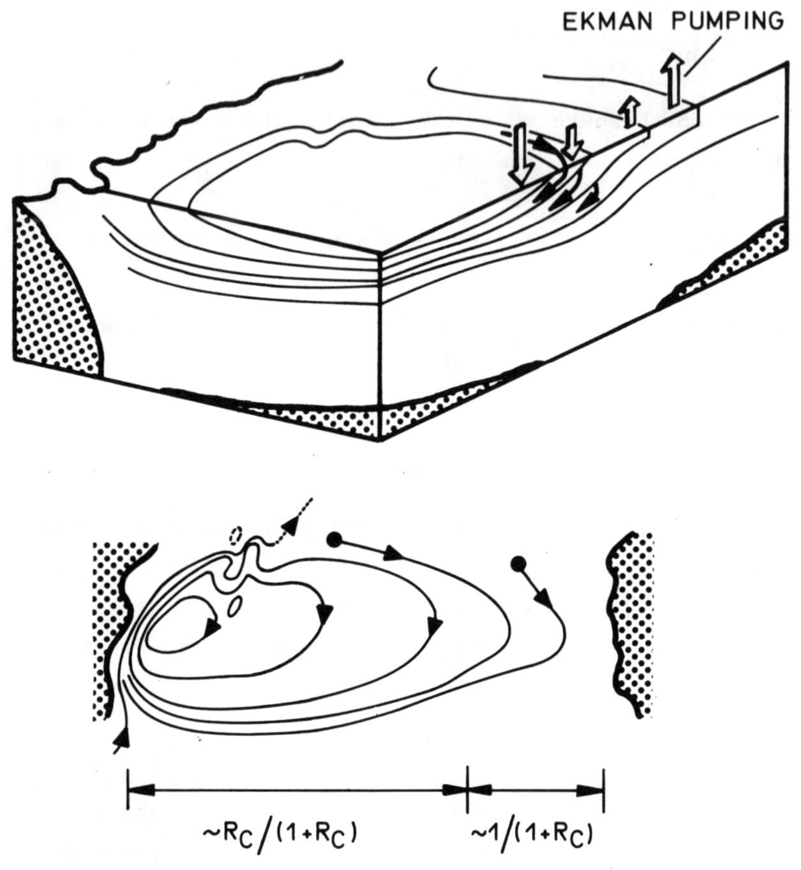

Fig. 1

Schematic diagram of the subtropical wind-gyre. Fluid pumped by wind downward from the mixed layer joins a larger stream from the western boundary, in eastward-flowing parts of the gyre.

In the experiments of Stommel, Arons and Faller (1958) and the supporting theory, it was assumed that small-sized sources and sinks on a β-plane connect with one another via narrow zonal currents and boundary layers. Yet our remarks imply that one should find in that circumstance narrow (in the north-south direction) gyres superimposed on the mass-carrying flow. As the size L of the source and sink is reduced, the recirculation should increase. The limit $R_c \to \infty$ is singular until friction arises to limit the strength of the gyres. Indeed, in the photographs published by Stommel **et al.**(1958) great clouds of dye occurred instead of thin zonal jets, and this seems to be evidence of such recirculation. The other examples discussed by them, involving broadly distributed sinks, show the strong recirculation we describe here; indeed, that was the central message of their work.

The implication to observational oceanography is that the thermohaline circulation drives gyre-like motions which tend to obscure the simple arterial flow of the deep ocean. There is evidence from tracer studies that this is so. Consider, for example, the salinity, nutrients and oxygen concentration in the South Atlantic, at the level of Antarctic Intermediate Water $\sigma_\theta \sim 27.3$, 750m depth). The distributions (Wüst, 1935) are to a first approximation a function of latitude only. One sees the northward flowing intermediate water as a broad, yet relatively weak western boundary intensification of these properties. Intermediate scale horizontal gyres may have redistributed the western boundary current-borne tracers across the basin.

(ii),**The wind-driven circulation.** The central issue here is the dynamics of the wind-driven flow of the upper km. or so of the oceans. The above argument shows that the gyres have much more transport than the inward flow of fluid from the upper mixed layer, which drives them. Let us look in more detail at the constraints felt by the incoming fluid. For fluid governed by the Sverdrup relation (6) we integrate to find the ratio of transport moving south across a latitude y, to the total downward Ekman pumping poleward of y. This gives a y-dependent generalization of the recirculation index:

$$R_c(y) = \frac{\int\int v \, dzdx}{\int\int w_E \, dydx} = \frac{(f/\beta)\int w_E \, dx}{\int\int w_E \, dydx}$$

where the x-integration is over the Sverdrup interior, the z-integration is over the water column, and the y integration carries poleward to the line $y=y_0$ of vanishing Ekman pumping. Of course the estimate applies locally in x, as well, without the x-integrations in either denominator or numerator, but the x-integration assures us that we are accounting for all the down-pumped fluid that might cross the latitude of observation. A scale estimate of R_c is just

$$R_c(y) \sim a/(y-y_0)$$

As an example suppose

$$w_E = \alpha(y-y_o)$$

The net downward transport of fluid between latitude y and latitude y_o is then $(\alpha/2)(y-y_o)^2$. Then

$$R_c(y) = 2a/(y-y_o)$$

where a=(Earth radius) x tanλ, and the x-integration is over the Sverdrup interior. As one approaches the zero wind-curl line the meridional circulation is entirely dominated by recirculating fluid. Even at the gyre center, typically 1500 km or so south of y_o, the appropriate generalization of R(y) is moderately large, say 3 to 5. The model of Pedlosky and Young (1983) is an example that bears out this estimate. The wind-driven theory of Rhines and Young (1982) was based on the largeness of R_c, and it emphasized the persistent shaping of the interior potential vorticity field by lateral eddy-stirring. Any model that obeys Sverdrup dynamics must draw in boundary current fluid in excess of the Ekman pumping according to this ratio.

This index of recirculation intensity is a lower bound, for it neglects the inertial enhancement of the gyres in the western sides of the oceans. Typically this leads to local recirculation indices of order 10 or more.

The pinching of the circulation. These arguments show that the injection of new water into the geostrophic interior must be accompanied by a greater recirculating flow. With the Sverdrup meridional transport smoothly varying in x, this inequality applies throughout the span of the basin. The consequence is that, in regions of eastward surface flow, the incoming fluid is pressed into a narrow stream on the eastern side of the basin. The above scale-analysis shows that the width (east and west) of the entering plume of new fluid is a fraction $\sim(y-y_0)/2a$ of the basin width.

Luyten, Pedlosky and Stommel (1983a) (LPS) provide a model of the entry of 'new' fluid into this system, and indeed show that Sverdrup dynamics can, in a layered model with a few isopycnal layers, constrain many of the downstream properties of the flow. But why do the explorations with the LPS model frequently show a very small pool of recirculating water, surrounded by extensive ventilation and shadow zones? It is simply that much of the fluid that they label as 'ventilated from the surface' has in fact entered the domain from the western boundary current. If we take their model to have literally small diffusion, then this entering fluid does not take on any properties of the mixed layer; with respect to passive tracers it remains distinct, and the pool of recirculating waters dominates the domain. The potential vorticity of the inflowing boundary current waters is in fact altered throughout each exposed region of a given constant-density layer; this alteration may or may not be significant, depending on the intensity of mixing (see appendix). The alternate interpretation of the model is that it implies extremely strong and selective density mixing. The deep mixed layers, thus created, reach from the surface to 900m depth near the gyre boundary, in the case they study. Then, indeed, tracers and potential vorticity will be altered throughout their ventilated zone by massive mixing.

It may be that the layered ventilation model approximates to some degree the circulation of a weakly diffusive continuously stratified ocean. In the model without diffusion, however, the ingestion of a passive tracer from the mixed layer will vary strongly with the number of layers in the model, since this number determines the effective mixed layer depth. Also greatly altered when more isopycnal layers are added to improve the representation of the stratification will be the size of the recirculation zone at densities near those near the zero Ekman pumping line. A model with few layers, though meant to describe the injection of fluid into the geostrophic interior from 'above', more resembles an injection from the 'side' (owing to the very deep mixed layers implicit in such a model)[†]. Thus there is a global aspect that keeps asserting itself: by incorporating more layers, in order to simulate better a weakly diffusive, continuously stratified ocean, more and more of the deep water is shielded from the **direct** modification of its potential vorticity and tracer profiles. The incoming western boundary current begins to dominate the potential vorticity of the gyre, and determination of its potential vorticity requires more physics than is currently in the hyperbolic calculation. Talley (1985) applies the LPS model to the North Pacific, and suggests that, for outcrops far from

---

[†]. Hendershott (1986) has constructed just such a model using quasi-geostrophic dynamics, in which the outcropping of density surfaces is simulated by applying atmospheric forcing to the deeper layers directly on a 'stair-case' poleward boundary of the layer model. It is important to learn whether such a model, which resembles one constructed with ρ as the vertical coordinate, contains essentially the same physics as the LPS model.

the zero pumping line, there is only small sensitivity to the number of layers chosen to represent the true continuous density field.

The configuration of a few homogeneous-density layers is thus equivalent to assuming a special distribution of large mixing coefficients. The water emanating from the western boundary in a thin jet, which feeds the recirculation can then have its properties modified by mixing, during its close encounter with the surface outcrop. But, it is important to recognize that in the limit of vanishing diffusivity of density and momentum, these arguments suggest that the ventilated fraction of the ocean (in the sense of tracer ingestion from the mixed layer or potential vorticity modification) becomes small, essentially zero at the zero wind curl line, increasing to R_c^{-1} at the gyre center.[+]

Rhines and Young (1982) argue that 'edge-' mixing of the potential vorticity at the rim of the gyre ultimately determines the circulation, and this points to the need of evaluating the contribution of the western boundary current with new observations. Tracer properties of the boundary currents should be particularly valuable as proxies for potential vorticity mixing. Sarmiento (private communication) reports that Radium-228, which comes from the sediments on the continental rise, has entered the North Atlantic gyre so efficiently that strong 'edge-' mixing must be a significant part of thermocline ventilation.

[+] The thermocline equation of Welander (1971) shows this more directly:
$$M_{zzt} + J(M_z, M_{zz}) + (\beta/f)M_x M_{zzz}) = \kappa M_{zzzz}$$
where M_z is the pressure, and κ is the density vertical mixing coefficient. If κ=0, the nonlinear terms have scale amplitudes of

$$M^2/(L_x L_y d^3) \text{ and } M^2/(L_x d^3 a) \quad (a=f/\beta)$$

whose ratio is just the recirculation index, a/L_y. This suggests that these horizontal and vertical advective terms cannot balance unless the solution can contrive to have a north-south scale as great as the Earth's radius. Instead, a diffusive transition layer is indicated,

The problem of wind gyres, then, rather than being a simple inpumping of surface boundary conditions, more resembles forced ventilation ('blowing on a radiator' rather than letting it convect by itself). This view is supported by the great eastward excursions that one estimates for fluid particles near the sea surface (see, e.g., Sarmiento (1983) whose reconstruction of one-year trajectories demonstrates the 'pinch effect' despite the use of time average data to estimate velocity). Surface drifters (e.g.,McNally **et al.**, 1983) illustrate this rapid east-west communication. Ventilation must eventually be studied as a Lagrangian problem although the vertical structure is very complex: North Pacific drifters complete a circuit about the subtropical gyre in about 4 years, moving at a speed larger than that of the underlying geostrophic flow.

The numerical model of Cox and Bryan (1984) and Cox (1985) well illustrates these competing issues. Their steady, stratified subtropical gyre experiences water-mass transformation where a concentrated stream of boundary layer fluid separates from the coast and flows eastward into mid-ocean. Just as in LPS, this simulation exhibits a relatively small pool of isolated properties, because of the large mixing between neighboring density levels (and between neighboring streamlines) along the outcrop. The region occupied by recirculating streamlines is far larger than the 'tracer' pool, and increases nearly to fill the basin at the deeper levels. Mode waters with low potential vorticity form at the deeply convecting outcrop lines, yielding permanent tongues of low Q. Though very diffusive and lacking explicit eddies, the model is the first to exhibit both ventilation and recirculation in a simple context. Cox and Bryans's Lagrangian analysis of the density and potential vorticity variations following a fluid parcel reveals the dramatic, irreversible changes that occur in the western boundary current and in the outcrop zones.

---

of thickness $(\kappa a/V)^{1/2}$, as suggested by Welander (**op.cit.**).

Subsequent high-resolution studies of Cox (1985), using a 170x170x18 level model, add significantly to the picture. The $1/3^0$ lateral resolution allows mesoscale eddies to develop, and they mix the potential vorticity of the gyre rapidly. In addition dramatic convective mixing occurs where the boundary current enters the interior. Both buoyancy flux and tracer influx from the atmosphere are greatly enhanced by this 'forced advection', as we have outlined above. The strong Cartesian mixing in the western boundary current still colors the interior potential vorticity of the gyre, but lateral eddy mixing in the North Equatorial Current becomes a central effect in shaping the mean gyre.

What is the geometry of the isopycnal prisms? On average the meridional velocity in the oceans far exceeds that required by mass conservation alone: hence, what is the nature of the inflow? Do the flow lines converge horizontally or converge vertically in order to obey mass conservation? The argument given above suggests that on average there must be much **horizontal** convergence and hence much 'pinching' of the circulation into narrow jets of flow. This produced a plausible picture for those surfaces outcropping in regions of eastward surface flow. But, where the upper geostrophic flow is westward, it is difficult to imagine how this lateral pinching could occur. It is plausible that in these regions of westward wpper-level flow the required convergence should be vertical rather than horizontal. Below, we show that this indeed occurs in the North Atlantic, and that the effect is closely tied to the existence of the thermocline.

The North Atlantic. We can examine this question by looking at the relation between the typical thickness of an isopycnal sheet in mid-gyre (the 'throat'), and comparing it with the area at the sea surface which, in late winter, has fluid density within this interval (the 'mouth'). Fig. 3a is a plot of outcrop area ΔA against area of the 'throat', $L_x \Delta d$,

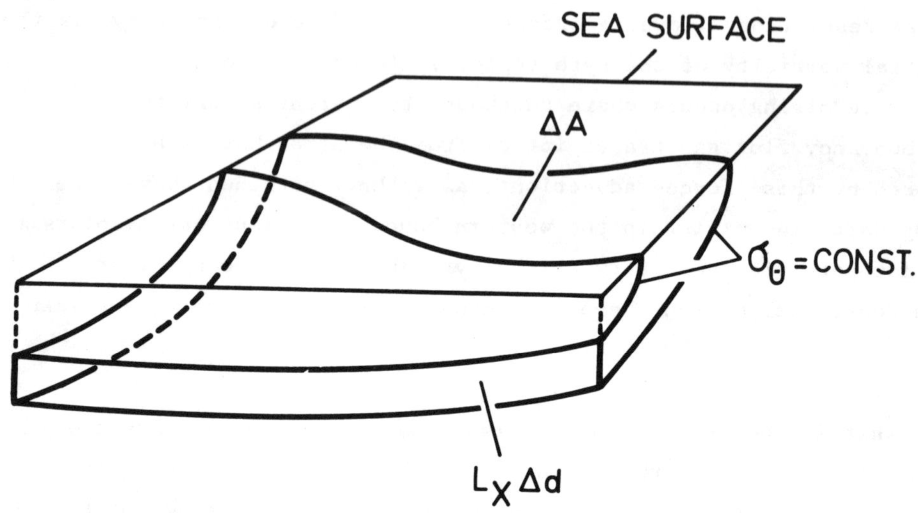

Fig. 2

Definition sketch showing the lateral convergence of the entering fluid. During the close passage of fluid from the west 'pool' water can be ventilated by mixing. With weak mixing, the prism of inflowing fluid remains distinct. The 'mouth' and 'throat' of the circulation are shown.

(see fig. 2) with potential density as a parameter. L_x is taken to be the east-west scale of the gyre. The data has been divided into increments of 0.1 σ_θ , and is taken from Sarmiento's (1983) tabulation of outcrop areas and average thicknesses, which in turn are based on the Levitus (1982) 1° smoothed hydrographic atlas. We see that from σ_θ of 26.2 down to 27.2 the outcrop windows have nearly constant surface area, while the average (vertical) thickness of the layers increases steadily. Below 27.2, the areas decrease rapidly. An equally interesting plot is the ratio of outcrop area to the prism volume contained within the gyre.

The ratio of areas of the horizontal 'mouth' at the base of the mixed layer to the vertical 'throat' at some constant latitude downstream is $\Delta A/L_x \Delta d$ (fig 3a), where L_x is taken to be 3.3×10^3 km. If the flow were uniform across the prism, the ratio would give the average vertical convergence of the flow, in passing from mouth to throat. But we have shown that, on average, vertical convergence alone cannot satisfy Sverdrup dynamics.

This ratio, fig 3b, shows three distinct regimes. First, in the deep water below σ_θ = 27.2 it takes small values. The boundary at 27.2 corresponds to the isopycnal surface which outcrops near the zero Ekman-pumping line. $\Delta A/L_x \Delta d$ is nearly constant through much of the wind gyre, up to σ_θ=26.4. Above this surface $\Delta A/L_x \Delta d$ takes on much larger values in the pycnocline where the isopycnal thickness Δd is small.

Suppose we compare this observed geometry with a model of the velocity field which obeys the Sverdrup relation. If the Ekman pumping has the form

$$w_E = \alpha g(x,y) z_0^2$$

then a model of the the vertical and meridional velocity field satisfying $\beta v = f w_z$ is

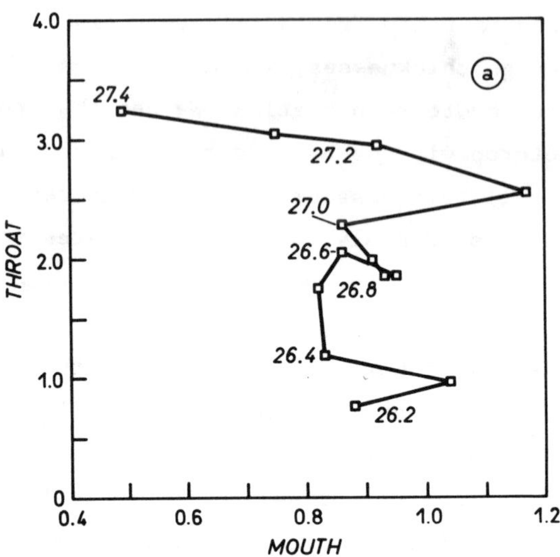

Fig. 3a

The area of the 'mouth' (i.e., the horizontal window at the sea surface between two outcrop lines separated in density by 0.1 σ_θ) plotted against the area of the 'throat' (i.e. the area of a vertical plane lying east and west, and bounded by the same two σ-surfaces), with potential density a parameter.

Units: 10^{12} m^2 (mouth) resp. 10^8 m^2 (throat).

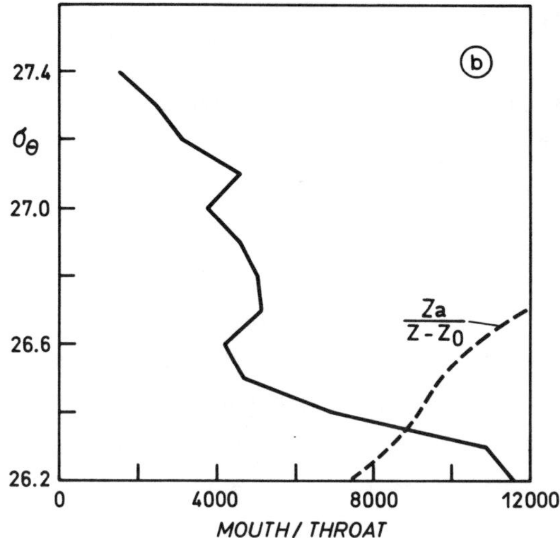

Fig. 3b

The ratio of area of the mouth and throat, $\Delta A/L_x\Delta d$, as a function of potential density, σ_θ. The range $27.2<\sigma_\theta<26.5$ is nearly uniform at about 4000. Deeper surfaces have relatively more lateral convergence, and shallower surfaces (in the pycnocline) relatively less. The dashed curve is the model for the ratio v/w from (3), with z_o=1000m. The coincidence of the curves above σ_θ=26.4 shows that lateral convergence is not required at those levels, and vertical convergence suffices there.

$$w = \alpha g(x,y)(z-z_0)^2$$

$$v = 2\alpha a g(x,y)(z-z_0)$$

where $a=f/\beta$ and z_0 is the base of the wind-gyre. These velocities have a ratio

$$v/w = 2a/(z-z_o) \qquad (8)$$

On fig. 3b we plot this reference curve for the choice z_o=1000m, using typical mid-gyre depths of σ-surfaces. In the upper waters, $\sigma_\theta < 26.4$, the agreement of the area ratio (mouth/throat) and the model v/w suggests that these density levels need not pinch laterally into narrow streams. They instead pinch vertically as they proceed southward and westward. Thus the flow on these uppermost density surfaces can proceed without great influx from other sources. This observation points to the strong interrelation between the wind-gyre dynamics and the existence of the thermocline (which is the region of small Δd's): in particular, the warm surface waters near the gyre center can uniformly flood the thermocline, whereas the cooler waters to the north move eastward to avoid the influx of boundary current water. Could this be the practical result of having an intense coherent Gulf Stream jet, rather than the distributed (in y-) influx predicted in viscous, linear circulation theory? If so, the western boundary current is exhibiting once again its potential to affect interior dynamics.

Sarmiento also calculates a total downward Ekman pumping of 8 Sverdrups for the range of outcrops, $26.2<\sigma_\theta<27.4$. By contrast, the Sverdrup geostrophic transport for the subtropical gyre in the same range of density is about 33 Sverdrups (e.g., Leetma **et al.** 1977) at 32°N. This figure is slightly larger than the average transport through the Florida Straits, which is also terminated below at a density of about $\sigma_\theta = 27.4$, and includes a thermohaline (meridional-cell) component. The recirculation index should equal the ratio of these transports, which is about 4.

North Pacific. The Pacific Ocean illustrates with apparent clarity the combined action of recirculation and injection of thermocline waters. The North Pacific, in contrast to the Atlantic, has no known production of deep or abyssal waters. According to climatological maps (a recent example being the Levitus atlas, 1982), the surface density is lighter (at corresponding latitudes) than in the Atlantic, and isopycnals that would reach to the surface due to air-sea buoyancy flux fail to do so in the Pacific. The excess precipitation in the Aleutian low and ICTZ have much to do with this (Warren, 1982), and the result is a stable halocline at ~ 150m depth that caps off the deep water, north of the polar front. Despite the lightness of the subpolar surface waters, the surface density still increases northward, so that we cannot have a buoyancy driven shallow thermohaline gyre. The upward Ekman pumping works against a much reduced gravitational burden, however, and seems to succeed creating an upward branch of circulation.

The greatest wintertime density to be found at the sea surface south of the zero Ekman pumping line in the North Pacific is about 26.25. The typical depth of this potential density surface further south is 300m, with a maximum of 450-500m in the western inertial recirculation. In the North Atlantic, the 'critical' density at the zero Ekman-pumping line is about 27.2, and that surface descends to fully 600m in the eastern part of the gyre, and 900m in the west.

The contrast between oceans is the more striking when one realizes how much deeper the North Pacific wind-circulation penetrates, than that in the North Atlantic. Maps of potential vorticity (Holland et al. 1984) show particularly clearly that the subtropical gyre at 1000m still extends across most of the North Pacific, although the circulation at 1000m may not simply mirror that of the upper-level gyres (Reid and Mantyla, 1978). In the Atlantic, by contrast, the 1000m level remnant of the wind gyre is very small in breadth, dominantly lying in the westward recirculation region. This occurs despite the shallow thermocline in the Pacific. One important difference that suggests this distinction is the inclined line of zero Ekman pumping in the North Atlantic, which in the theory tends strongly to shrink the mid-depth gyre.

The maps of Q for the North Pacific (Holland **et al**. 1984, Keffer 1984) illustrate vividly the entering plumes, typically with large values of Q derived from the subarctic gyre or from the shallow eastern tropical Pacific. These tongues (fig. 4) clearly diffuse away, and in doing so provide a measure of the destruction of the memory of Q for its upstream boundary conditions. The relative roles of vertical and lateral mixing in these tongues is not known, yet might be accessible to us, if enough is known about the velocity field from maps of Q and dynamic height. The recirculation of the gyres is the dominant feature of these maps, with plateaus of nearly uniform Q of great dimension. Gradients of Q occur in the entry plumes and at shallow levels where direct convective forcing may be active. The diagnostic equation (10) below for $Q(\psi)$ shows how 'islands' of Q are essential to balance interior sources of Q upon a given σ_θ-surface. The subarctic gyre of the North Pacific is an example of such islands. There is a shallow halocline maintained by excess precipitation, which dominates the map of Q on the σ_θ=26.5 surface. This, and other buoyancy-driven regions (the 18° water of the North Atlantic, the northern gyre of the North Atlantic) are well-known features of water-mass analysis, yet the potential vorticity analysis reflects the potent dynamics of these water masses.

Ventilation bottlenecks. Important entry points for tracers appear thus to be principally the regions of large gradients in the poleward-westward rim of the gyre. The gyres themselves adjust rapidly, assuming a nearly homogeneous tracer value which is the average of the boundary conditions around their rim. Musgrave (1984) has carried out

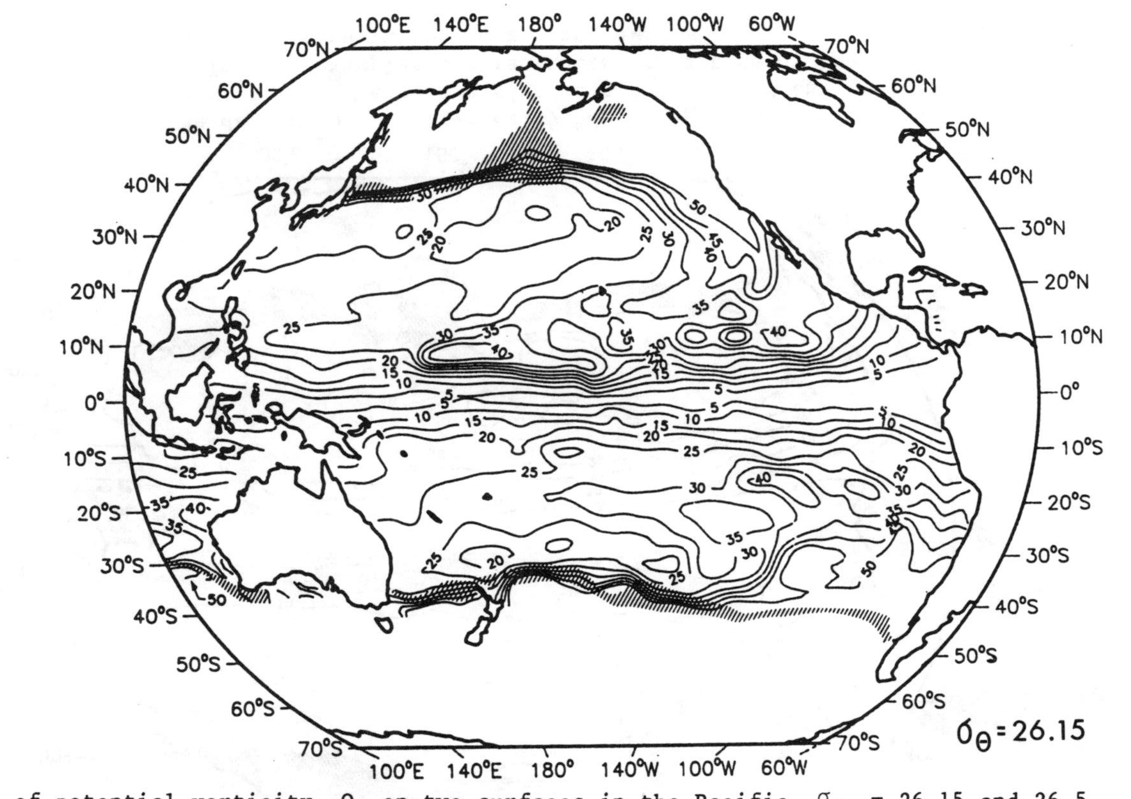

Fig. 4: Maps of potential vorticity, Q, on two surfaces in the Pacific, σ_θ = 26.15 and 26.5 from Holland et al., 1984 and Keffer, 1984. On the shallower surfaces the massive source of large values of Q is traceable both to the large salinity gradient along the polar front, and also the salinity contrast provided by tropical rainfall, mixed vertically downward. In the South Pacific the interaction with the polar front/circumpolar current appears to be strong.

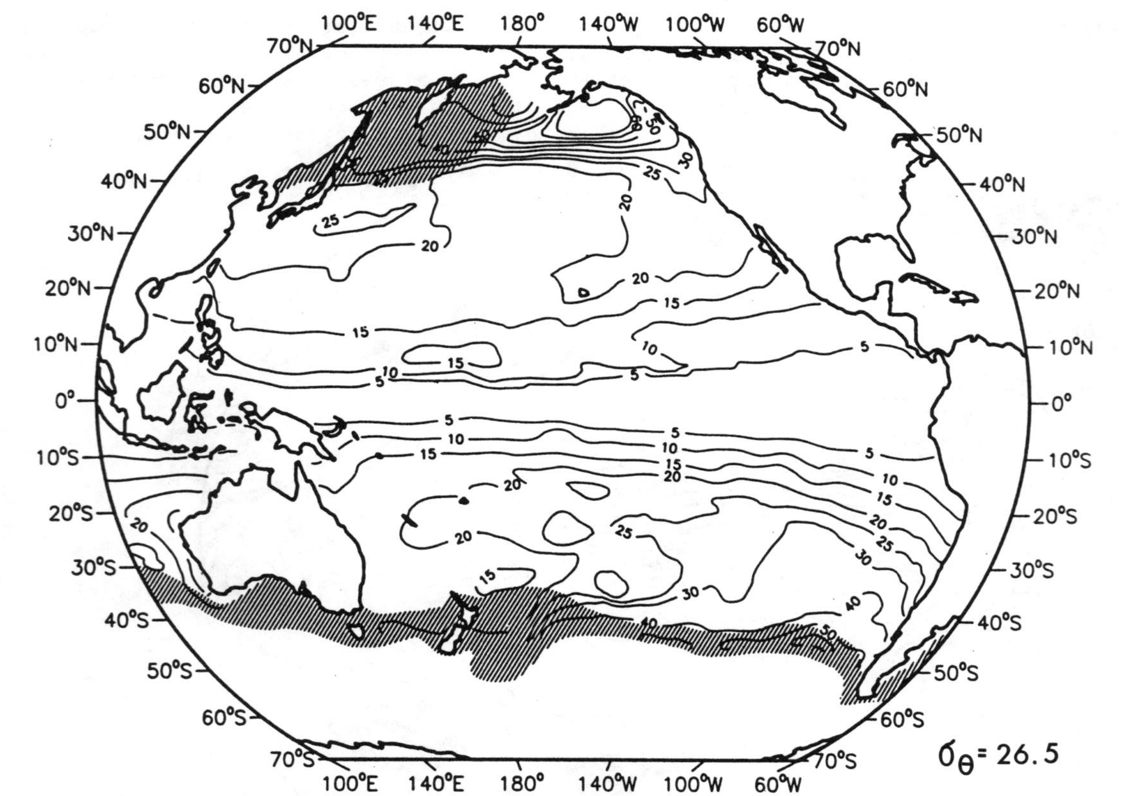

Fig. 4 (continued): On the deeper surface (typically 500 m depth in mid-gyre) a tongue of large Q circulates in from the polar front. There is a sign of a source of Q the creates in 'island' in the western subtropical gyre. In both examples the recirculatory nature of the flow is evident, as is the dissipation of the contrast of Q seen in the entry tongue.

simulations of tracer evolution in gyres, and finds in an idealized situation that the influx rate is almost entirely governed by the diffusion layer at the gyre rim. He also finds that the homogenized value of the tracer within the gyre is an interesting function of the degree of boundary current intensification and Peclet number (the ratio of inertial to diffusive terms). We postulate a simple one-dimensional advection diffusion model of the form

$$U\theta_x = \kappa\theta_{yy}$$

where U is the eastward velocity in the vicinity of the outcrop line. x is a coordinate along the bounding surface and y is normal to x. The result is a net influx of tracer of order $U(\theta_n-\theta_s)(\kappa L_x/U)^{1/2}$ where L_x is the east west width of the domain and $(\theta_n-\theta_s)$ is the tracer contrast between northern and southern boundaries. This predicts that the Nusselt number of a gyre with prescribed tracer difference imposed from its poleward side to its equatorward side (that is, the ratio of meridional flux of tracer to the purely diffusive flux, were there no circulation) is proportional to the square root of the Peclet number, UL_x/κ. This illustrates the role of 'forced advection' in ingesting tracers.

In experiments which combine gyre recirculation with a weak ventilation, Musgrave finds that the ventilation can pull the value of the homogenized tracer close to that of the inflow boundary, whereas without convective ventilation that homogenized value lies intermediate between poleward and equatorward extreme boundary values of the tracer. The homogenized structure of the gyre is otherwise unaltered, however.

These remarks are meant to stress the difference between the actual forced ventilation and the weaker ventilation that would occur due only to Ekman pumping downward of the surface boundary values. A good illustration is Bunker and Worthington's (1976) map of heat loss from the sea to the atmosphere; the northwest corner of the subtropical gyre is the most intense region of surface flux. It is the forced ventilation by warm Gulf Stream and Sargasso waters that make it so.

Integral balance of the potential vorticity. Earlier, we have used dissipation integrals to discuss the net vorticity balance of the gyres (Rhines and Young, 1982a). In a nearly cyclic flow, where the pumping of fluid through the system is weak, we can approximately write

$$\int \mathbf{F}_Q \cdot \mathbf{n} ds = \iint S \, dxdy, \qquad (9)$$

where $\mathbf{F}_Q$ is the eddy flux of potential vorticity, $\mathbf{n}$ is an outward unit vector normal to a mean streamline, and ds is an arc length element long the streamline, S is the net forcing of potential vorticity, minus small-scale dissipation.

If the isopycnal eddy transport of Q, $\mathbf{F}_Q$, is modelled as a downgradient flux of Q, then

$$\int \kappa \nabla Q \cdot ds = -\iint S \, dxdy, \qquad (10)$$

κ being the Lagrangian diffusivity due to mesoscale motions. In absence of forcing or other modes of dissipation, S=0 and we recover homogenization, $\nabla Q = 0$. But more generally, there must be a gradient of Q to remove the source-sink effect in the domain encircled by the streamline. If we make the marginally valid assumption that κ is small (in the sense described by Rhines and Young, 1982a) we have nearly that

$$Q \approx Q(\psi)$$
$$\nabla Q \approx (\partial Q/\partial \psi) \nabla \psi$$

and hence from (10)

$$\partial Q/\partial \psi = -\iint S \, dxdy / \int \kappa u.ds. \qquad (11)$$

This formula explicitly relates the variation of potential vorticity within the gyre to sources and sinks of Q enclosed (upon a given isopycnal surface) by the streamline in question. It can yield explicit solutions for buoyancy forced problems (currently being carried out by Young, Dewar, and others). The formula holds for a model of the LPS type and it emphasizes that jet-like flow regions, where U is large, dominate the denominator and tend to bring down the magnitude of the variation of Q, despite the presence of sources, S. It is just these jet-like regions which are ignored in LPS calculations, and this is another way of concretely emphasizing the crucial role that such regions play in determining the structure of the potential vorticity. Western boundary currents are one source of large contributions to the denominator of (6), but the greater freedom of the jet to meander once it has separated from the coast suggests that there, near the encounter with the outcrops, the greatest contributions to $\int \mathbf{F}_Q \cdot \underline{n} ds$, and hence to $\partial Q/\partial\psi$, will occur. In their analysis of the average flux of Q in an eddy resolving numerical model, Rhines and Holland (1980) found these regions, and the inertial recirculation zone, to dominate the eddy-induced form drag, and hence the dissipation integrals. The role of western boundary currents in this class of problem has begun to be analyzed by Young and Rhines (1983), Ierly and Young (1984), Young (1984), Luyten and Stommel (1984), Welander(1984), and Haidvogel (personal communication: an eddy resolving circulation calculation with stretched finite difference grid to better resolve the boundary currents).

Pedlosky (1983) gives a scale analysis which focuses on the brief encounters of a recirculating flow with an upper mixed layer. He suggests that these can defeat the homogenization tendency desribed by Rhines and Young (1982). The variations of Q would occur on the strip of fluid that passes through the mixed layer (but the entire region within the innermost streamline to contact the mixed layer would even so remain homogenized). A re-analysis of this question is given in the appendix.

In the case of strong mixing, the gyre paradoxically is far less homogeneous in properties, because of the incursion of non-constant property boundary conditions. Of greatest importance is to make observations that will tell us how deep the net mixing reaches in the outcrop regions; in the limit of weak mixing, only a small fraction of the volume of gyre fluid will be susceptible to this alteration of properties. Tritium data (Sarmiento 1983) suggests fairly vigourous mixing in the North Atlantic, however. Cox's (1985) high-resolution study represents a fortuitous situation in which the eddy mixing is strong, yet it tends to drive the gyre in the direction of homogenization. This happens because the strong eddy-mixing occurs interior to the gyre, without reaching out to the boundaries. The assumptions of the homogenization theory are then valid even though the Peclet number is not large.

Conclusion. We have here and earlier advanced the idea that the dynamics and tracer response of the wind-driven circulation involves the 'spinning of a flywheel' as well as 'squeezing toothpaste out of its tube', the latter dynamics being recently much in fashion. The collaboration between recirculation and direct injection, interacting through shear dispersion, determines the ventilation of the ocean interior. This continues in the spirit of Montgomery (1938), who studied both dynamics and kinematics of streamtubes entering the geostrophic interior from the mixed layer.

The simple Sverdrup constraint assures that typically 3/4 (as a lower limit) of the transport is recirculating fluid, and hence that in a weakly diffusive ocean the majority of the gyre is a 'pool', which communicates with the incoming stream of fluid through lateral mixing. Shear-dispersion theory points to active areas in which this mixing may occur: in particular where the flow is concentrated into a jet, and encounters a strong boundary condition (for example, at the upper mixed layer). The LPS model at first sight is a non-diffusive solution of a layered-density fluid, in which prescription of surface density and Ekman vertical velocity, plus plausible eastern boundary conditions determine a great deal of the structure of the wind-driven irculation and the main pycnocline, itself. We emphasize here that the model in fact presumes very large mixing, and implicitly assumes a special western boundary current model. The creation of the basic stratification appears more of a global problem, with the global aspect asserting itself through eastern boundary conditions, through the incursion of the separated western boundary current into the mid-ocean region of ventilation, and through prescription of the surface density field.

The homogenization predicted by Rhines and Young (**op.cit**) can account for the structure of the large 'pool', provided mixing is sufficiently weak. Even when stronger mixing or active sources of Q are present the recirculation effects still occur, and average potential vorticity budgets like eqns. (9) - (11) determine the (non-vanishing) variation of Q across streamlines; they do so in a fundamentally different way than simply asserting, as does the LPS model, the conservation of tracer or Q as fluid enters from the mixed layer.

Holland **et al.** (1984) suggest that the North Atlantic and North Pacific provide us with examples of both strong and weak ventilation, respectively. In the North Atlantic the down-pumping regions have direct 'windows' to the deep thermocline along isopycnal surfaces, while in the Pacific only the top 250m or so of the fluid has an isopycnal pathway to the surface in the down pumping region. Yet, direct ventilation clearly dominates in certain regions, particularly near the sea surface, and in the entry regions for plumes of mode-water. It is likely that after entry these waters are greatly affected by shear dispersion, by which the gyre mixes the tracer along mean streamlines. Later, western boundary currents (and more so, the intense-eddy regions of the boundary jets after separation from the coast) affect mixing **across** mean streamlines.

The invasion of tritium into the North Atlantic gyre has occurred at such a great rate, seemingly much faster than the simple product of vertical velocity and surface concentration would permit. (Sarmiento 1983). The great overburden of tritium, particularly on the deeper density surfaces, strongly suggests that special diapycnal mixing is having an impact on the circulation. If this mixing occurs in the most likely place, where the eastward flow from the boundary current moves out to sea, then the intense, deep mixing implicit in the LPS model finds some justification.

The North Pacific, however, exhibits instead tongues of injected tritium which are narrow and intense, reminiscent of the pinching of the circulation and shear dispersion described above. Given so little contact to the sea surface, recirculation seems to be a dominant influence (see also Keffer, 1984).

An observational investigation that managed to capture the movement of mixed layer fluid downward into the geostrophic interior would be valuable indeed. The horizontal excursions experienced by upper ocean fluid in a year's time are so large that one winter's active deepening may be carried to a far different, perhaps quieter ocean region before the next winter. In this way new layers with small Q could be 'buried' by the succeeding summers' heating. Again, the 3-dimensional nature of ventilation is evident. Perhaps the modern suite of quasi-Lagrangian tools (tracers, drifters, neutrally buoyant floats) can capture this process of water mass creation and submergence.

Postscript. During this fine study conference we attempted to make a change to the field which would lighten the burden for all future ocean-atmosphere dynamicists; the term potential vorticity is unwieldy and fails to convey the spirit of the quantity. It was observed that physicists, in whose quarters we were housed, had successfully captured the whimsy of scientific research with terms like 'charm' and 'color'. A contest was held, and the participants contributed many suggestions as a replacement for 'potential vorticity', among them:

strophity, *spin*, *spingity*, *dervishity* and *Sonja*,

the latter in honor of a fine Norwegian exponent of angular momentum conservation. Prof. Hide's suggestion, *piety*, was apparently motiviated by the Stommel-Arons-Faller experiments. We invite further entries to this competition in hopes that we will one day be freed of onorous terminology.

I would like to thank Bill Young for useful input to the discussion of dissipation integrals. The work was supported by the National Science Foundation under grant OCE 82-19780/OCE 84-45194.

Appendix: Scale analysis of the circulation integral.

Despite several interesting theories, we do not yet have a clear picture of the bulk potential vorticity balance of the wind-driven gyres. Holland and Rhines (1980) made diagnostic calculations using a numerical model of the circulation. The eddy potential vorticity flux that forms the integrand of (11) takes on particularly large values in the region of the free separated jet. The possible dominance of small regions in the integral balances is a worrisome feature that occurs in much of ocean circulation modelling, whether in global energy or vorticity budgets. The behaviour of the integrand in the western boundary current is highly model dependent. There were some seemingly robust results, in that investigation, however: form drag provides a strong driving of the abyssal recirculation, and relative-vorticity flux (a.k.a. lateral Reynolds stress) communicates the gyre momentum between adjacent streamlines, while lateral flux of q across the separated western boundary jet carries potential vorticity between the gyres, hence between regions of opposing generation by wind.

Eqn. (11) provides the relevant diagnostic relation. If we consider an isopycnal layer of thickness H (layered density profile), the source function for potential vorticity is

$$S \sim fw_E/H^2$$

where w_E is the Ekman velocity. Assuming κ to be constant the scale estimate of (4) is

$$\kappa\int_{\mathcal{C}_1}\nabla Q\cdot\mathbf{n}\,ds - \kappa\int_{\mathcal{C}_2}\nabla Q\cdot\mathbf{n}ds = \iint S\,dxdy \qquad (12)$$

$$\sim \quad \kappa\,\nabla Q\,2\pi L \qquad 0 \qquad \delta nfw_E/H^2$$

where $\mathcal{C}_1$ and $\mathcal{C}_2$ are the bounding contours, the former being the limiting mean streamline passing along the 'apex' of the outcrop, and the latter the streamline at the edge of the shielded fluid beneath the next isopycnal outcrop, fig.5, $2\pi L$ is the circumference of the gyre, n is the length along the outcrop of the forcing, δ is the width of the 'mouth'. Here $Q=(f+\zeta)/h$. Note that the second integral vanishes identically, provided that the forcing S is restricted to the outcrop zone. Now, if the flow does not pass through a concentrated boundary current, this yields for the potential vorticity gradient (non-dimensionalized with respect to β/H) :†

$$\nabla Q/\beta H \sim (1/2\pi)P\ (\delta/H)(d/L)(n/L)$$

where $P=UL/\kappa$ is the Peclet number, and $U \sim fw_E/\beta d$ is the average strength of the Sverdrup circulation.

Typical estimates for the North Atlantic are P=10 to 50, n~1000 km, L~2000km. The product $(\delta/H)(d/L)$ (~0.1 or less) describes the ratio of the large-scale-average slope of an isopycnal surface to the slope of the isopycnal near the outcrop.

We find thus that $\nabla Q/(\beta/H) \sim 0.1$ or smaller to 0.5 for Peclet numbers that are reasonably large (10 to 50); thus the homogenization result may be satisfied, at least marginally in the thin band of fluid that encounters the outcrop, as well as the much larger interior region.

† Pedlosky (1983) failed to note that the second integral vanishes, and thus estimated the LHS (left-hand-side) of (12) as $O(\kappa \Delta Q \delta/L)$ where ΔQ is the 'typical potential vorticity anomaly'. But, to the contrary, Q must have its entire variation in the ribbon of fluid passing through the outcrop, yielding our estimate for the LHS, which is smaller by a factor δ/L than Pedlosky's. Regardless of the variation of Q on that ribbon, homogenized Q appears everywhere in the interior (i.e., within $\mathcal{F}_2$), in this large-Peclet number idealization.

However, in addition, Young (1983) points out the major effect that western boundary currents may have in speeding up mixing across streamlines. If we include this intensification in the scale estimate above we find instead,

$$\nabla Q/(\beta/H) \sim (\ell/L)\ P\ (\delta/H)\ (d/L)\ (n/L).$$

where ℓ is the width of the western boundary current, typically $\ell \sim 100$ km. Now the homogenization result seems more likely, with the non-dimensional gradient of Q ranging from 0.03 or smaller to perhaps 0.2 within the outcropping band.

Remarks: First, the scale analysis shows some insensitivity to model configuration: as we increase the number of layers (thus by assumption reducing the implicit mixing, as seen in the mixed layer depth H), the estimate of ∇Q is not affected. However, the width of the ribbon of possibly inhomogeneous Q is just δ, which is sensitive to the model stratification. With small mixing these ribbons are narrow features on the rims of the gyre, at each isopynal level.

Second, the value (10-50) assigned to the Peclet number represents a strong assumption about mixing in the western boundary current (where the local Peclet number, $U\ell/\kappa$, brackets unity, here).

Third, these considerations are relevant to the ventilated thermocline models, owing to the implicit assumption of western boundary current dynamics in those models. The properties of boundary current fluid rising and flowing along the outcrop zones will affect the entry conditions into the gyre. The question is, does this fluid remember its past transit through the gyre? Or does the total boundary current in some average sense, remember that transit? It is interesting that by choosing a homogenized interior, the gyre can have a strongly mixing boundary current and yet have no loss of information or structure as fluid passes through it (except for a viscous sublayer on the inshore side).

Although there are many levels at which homogenization fails to occur in gyres, we suggest that the global constraint (11) is the most likely to be relevant.

The best way to resolve these matters would be to have extensive tracer- and Lagrangian flow measurements in the subtropical gyres of the world ocean. The modification processes at work in the western boundary currents could be examined with fine- scale surveys, while the broader sections across the gyre would give a second view of the effect of boundary currents on the distributions.

The deliberate events that ventilate the gyre in the intense 'forced advection' region must be examined with detailed observations. Particular elements, like warm Gulf Stream rings, might play a significant role in the ventilation (Joyce and Schmitt, private communication; warm rings in winter suffer deep convection, and are later reabsorbed by the Gulf Stream, implying a large conversion of water properties).

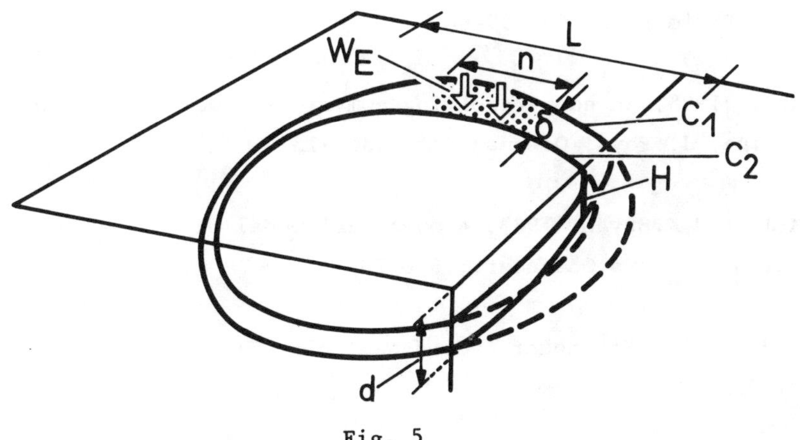

Fig. 5

Definition sketch for the analysis of potential vorticity near an outcrop line.

REFERENCES

Batchelor, G.K., 1969, Computation of the energy spectrum in homogeneous two-dimensional turbulence, Phys. Fluids **12**, 233-238.

Bunker, A.F.and L.V. Worthington, 1976, Energy exchange charts of the North Atlantic Ocean. Bull. Amer. Met.Soc. **57**, 670-678.

Bretherton, F.P., 1966, Critical-later instability in baroclinic flows, Q.Journ.Roy.Met.Soc. **92**, 325-334.

Brown, E., W.B. Owens and H.L. Bryden, 1986, Eddy potential vorticity fluxes in the Gulf Stream recirculation, J. Phys. Oceanog. **16**, 523-531.

Bryden, H.L., 1983, Sources of eddy energy in the Gulf Stream recirculation region. J. Marine Res. **40**, 1047-1068.

Colin de Verdiere, A., 1977, Ph.D thesis, WHOI-MIT Joint Program in Oceanography.

Colin de Verdiere, A., 1979, Mean flow generation by topographic Rossby waves, J. Fluid Mech. **94**, 39-64.

Cox, M.J., 1985, An eddy-resolving numerical model of the ventilated thermocline, J. Phys. Oceanog. **15**, 1312-1324.

Cox, M.J. and K.Bryan, 1983, A numerical model of the ventilated thermocline, J.Phys.Oceanogr., 698-705.

Ertel, H., 1942, Ein neuer hydrodynamischer Wirbelsatz, Meteorol.Zeit. **59**, 277-281.

Fjortoft, R., 1953, On the changes in the spectral distribution of kinetic energy for two-dimensional nondivergent flow, Tellus 5, 225-230.

Gargett, A. 1984, Vertical eddy diffusivity in the ocean interior, J.Marine Res. **42**, 359-393.

Green, J.S.A., 1970, Transfer properties of the large-scale eddies and the general circulation, Q.Journal Roy.Met.Soc., **96**, 157-185.

Haidvogel, D.P. and P.B. Rhines, 1983, Waves and circulation driven by oscillatory winds in an idealized ocean basin, Geophys. Astrophys. Fluid Dyn. **25**, 1-63.

Hendershott, M. 1986, The ventilated thermocline in quasigeostrophic approximation, submitted to J. Phys. Oceanog.

Hogg, N.G., 1983, A note on the deep circulation fo the western North Atlantic; its nature and causes. Deep-Sea Res., **30**, 945-961.

Holland, W.R., T. Keffer, and P. Rhines, 1984, The dynamics of the oceanic circulation: the potential vorticity field. Nature, **308**, 698-705.

Holland, W.R., and P.B. Rhines, 1980, An example of eddy-driven circulation, J.Phys.Oceanogr. **10**, 1010-1031.

Hoskins, B.J., 1983, Modelling of the transient eddies and their feedback on the mean flow, in Large-scale dynamic processes in the atmosphere, Hoskins and Pearce eds., Academic Press, London.

Hoskins, B.J., M.E. McIntyre and A.W. Robertson, 1986, On the use and significance of isentropic potential vorticity maps, Q.Journal Royal Met. Soc. **111**, 877-946.

Ierly, G. and W.R. Young, 1983, Can the western boundary layer affect the potential vorticity distribution in the Sverdrup interior of the wind gyre?, J.Phys.Oceanogr. **13**, 1753-1763.

Keffer, T.,1984, The ventilation of the world ocean: maps of potential vorticity. J. Phys. Oceanog. **15**, 509-523.

Leetma, A., P.P. Niiler, and H. Stommel,1977, Does the Sverdrup relation account for the wind-driven ocean circulation?, J.Marine Res. **35**, 1-9.

Levitus, S., 1982, Climatological atlas of the world ocean, NOAA Professional Paper 13, U.S.Dept. of Commerce.

Luyten, J., J. Pedlosky and H. Stommel, 1983a, The ventilated thermocline, J.Phys. Oceanogr **13**, 292-309.

Luyten, J., J. Pedlosky and H. Stommel, 1983b, Climatic inferences form the ventilated thermocline, Climatic Change, **5**, 183-191.

McNally, G.J., W.C. Patzert and A.D. Kirwin, and A.C. Vastano, 1983, The near-surface circulation of the North Pacific using satellite tracked drifting buoys, J.Geophys. Res.**88**, 7507-7518

Montgomery, R.B., 1938, Circulation in the upper layers of the southern North Atlantic deduced with the use of isentropic analysis. Papers in Phys. Oceanogr. and Meteorol.**6**, 55 pp.

Musgrave, D. 1984, A numerical study of the roles of sub-gyre scale mixing and the western bondary current on homogenization of a passive tracer. J. Geophys.Res., **90**, 7037-7043.

Pedlosky, J. 1983, On the relative importance of ventilation and mixing of potential vorticity in mid-ocean gyres, J.Phys.Oceanog. **13**, 2121-2122.

Pedlosky, J. and W.R.Young, 1983, Ventilation, potential vorticity homogenization and the structure of the ocean circulation. J.Phys Oceanog. **13**, 2020-2037.

Plumb, R.A., 1979, Eddy fluxes of conserved quantities by small-amplitude waves, J. Atmos. Sci. **36**, 1699-1704.

Reid, J.L. and A. Mantyla, 1978, On the mid-depth circulation of the North Pacific Ocean, J.Phys. Oceanogr.,**8**, 46-951.

Rhines, P.B., 1977, The dynamics of unsteady currents, in The Sea, vol.6,E.D. Goldberg ed., Wiley Interscience, N.Y., 189-319.

Rhines, P.B., 1986, Vorticity dynamics of the oceanic general circulation, Ann. Revs. Fluid Mech. **18**, 433-497.

Rhines, P.B., W.R. Holland and J.C. Chow 1985, Experiments with buoyancy-driven ocean circulation, NCAR Technical Note TN-260+STR, National Center for Atmospheric Research.

Rhines, P.B. and W.R. Young, 1982a, Homogenization of potential vorticity in planetary gyres, J.Fluid Mech. **122**, 347-368.

Rhines, P.B. and W.R. Young, 1982b, A theoretical study of the wind driven circulation, I.,mid-ocean gyres, J.Marine Res., **40**, 559-596.

Sarmiento, J.L., 1983, A tritium box model of the North Atlantic thermocline, J.Phys. Oceanogr. **13**, 1269-1274.

Sarmiento, J.L.,1983b, Simulation of bomb tritium entry into the Atlantic Ocean, J. Phys. Oceanogr. **13**, 24-1939.

Schmitz, W., W. Holland and J.F. Price, 1983, Mid-latitude mesoscale variability, Rev. Geophys. Space Phys. **21**, 1109-1119.

Stommel, H., A. Arons, and A. Faller, 1958, Some examples of stationary planetary flow patterns in bounded basins, Tellus **10**, 179-187.

Talley, L. 1985, Ventilation of the subtropical North Pacific: the shallow salinity minimum. J. Phys. Oceanog. **15**, 633-649.

Taylor, G.I., 1917, Observations and speculations on the nature of turbulent motions, in G.I. Taylor, Scientific Papers, G.K. Batchelor ed., Cambridge Unversity Press, 1960.

Taylor, G.I., 1921, Diffusion by continuous movements, in G.I. Taylor, Scientific Papers, G.K. Batchelor ed., Cambridge University Press, 1960.

Vasholz, D.P., and L.J. Crawford, 1985, Dye dispersion in the seasonal thermocline, J. Phys. Oceanog. **15**, 695-712.

Wallace, M., 1978, Trajectory slopes, conuntergradient heat fluxes and mixing by lower stratospheric waves, J. Atmos. Sci. **35**, 554-558.

Warren, B., 1983, Why is no deep water formed in the North Pacific? J.Marine Res., **42**, 327-347.

Welander, P., 1971, The thermocline problem. Phil.Trans.Roy. Soc. **A270**, 69-73

Whitehead, J.A., 1975, Mean flow dirven by circulaiton on a β-plane, Tellus **27**, 358-364.

Wunsch, C. and D. Roemmich, 1985, Is the North Atlantic in Sverdrup Balance?, J. Phys. Oceanog. **15**, 1876-1879.

Wüst,G., 1935, Schichtung und Zirkulation des Atlantischen Ozeans. Die Stratosphare. Reports of the Meteor expedition, repub. by W.J.Emery 1978, Amerind, New Delhi, 112pp.

Young, W.R., 1984, The role of western boundary layers in gyre-scale ocean mixing, J.Phys. Oceanogr. **14**, 478-483.

Young W.R. and P. Rhines, 1983, A theory of the wind-driven circulation, II. The role of western boundary currents. J.Marine Res. **40**, 849-872.

Young, W.R., P. Rhines and C.J.R. Garrett, 1983, Shear-flow dispersion, internal waves and horizontal mixing in the ocean, J. Phys. Oceanog. **12**, 515-527.

THERMOHALINE EFFECTS IN THE OCEAN CIRCULATION AND RELATED SIMPLE MODELS

PIERRE WELANDER

School of Oceanography
University of Washington
Seattle, Washington 98195

1. HISTORIC BACKGROUND

The discovery at the end of last century that small year-to-year variations in the temperature and salinity affected the "herring periods" played an important role for the development of physical oceanography in Scandinavia. Helland-Hansen, Nansen, Ekman and their collaborators in the "Bergen School" developed the instruments needed to measure these and other oceanographic parameters of interest with the required accuracy. They also developed the theoretical tools by help of which the ocean currents could be calculated from the measured temperature and salinity fields, in particular the so-called "dynamic method".

These oceanographers were thus forced to consider the equation of state for the sea water in detail. They determined with remarkable precision such coefficients as the thermal expansion and the compressibility, as functions of pressure, temperature and salinity, and introduced new practical useful density variables, such as the well known σ_t and σ_θ . Temperature and salinity, as well as oxygen concentration, were used to define water masses, and to track the spreading of such masses in the oceans. Such studies were developed further, and took on the form of almost an art in the later "Kiel School", with Wüst and Defant in the forefront. Today, with a large hydrographic data material available, it is posssible to study water masses and their spreading in much more detail all over the World Ocean, and practical use has also been made of observed distributions of

J. Willebrand and D. L. T. Anderson (eds.), Large-Scale Transport Processes in Oceans and Atmosphere, 163–200.

many tracers, including such radioactive tracers as radiocarbon and tritium. Still, much of the present work is based on ideas and methods developed by the pioneers in the Bergen and Kiel Schools.

The first real theory of the ocean circulation, relating the currents to the forcing at the sea surface, was advanced by Ekman (1905, 1923). He dealt with a wind-forced homogeneous ocean and, besides his "spiral", discussed the role of topography and the beta-effect (his "Planetarische Wirbel"). The theory was later extended to a stratified ocean, in a little known paper (Ekman, 1924). After Ekman's promising start, the ocean circulation theory advanced in a sluggish way. It was not until the late 1940's that the absolutely critical role of the beta-effect on the oceanic transport field became clear, through the two monumental papers by Sverdrup (1947) and Stommel (1948). Sverdrup pointed out that the mass transport components $M^x = \int \rho u dz$ and $M^y = \int \rho v dz$, in the zonal and meridional directions, respectively, satisfied the two equations[1]

$$\beta M^y = \text{curl}_z \vec{\tau}^w , \quad \frac{\partial M^x}{\partial x} + \frac{1}{\cos \phi} \frac{\partial}{\partial y} (M^y \cos \phi) = 0 \tag{1.2}$$

allowing a full determination of the transport field when one boundary condition was added. In particular, he applied the condition $M^x = 0$ at an eastern oceanic boundary. The theory assumed that the geostrophic forces and the wind force were in integrated balance, and that the motions were negligibly small below the depth to which the vertical integration was carried. These assumptions appear reasonable, and the Sverdrup equations are even today accepted by most oceanographers as the proper zeroth order theory for the interior oceanic transport. Why the boundary condition in Sverdrup's theory should be placed at the eastern side of the oceans,

---

[1] Here ϕ is latitude, ∂x and ∂y displacements northward and eastward, respectively, β the y-derivative of the Coriolis parameter, and $\vec{\tau}^w$ the wind-stress vector.

rather than at the western, was explained by Stommel (loc. cit.): higher order frictional or internal effects are needed to close the problem, and the nature of the overall vorticity balance is such that these added affects must be concentrated to the western side, producing such well known phenomena as the Gulf Stream in the Atlantic Ocean and the Kuroshio in the Pacific Ocean. After the appearance of these two papers followed a number of studies of the western boundary current dynamics; for a review of these work the reader is referred to the book "The Gulf Stream" by Stommel (1965). In spite of these progresses, the "Grand Theory", which could explain all currents in terms of only the applied surface forcing (wind stress, heating/cooling, evaporation/precipitation) was still far away. There was an understandable reluctance to give up the simple Sverdrup theory, and tackle the problem of predicting the interior mass field of the oceans, requiring explicit modeling of such unpleasant processes (from the theoreticians viewpoint) as non-linear advection and turbulence. In the 1950's some progress on the larger problems was nevertheless made, along the line of a similarity theory. This theory assumed that non-linear advection and vertical (that is, practically, cross-isopycnal) diffusion of density are in balance, that the vertical profile shape is universal (the amplitude and scale-depth may vary with horizontal position). As is usual in ocean models, a Boussinesq approximation was employed, allowing temperature and salinity to be combined into a single density variable. For a review of the similarity theory the readers are referred to Veronis (1969).

The incompleteness of the similarity theory was obvious from the beginning, since the boundary conditions had to be chosen very specially. Further, boundary currents could not be added to close the problem. It also turned out that the role of the vertical diffusion was mystic; for example, the first two versions of the theory, by Robinson and Stommel (1959) and Welander (1959) gave practically identical solutions, although the former version assumed diffusion to be critical while the second version asssumed it negligible.

In the early 1960's a new powerful tool was put to practical use in ocean circulation theory: the fast electronic computers. The pioneering work in developing three-dimensional numerical models for the ocean circulation was done by Sarkisyan (1962) in U.S.S.R. and Bryan (1963) in U.S.A. From that time on numerical models have been playing an ever growing role in theoretical studies of the ocean circulation. Only numerical models can deal effectively with the complications of irregular topography and coastline, non-linear processes and realistically described surface boundary conditions. The most advanced models of eddy-resolving type generate their own turbulence (at the meso-scales), deal with effects of seasonal variation, and consider active coupling to the atmosphere. It is not possible to discuss here all the interesting numerical model works which have been published; the interested readers may consult some recent papers on the subject, for example, Bryan, Manabe and Pacanowski (1975), Bryan and Lewis (1979), Meehl, Washington and Semtner (1982), and Hasselmann (1982).

Modern theoretical oceanographers do also, of course, pursue model studies along analytical lines. It has been found possible to bypass the restrictions of the similarity theory mentioned previously by going to layered models. Interesting new models of such type have been developed by Luyten, Pedlosky and Stommel (1983) in a so-called "ventilated" version (the flow is directly forced by Ekman pumping), and by Rhines and Young (1983a, b) in an "unventilated" version (streamlines do not connect to the Ekman layer). These models have predicted important new phenomena in the oceans, such as the "shadow zones" at the eastern sides, and the layers of homogenized potential vorticity; these apparently are seen in the observed oceanic fields.

The present article deals with two special aspects of the ocean circulation problem: the vertical structure of the main thermocline, and phenomena associated with so-called "mixed boundary conditions", meaning boundary conditions for temperature and salinity which separate these variables (they cannot be lumped into a single density variable). These latter phenomena are discussed in the context of highly simplified models, of the "box type" and "loop type". Such models may seem removed from the real oceans; however, they are very useful for exploring the physical nature of phenomena which may appear in very complex form in reality. It may also happen that a new phenomenon is discovered in a very simple model, leading to a search for and the successful discovery of a similar phenomenon in a larger numerical model. We will find such an example in the "Rooth-Bryan flip", to be discussed further in Section 4 of this article.

At the suggestion of one of the editors (J.W.) a short section on thermal and thermohaline oscillators has been added at the end of the article. This subject has interested this author through a number of years, and the readers may at least have interest to look to the "oscillator gallery" which I have put together in Figure 10. Short comments on the physical nature of these specimens are given, and the readers can learn more from the references given. Of course, most fun comes from actually building some of these oscillators yourself; in many cases this is possible in the "kitchen laboratory".

2. THE BOUSSINESQ APPROXIMATION, AND THE DENSITY FORM MODELS

In classical hydrodynamic models, dealing with a one-component stratified fluid at laboratory scales, the Boussinesq approximation is commonly employed. The full approximation involves several steps:

(i) the density is replaced by a constant value, except when it is coupled to gravity,

(ii) the continuity equation is replaced by the incompressibility condition,

(iii) the equation of state is simplified, to express a linear relation between density and temperature, and

(iv) dissipation of heat is neglected.

For consistency such parameters as specific heat, conductivity, etc., should be treated as constants. For a detailed discussion of the Boussinesq approximation see, for example, Spiegel and Veronis (1960).

The Boussinesq approximation will only work when the vertical scale of the phenomenon studied is small compared to the scale height of the fluid or gas (the height over which the density falls by a factor e^{-1}), and it is also necessary that local pressure variations are small (to avoid large density changes) and slow (to avoid sound waves). Studying the circulation of the lake the Boussinesq approximation seems justified, except for the linearization of the equation of state, which obviously is a bad approximation at low temperature. It is, however, possible to use a "quasi-Boussinesq" approximation, introducing a non-linear equation of state, still involving only density and temperature. One may think that the approximation can be generalized to include salinity as well, when dealing with oceanic water, but here special complications arise. When we apply the Boussinesq model to the real world it is obvious that the model temperature and model density represent not the in situ values but the potential values. For fresh water we have a uniquely defined potential density, producing equivalent dynamic effects independent of the reference level (the depth or the pressure at which we identify potential and in situ density). For salt water this is not the case, and ambiguous results are obtained when larger pressure variations are in play. As an example, the potential density field for a section along the western trough of the Atlantic Ocean is shown in figure 1a; this potential density is referenced to the surface (zero decibar). If we use 4000 m (4000 decibars) as reference we produce a different picture, shown in Figure 1b. In Figure 1a, we see static instability near the bottom, in the corresponding portion of Figure 1b we see static stability. Calculations by use of a Boussinesq model would have predicted convective overturning in the first case, and no overturning in the second case. In the example given the difference between the two pictures can be understood in terms of the compressibility effect. Cold water is more compressible than warm water. In the region where there is a static instability in Figure 1a warmer and saltier water (North Atlantic Deep Water) overlies colder and fresher water (Antarctic Bottom Water). When we bring these waters to zero decibars the larger expansion coefficient for the colder water makes this become lighter, and a

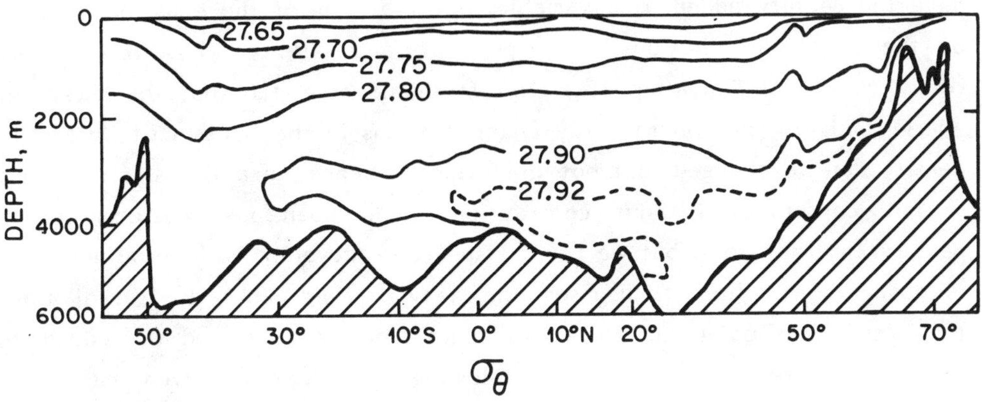

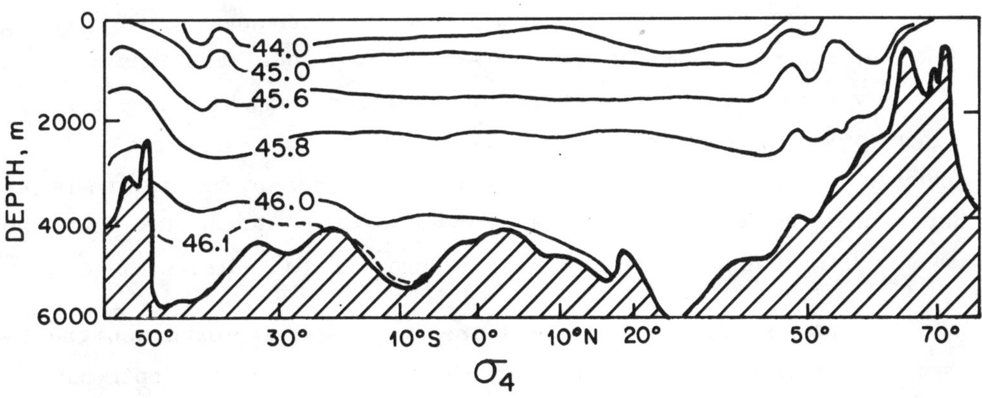

Fig. 1 (a) Isolines for potential density σ_θ referred to zero decibars, and (b) for potential density σ_4 referred to 4000 decibars, along the western trough of the Atlantic Ocean. Redrawn from GEOSECS Atlantic Expedition, Sections and Profiles (1980).

statically unstable situation is created. The in situ situation is actually a statically stable one, as indicated by Figure 1b. It is only by the presence of salt that we can produce these two different results; in a freshwater case the potential densities, referred to the same two levels, would give identical predictions for the static stability. The correct Boussinesq-approximated equation of state for sea water does not merely have the salinity added as a variable, but the form of the equation depends on the actual distributions of temperature and salinity. Thus, it does not really qualify as "equation of state". It should be noted that the mass and energy budgets take on radically changed forms in the Boussinesq fluid. The density can be changed, but not the volumes, hence, in a formal sense mass is not conserved. Without contractions and expansions pressure cannot perform internal work, but we find an apparent buoyancy work term: net work is done when lighter fluid rises and heavier fluid sinks. In a real fluid the pressure is doing all the work, apart from friction, and the buoyancy work is net zero (as is seen by following a particle of given mass in a closed loop). The gravity is important for the creation of mechanical work in the real oceans, but the effect is an indirect one, through the setup of hydrostatic pressure variations against which expanding and contracting fluid elements do work. It should also be noted that heat diffusion is a necessary ingredient: to get net positive work expansion must take place at higher pressure, contraction at lower pressure, and this can only be achieved if heat is diffused down from the surface. In the atmosphere we have a more effective "engine" for here the heating naturally takes place at lower levels, cooling at higher levels, through radiative effects.

In spite of all these problems the Boussinesq approximation can be used to model, to the zero'th order, the circulation in the upper part of the oceans, where the stratification is strong and the compressibility effects are not too large. With the Boussinesq approximation employed it may further become possible to formally eliminate either the temperature or the salinity from the problem, or equivalently, replace these by the density variable. If we have conservation of the temperature and salinity in the model: $\frac{dT}{dt} = 0$, $\frac{dS}{dt} = 0$, and an equation of state of the form $\rho = \rho_0\,(1-\alpha T+\gamma S)$, we obviously have $\frac{d\rho}{dt} = 0$ as well, and the equation, together with the dynamic equations and the incompressibility relation, will close the problem. We obviously have a different situation if molecular effects are important, as in the double-diffusive regimes. The corresponding molecular form equations are $\frac{dT}{dt} = \kappa\nabla^2 T$, $\frac{dS}{dt} = \kappa_S\nabla^2 S$, and these cannot be combined to a corresponding equation for the density when the diffusivities κ and κ_S are different (diffusivity for salt is about 1% of that for heat). For a discussion of the double-diffusive regimes, which are believed to be important at least locally in the oceans, the readers are referred to Turner (1973).

The above "density form" model may also fail for another reason. The boundary conditions applied at the sea surface may be a fixed temperature and fixed salinity, or a fixed temperature flux and fixed salinity flux (proportional to the applied heat flux and the evaporation/precipitation, respectively). These conditions fix the density, in the first case, and the density or buoyancy flux, in the second case. The more general Rayleigh type conditions, giving the turbulent surface fluxes proportional to the difference between a forcing value and the local surface value, $-\overline{w'T'} = \sigma(\tilde{T}-T)$, $-\overline{w'S'} = \sigma(\tilde{S}-S)$, where $\tilde{T}$ and $\tilde{S}$ are the forcing values will work as well. Such conditions are sometimes used in numerical models to "nudge" the solution towards a wanted climatological mean state; in such a case $\tilde{T}$ and $\tilde{S}$ are observed mean values, given on a function of horizontal coordinates. These Rayleigh type conditions are, however, only consistent with a "density form" if the transfer coefficient σ is the same in the conditions for temperature and for salt, and it can be argued that this is an unrealistic assumption. Firstly, note that the coefficient has the dimension of the velocity, and that its division into a depth H (which could be the mixed layer depth, or some other suitable model depth) gives us a time constant, $\tau = H/\sigma$, or response time, for the layer considered. The response times for temperature and salinity are generally much different: the temperature responds fast, the salinity slowly. An alternative statement is that the thermal feedback is strong (the flux is strongly influenced by T) and the haline feedback is weak (the flux is practically independent of S). When we apply different values for the coefficient in the Rayleigh surface conditions for T and S we can no longer transform the problem to a "density form". Such cases of "mixed boundary" conditions are discussed further in Section 4.

3. MAIN THERMOCLINE REGIMES, AND THE ROLE OF VERTICAL DIFFUSION

We consider the possible regimes in the main thermocline, as we proceed down from the bottom of the top Ekman layer/mixed layer to the top of the deep water. The basic equations are assumed to be the same as in classical similarity theory for the thermocline, that is, "density form" equations, with density determined by a balance of three-dimensional advection and vertical diffusion, and geostrophic-hydrostatic dynamics. The diffusivity κ may be variable. Using standard beta-plane coordinates we have thus for the model density

$$u \frac{\partial \rho}{\partial x} + v \frac{\partial \rho}{\partial y} + w \frac{\partial \rho}{\partial z} = \frac{\partial}{\partial z} (\kappa \frac{\partial \rho}{\partial z}) \tag{3}$$

The diffusivity which really represents a cross-isopycnal mixing (mixing of density along its own iso-surfaces is a trivial process), is assumed to lie in the range 0.01 to 0.1 cm^2s^{-1} in the strongly stratified upper part of the main thermocline, and have a large value, of order 1 cm^2s^{-1}, in the lower part representing the upper deep sea. Justification for such values comes from studies of tracers as well as measurements of certain turbulent parameters, such as the variance of the small scale temperature gradient, the dissipation, etc. For reviews of these studies, see Broecker and Peng (1982), Gargett (1984), and Gregg (1985). The diffusivity typically increases with the depth, due to an inverse relation to the Brunt-Väisälä frequence N (the law $\kappa = \kappa_0 (N/N_0)^n$, with n in the range -1 to -2 is suggested by several investigators). It may be noted that the value 1 cm^2s^{-1} agrees with the earlier estimate for the deep sea by Munk (1966).

We introduce the following scaling constants: the horizontal and vertical density variation, $\Delta\rho$; the horizontal scale, L; the vertical scale, D; the current, U; and the vertical velocity, W. The scale L is assumed to be planetary, of the order of earth's radius, R. The vertically variable diffusivity is replaced by a constant, κ_0 , the latitudinally variable Coriolis parameter by a constant, f_0. The constants D, U, and W are unknown, the other constants are given. From the assumed geostrophic and hydrostatic balances we get a thermal wind equation, for example, $f\frac{\partial u}{\partial z} = \frac{g}{\rho_o}\frac{\partial \rho}{\partial y}$ for the zonal current. Assuming that the current is weak at large depths, the constant U can also represent the current variation vertically, and we have the scaling relation $U \sim g\frac{\Delta\rho}{\rho_o f_o}\frac{D}{L}$. Using the incompressibility relation we find $W \sim \frac{UD}{L} \sim g\frac{\Delta\rho}{\rho_o f_o}\frac{D^2}{L^2}$, representing one relation between D and W. A second such relation is obtained from the scaling of equation (3). The advection terms are of the same order, as found from the incompressibility relation, and if diffusion is included we must therefore have $\frac{W\Delta\rho}{D} \sim \frac{\kappa_o\Delta\rho}{D^2}$. Solving for D and W

$$D \sim \left[\frac{\kappa_o f_o L^2}{g\Delta\rho/\rho_o}\right]^{1/3}, \quad W \sim \left[\frac{\kappa_o^2 g\Delta\rho/\rho_o}{f_o L^2}\right]^{1/3} \tag{4}$$

Using the realistic values $\Delta\rho/\rho_o = 1/500$, $g = 10^3 \text{cm s}^{-2}$, $f_o = 10^{-4} \text{ s}^{-1}$, $L = 6.10^8$ cm, and κ_o in the assumed range 0.01 to 0.1 cm^2s^{-1}, we find that D is in the range 55 to 120 m, and W in the range $1.8 \cdot 10^{-6}$ to $8 \cdot 10^{-6}$ cm s^{-1}. The estimate applies to the upper strongly stratified part of the main thermocline. To be able to match the thermocline solution to that of the overlying Ekman layer, the vertical velocity must be comparable to the Ekman vertical velocity, which is determined by the formula $w_e = \text{curl}_z(\vec{\tau}^W/\rho_o f) \sim \frac{\tau_o}{\rho_o f_o L}$, where τ_o is the wind stress amplitude. Choosing a value $\tau_o = 1$ dyne cm^{-2} gives $w_e \sim 1.8 \cdot 10^{-5}$ cm s^{-1}. This estimate is actually on the low side; the scale L should in this context be less than the planetary scale, for example, $L = 2 \cdot 10^8$ cm. In the latter case we get the higher value $w_e \sim 5 \cdot 10^{-5}$ cm s^{-1}. In any case, it is found impossible to match the thermocline solution to the Ekman one. Since the matching of vertical velocities is necessary, we must add the scaling relation $W \sim w_e$, and consequently drop one previous scaling relation. We cannot give up the thermal wind relation or incompressibility, but terms can be dropped in equation (3) for the density. Since the advection terms are of the same order, our choice is to drop all advection, or diffusion. We cannot have vertical diffusion alone, if we want a thermocline, but it is possible to drop diffusion, and have an advective or ideal fluid thermocline, driven directly by the Ekman pumping. The new scalings give

$$D \sim \left[\frac{w_e f_o L^2}{g\Delta\rho/\rho_o}\right]^{1/2}, \quad W \sim w_e \tag{5}$$

With the parameters given earlier (the larger w_e -value) we now find $D \sim 300$ m, which looks better than the depth found for the diffusive thermocline. We can verify that the diffusion term indeed is smaller with this scaling; the estimate is $\frac{\partial}{\partial z}(\kappa \frac{\partial\rho}{\partial z})/w \frac{\partial\rho}{\partial z} \sim \frac{\kappa_o}{w_e D} = 0.007$ to 0.07, using κ_o -values in the assumed range. For a more detailed discussion of these thermocline scalings see Welander (1971).

As we proceed into the deep portion of the main thermocline, or the upper portion of the deep sea, the scaling is expected to change. The density gradients become small, and the current decreases to the same extent, by the thermal wind relation. The advection terms, proportional to both, will therefore decay in a quadratic sense. This is not true for the diffusion term: it is linear in the density gradient, and multiplied by a diffusivity which increases rather than decreases with the depth. At some larger depth we must therefore assume that diffusion and advection can balance, that is, a deep diffusive thermocline will develop. It is difficult to estimate the values of D and W associated with such a regime, with the parameters $\Delta\rho$ small and badly defined, and we do not expect this thermocline to stand out, in measured profiles of temperature or density. It should be noted that another change of balance may take place in this regime. If the density gradients drop we can only conclude that the current drops at a corresponding rate, and that the <u>horizontal</u> advection terms drop quadratically. It is necessary that the <u>vertical</u> velocity drops to zero as we approach a nearly homogeneous deep sea. If the current is zero we can only conclude from the incompressibility relation that $\frac{\partial w}{\partial z}$ is zero; w itself may have a non-zero value[2]. In fact, the similarity thermocline solutions all give a constant, non-zero w at large depths. If horizontal and vertical density gradients decay at comparable rates, the conclusions will be that vertical advection becomes dominating over horizontal advection. The correspondingly simplified equation $w \frac{\partial \rho}{\partial z} = \frac{\partial}{\partial z} (\kappa \frac{\partial \rho}{\partial z})$, expressing a local vertical balance, was actually used long ago by the geochemists in studying tracer balances in the deep sea.

It may also be argued that topography tends to create such a "geochemist's thermocline". The topography introduces a horizontal scale L_{top} which is much smaller than the planetary one, and it amplifies the vertical velocity and the vertical advection term in a corresponding way, by a factor L/L_{top}. The necessity for strong vertical velocities crossing isopycnal surfaces becomes intuitively obvious when one looks at data in the deep sea: one sees jagged topography in the contrast to a smooth, large-scale density field. An overview of the different regimes suggested by the previous discussion is given in Figure 2.

Some comments about the ideal fluid thermocline regime should be added. It is well known that for an ideal fluid boundary values in the form of densities and normal velocities can be prescribed, with the restriction that density can be prescribed only at inflow points. Our ideal fluid thermocline has the bottom of the Ekman layer/mixed layer as upper boundary; there may also be free lateral boundaries in some model calculations. At these boundaries we must prescribe the density at inflow points, and calculate outflow points from the solution. At the transition from inflow to outflow we will generally find a discontinuity in density; this discontinuity obviously will propagate along the streamline, producing an interior discontinuity surface. To remove such discontinuities diffusion

---

2 Because of the bottom a constant w cannot prevail throughout the deep sea. In a proper solution carried into a homogeneous deep sea there must be a small depth-independent current, and an associated small value of $\partial w/ \partial z$. The vertical velocity thus shows a weak linear variation with the depth. The velocity values must be adjusted to satisfy the boundary condition of zero normal motion at the bottom. From point of view of the thermocline regime we can, however, consider w constant asymptotically.

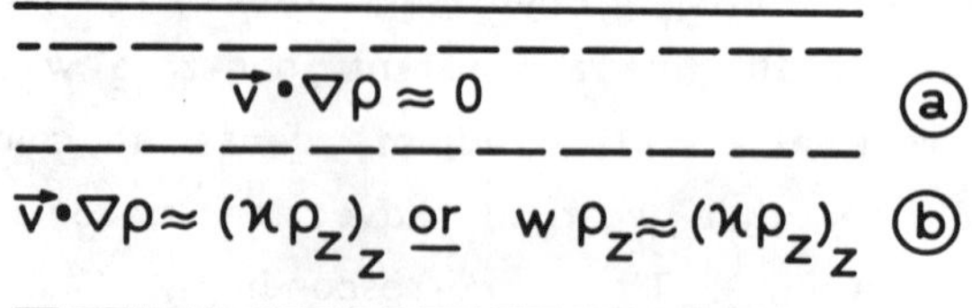

Fig. 2 Schematic picture of main thermocline regimes in a vertical section:
an ideal fluid thermocline (a) overlying an asymptotic diffusive thermocline (b); the latter may degenerate to a "geochemist's thermocline", with vertical advection and diffusion balancing to zero'th order.

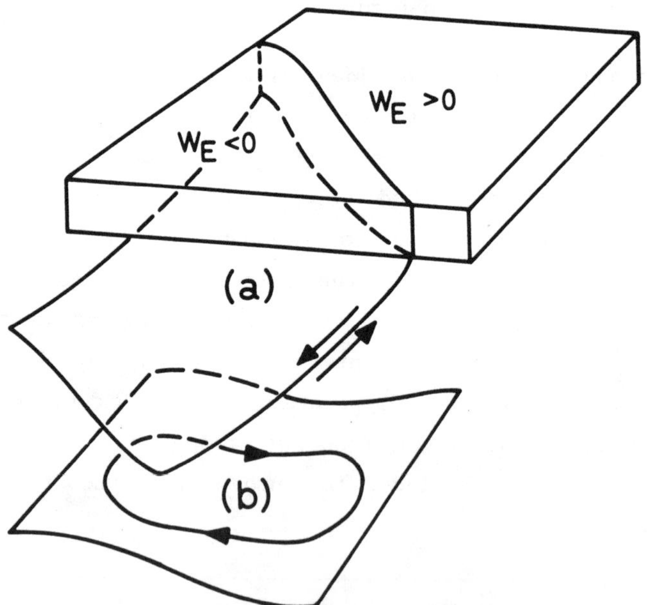

Fig. 3 Singular regimes in the ideal fluid thermocline:
(a) discontinuity surfaces joining the Ekman layer where $w_E = 0$,
(b) regions with closed streamlines, which do not intersect the Ekman layer, or lateral boundary layers.

is needed. We see here the ideal fluid representation of oceanic fronts; such fronts are a required feature in any realistic ocean circulation model.

Another problem arises when the streamlines of the ideal fluid thermocline close internally, without intersecting any boundaries. In such a case the solution becomes undetermined, and again we must add diffusion. See Figure 3. (Further discussion of this problem is found in another article in this volume by P. Rhines.)

4. PHENOMENA RELATED TO MIXED BOUNDARY CONDITIONS

As mentioned earlier, "mixed boundary conditions" refer to boundary conditions for temperature and salinity at the sea surface, which cannot be translated into an equivalent condition for a density or buoyancy variable. As an example, such conditions arise when we prescribe one of the variables, say, the temperature, and the flux of the other, say, the salinity flux.

We will, in particular consider boundary conditions of the Rayleigh type, discussed briefly in Section 2. In this case we will, however, assume that the transfer coefficient for temperature and salinity fluxes are different; and we write

$$\overline{-w'T'} = \sigma\ (\tilde{T}-T)\ ,\ \overline{-w'S'} = \sigma_S\ (\tilde{S}-S) \qquad (6a,\ b)$$

This gives us two response times, a thermal response time $\tau = H/\sigma$, and a haline response time $\tau_S = H/\sigma_S$, where H is a suitably defined layer depth. For the upper ocean down to, say 100 m, τ is of order weeks to a few months, while τ_S must be very large, of order hundreds to thousands of years, considering that there is practically no feedback from the surface salinity S onto the salinity flux. In fact, as a first approximation it seems reasonable to go to the limit $\sigma_S \to 0$, $\tilde{S} \to \infty$, with a finite product $\sigma_S \tilde{S}$, marking the salinity flux prescibed, $-\overline{w'S'} = \sigma_S \tilde{S} = Q_S$. If we want a still simpler case, we can take the other limit, $\sigma \to \infty$, for the thermal response, meaning a prescribed surface temperature, $T = \tilde{T}$.

Stommel (1961) apparently was the first to note the new phenomena which arise in simple thermohaline convection models when mixed boundary conditions are applied. He dealt with a system in the form of two well-mixed boxes, connected by upper and lower hydraulic links (flow proportional to the pressure difference at the respective levels). See Figure 4a. The equations for this system are

$$V\dot{T}_1 = \kappa(\tilde{T}_1 - T_1) + |q|(T_2 - T_1) \ , \quad V\dot{S}_1 = \kappa_S(\tilde{S}_1 - S_1) + |q|(S_2 - S_1) \qquad (7a,\ b)$$

$$V\dot{T}_2 = \kappa(\tilde{T}_2 - T_2) + |q|(T_1 - T_2) \ , \quad V\dot{S}_2 = \kappa_S(\tilde{S}_2 - S_2) + |q|(S_1 - S_2) \qquad (8a,\ b)$$

$$q = C(\rho_1 - \rho_2) \ , \quad \rho = \rho_0(1 - \alpha T + \gamma S) \qquad (9,\ 10)$$

where V is the box volume (equal boxes), q is the flow (volume transport) in the tubes, κ and κ_S constants proportional to the previous constants σ and σ_S, and C is a hydraulic constant. The hydrostatic and Boussinesq approximations are obviously applied, and it is assumed that the time-lag in the tube flow is negligible.

We can without restriction assume an antisymmetric forcing, $\tilde{T}_2 = -\tilde{T}_1$, $\tilde{S}_2 = -\tilde{S}_1$. Starting from an antisymmetric initial state the solution then stays antisymmetric; if there is an initial symmetric component this will die out in time. Choosing $2\tilde{T}_1$ and $2\tilde{S}_1$ as units for T and S, and $\frac{V}{\kappa}$ as unit for time, and introducing the variables $x = T_2 - T_1$, $y = S_2 - S_1$, we find the simple system

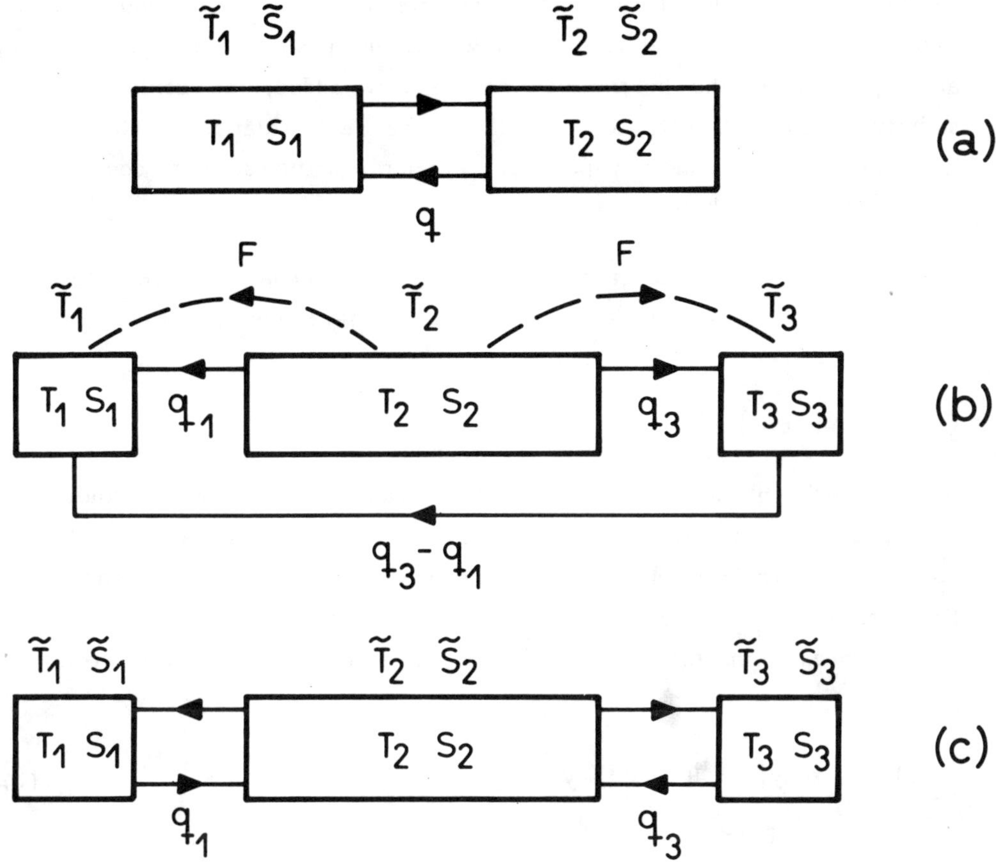

Fig. 4 (a) Stommel's two-box model. The forcing temperature $\tilde{T}$ and salinity $\tilde{S}$ act through a Rayleigh type transfer law.
(b) Rooth's three-box model. The center box represents the waters at low latitudes, the two outer boxes represent high latitude waters. The forcing temperatures $\tilde{T}_e$ and $\tilde{T}_p$ and the freshwater flux 2F from the center box, going equally into the outer boxes, are prescribed. A single deep pole-to-pole connection can transport deep water.
(c) The three-box model discussed in the present paper. Two deep pole-equator connections exist, allowing a circulating symmetric state. See the text for details.

$$\dot{x} = 1-x-|q|x , \quad \dot{y} = \lambda(1-y)-|q|y \tag{11a, b}$$

$$q = mx-ny \tag{12}$$

where λ is the ratio of the tranfer coefficients κ_S and κ ; and m, n are two constant proportional to α, λ . We assume positive values for both $\tilde{T}_1$ and $\tilde{S}_1$, that is, the heated box has salinity added, the cooled box has salinity subtracted. This corresponds to the real situation in the oceans, with net heating and evaporation at low latitudes, net cooling and precipitation at high latitudes.

Consider firstly the case $\lambda=1$, that means equal response time for T and S. For the density variable $z = mx-ny$ we have then

$$\dot{z} = m-n-z - |z| z \tag{13}$$

Setting the right hand zero, to get the steady states, only one solution is found, as the readers easily can verify. This state is associated with positive flow ($q>0$) when $m>n$, with negative flow ($q<0$) when $m<n$. In the first case we have a dominant thermal torque and in the second a dominant haline torque. In the case $\lambda \neq 1$ we have two steady state relations

$$1-x-|mx-ny|x = 0 \ , \quad \lambda(1-y) - |mx-ny|y = 0 \tag{14a, b}$$

representing two hyperbolas in the x-y-plane. It is found that these can intersect in one or three points, giving one or three steady states. If we have a single steady state it must obviously be stable; if we have three steady states an analysis shows that two are stable and one is unstable. It can be proven that no limit cycle is associated with the unstable steady state; thus the system cannot carry out self-sustained oscillations but must end in one of the two existing stable steady states. We are interested in the case $\lambda < 1$, meaning that the thermal response time is shorter than the haline one. In this case, with two stable steady states present, we find either a fast loop with dominant thermal torque, or a slow loop with reversed flow and dominant haline torque; see Figure 5a, b. We can understand the situation most simply by looking at the limiting case were the box temperature as well as the salinity fluxes are kept fixed;

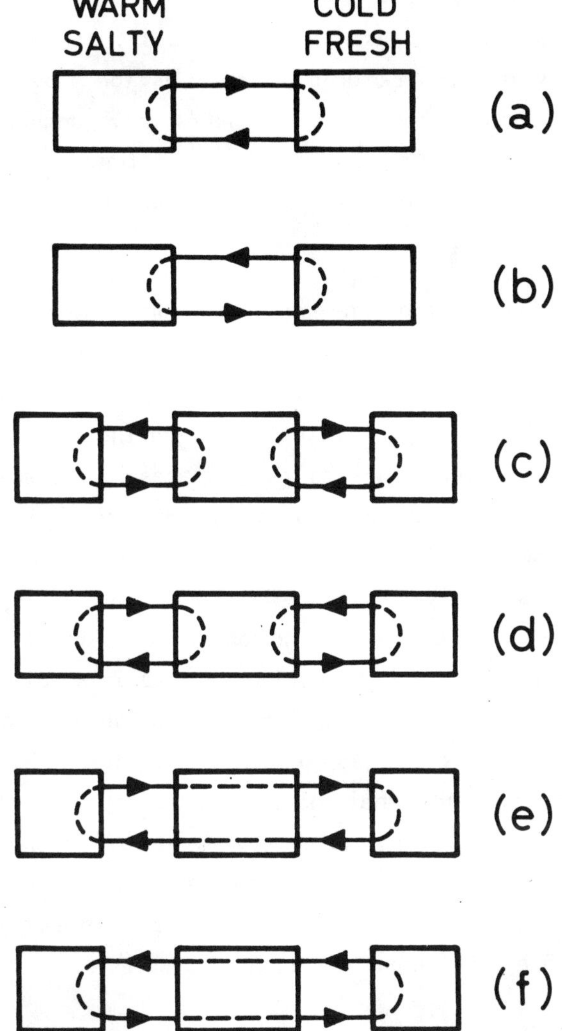

Fig. 5 (a) The stable fast-cell solution in Stommel's two-box model. The thermal torque dominates over a braking haline torque.
(b) The stable slow-cell solution in the same model. The haline torque dominates over a braking thermal torque, producing a reversed convection cell. Between these two cases exists a third, unstable solution. See the text for details.
(c), (d) Symmetric three-box solutions, corresponding to two solutions of type (a) and type (b), respectively, placed back to back.
(e), (f) Asymmetric solutions, corresponding to one solution of type (a) and one of type (b), placed back to back.

in this case there are always two stable steady states. The thermal torque dominates if the flow is fast, since it stays independent of the flow, while the salinity differences drop to zero as the flow is increased. For slow flows the haline torque must necessarily dominate; the salinity difference actually goes to infinity when the flow slows to zero. For a more detailed discussion of the solution the readers may consult the article by Stommel (1961).

The assumption of zero time-lag in the connections of the above model may be questioned. In the real oceans the movements of waters between equatorial and polar regions are associated with time-lag of decades to centuries. Although such lags have no effect on the steady states they may affect the transient approaches to these states. A suitable model for the exploration of time-lags is the so-called "Howard-Malkus loop" (Malkus, 1972), in our case formulated in a thermohaline version. The loop is shown in Figure 6. It is subjected to a Rayleigh type forcing as in the box model, but $\tilde{T}$ and $\tilde{S}$ are now given functions of the angle ψ. The temperature T and salinity S of the fluid in the loop are assumed to depend on ψ and time t only and the angular velocity $\omega=\dot{\psi}$ is assumed to depend only on t, under the assumed Boussinesq approximation. Inertial effects are neglected in our case, and we assume that the flow represented by the variable ω, is proportional to the buoyancy torque. The torque is proportional to the negative density multiplied by sin ω and integrated around the loop. We have three equations for T, S and ω, after expressing the density in terms of T and S by the linear equation of state:

$$\frac{dT}{dt} = \frac{\partial T}{\partial t} + \omega \frac{\partial T}{\partial \Psi} = r[T(\psi)-T] \tag{15a}$$

$$\frac{dS}{dt} = \frac{\partial S}{\partial t} + \omega \frac{\partial S}{\partial \psi} = r_s[\tilde{S}(\psi)-S] \tag{15b}$$

$$\omega = \oint (mT-nS)\sin \psi \, d\psi \tag{16}$$

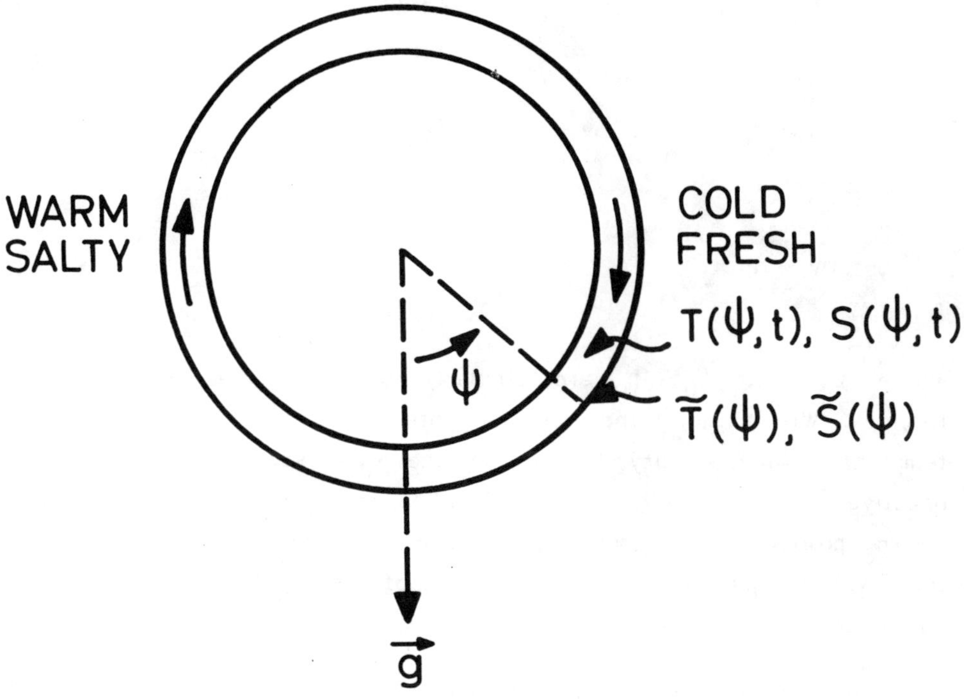

Fig. 6 The Howard-Malkus circular convection loop in a version with combined thermal and haline forcing. Forcing $\tilde{T}$ and $\tilde{S}$ are given along the loop. Inertia is neglected in the present version. See the text for details.

where r, r_S are two coefficients proportional to σ, σ_S ; m, n two coefficients proportional to α, γ . If we introduce the variables

$$x = \oint T \cos\psi \, d\psi \,, \qquad y = \oint T \sin\psi \, d\psi$$

$$\xi = \oint S \cos\psi \, d\psi \,, \qquad \eta = \oint S \sin\psi \, d\psi$$

we find, multiplying the equations (15a) and (15b) once with $\cos\psi$, and integrating around the loop, and once with $\sin\psi$ and integrating similarily,

$$\dot{x} + \omega y = r(x_o - x) \;, \quad \dot{y} - \omega x = r(y_o - y) \tag{17a, b}$$

$$\dot{\xi} + \omega\eta = r_s(\xi_o - \xi) \;, \quad \dot{\eta} - \omega\xi = r_s(\eta_o - \eta) \tag{18a, b}$$

$$\omega = my - n\eta \tag{19}$$

where x_o, y_o, ξ_o, η_o are similarly defined loop-integrals for the forcings. We consider the case of antisymmetric forcing, with positive temperature and salinity forcing on one side, say, the right side, and negative forcing on the other side. Thus we set $x_o = \xi_o = 0$ and choose ψ_o and η_o positive; with a suitable choice of units for T and S we can make these values equal to +1. After insertion of the ω-value the equations take on the form

$$\dot{x} + (my - n\eta)y = -rx \;, \quad \dot{y} - (my - n\eta)x = r(1 - y) \tag{20a,b}$$

$$\dot{\xi} + (my - n\eta)\eta = -r_s\xi \;, \quad \dot{\eta} - (my - n\eta)\xi = r_s(1 - \eta) \tag{21a,b}$$

We consider firstly the case $r_s = r$, equal response time for temperature and salinity, allowing a density form. With the new variables $z = mx - \eta\xi$, $\zeta = my - n\eta$ we get the two equations

$$\dot{z} + \zeta^2 = -rz \;, \quad \dot{\zeta} - z\zeta = r(m - n - \zeta) \tag{22,23}$$

The corresponding steady state equations, $\zeta^2 = rz$ and $-z\zeta = r(m-n-\zeta)$, represent a parabola and a hyperbola, respectively, in the z-ζ-plane. It is easy to verify that these can intersect in only one point, giving a single, and stable, steady state.

In the case where $r_s \neq r$ the situation becomes more complex, with the steady states represented by the intersection of four quadratics in the

x-y-ζ-η-space. It is found possible to obtain three steady states, as in the previous box model, but the analysis is too complicated to be reproduced here. An interesting new feature is the existence of self-sustained oscillations, in certain parameter ranges. These are associated with unstable steady states, and are generally of the single-periodic type, that is, they correspond to limit cycles in the x-y-ζ-η-space (numerical examples have only produced the limit cycle case, but it has not been proved that this is the only possibility). One numerical example, calculated by time-stepping the above equations on a programmable calculator (HP 41C), is shown in Figure 7; in this case a limit cycle is obviously present.

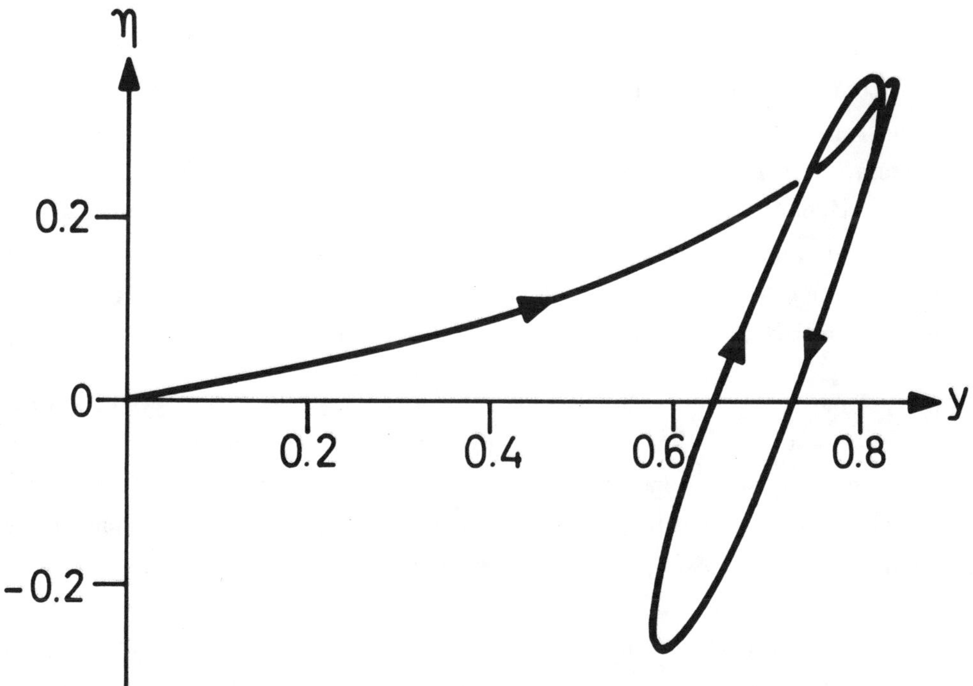

Fig. 7 Example of a numerical solution to the equations (17, 18, 19) governing the thermohaline, non-inertial Malkus-Howard loop. While the purely thermal version must go into a stable steady state, this version allows self-sustained oscillations, which generally takes the form of a limit cycle. Parameter values are $\lambda = 0.1$, $\eta_0 = 1.9$, $y_0 = 1.5$, initial state $x = y = \varepsilon = \eta = 0$.

From Stommel's two-box model one can go to a three-box version, to allow a "global" representation of the oceans: one "equatorial box" and two "polar boxes". Rooth (1982) considered such a model, with two upper connections, and a single deep pole-to-pole connection (Figure 4b). He applied the forcing asymmetrically, with a Rayleigh law for the temperature, and a given freshwater flux from the equatorial to the polar boxes (equivalent to a given salinity flux if the salinity variations are small). The prediction was that the symmetric state would be unstable in a certain parameter range, and that a steady single cell, of the pole-to-pole type, would then develop. Rooth believed that a non-linear equation of state was needed to get the unstable development, but a later analysis has shown that such a non-linearity is not required. This general idea turned out to be a fruitful one, Bryan (1986) verified the presence of such an instability, and the following "flip" to a pole-to-pole cell, in a numerical experiment using a three-dimensional model, with symmetrized geometry, and forcings corresponding to those of Rooth. In Bryan's calculation the "flip" was triggered by a moderate change (+2°/oo in one case, -1°/oo in another) of the salinity in a 50 m thick surface layer, in one polar region (poleward of 45° lat.). The switch from the original two-cell solution to a one-cell solution was, as one would expect, associated with dramatic changes in the poleward heat flux.

This interesting phenomenon obviously deserves further exploration, and this author has attempted to give some contribution, by looking at two specific models in some detail. One model is a version of Rooth's 3-box model which has two deep connections (Figure 4c), allowing a symmetric flow solution (in Rooth's case, with only one deep connection, the symmetric state must be one of zero motion). The single pole-to-pole connection is left out; putting it in will not change the model results in any qualitative way. We will consider here only the simplest case, where the temperatures of the boxes, and the salinity fluxes, are given. The equations for the two polar salinities (see Figure 4c) are then

$$V_p\dot{S}_1 = -F + |q_1|\,(S_2-S_1)\quad,\quad V_p\dot{S}_3 = -F + |q_3|\,(S_2-S_3) \qquad (24a,b)$$

$$q_1 = a\Delta T + b(S_2-S_1) \;, \quad q_3 = a\Delta T + b(S_2-S_3) \tag{25a,b}$$

where V_p is the polar box volume, and $\Delta T=\tilde{T}_e-\tilde{T}_p$ is the applied temperature difference; the constants a,b are proportional to α,γ. The constant F is positive, to give the prescribed negative salinity flux into the polar boxes. The equatorial salinity represents an additional variable, but it can immediately be expressed in terms of S_1, S_3, since the mean salinity $\overline{S}$ is conserved:

$$V_p(S_1+S_3) + V_eS_2 = (2V_p+V_e)\overline{S} = \text{constant} \tag{26}$$

where V_e is the equatorial box volume. We choose the mean salinity $\overline{S}$ as unit for S_1,S_3, and $\frac{V_p\overline{S}}{F}$ as unit for time. After insertion of the value for S_2, q_1, and q_3 the equations become

$$\dot{S}_1 = -1 + d|1-\varepsilon-mS_1-nS_3|(1-mS_1-nS_3) \tag{27a}$$

$$\dot{S}_3 = -1 + d|1-\varepsilon-nS_1-ms_3|(1-nS_1-MS_3) \tag{27b}$$

where $d = q\overline{S}\left(\frac{2V_p+V_e}{V_e}\right)^2$, $\varepsilon = \frac{a\Delta T}{b\overline{S}}$, $m = \frac{V_p+V_e}{2V_p+V_e}$, $n = 1-m$.

Steady states are given by the two equations

$$-1 + d|x-\varepsilon|x = 0 \;, \quad -1 + d|y-\varepsilon|y = 0 \tag{28a, b}$$

where $x = 1-mS_1-nS_3$, $y = 1-nS_1-nS_3$ are variables representing two straight lines in the S_1-S_3-plane. The function $F(x) = -1 + d|x-\varepsilon|x$ can have one or three zeros (see Fig. 8); the latter case is found when $d\varepsilon^2 > 4$, that is, when the thermal forcing is sufficiently strong. The steady state equation $F(x) = 0$ is represented by three parallel lines in the S_1-S_3-plane, and the corresponding equation $F(y) = 0$ is represented by three parallel lines, with a different slope. These six lines obviously intersect at a total of

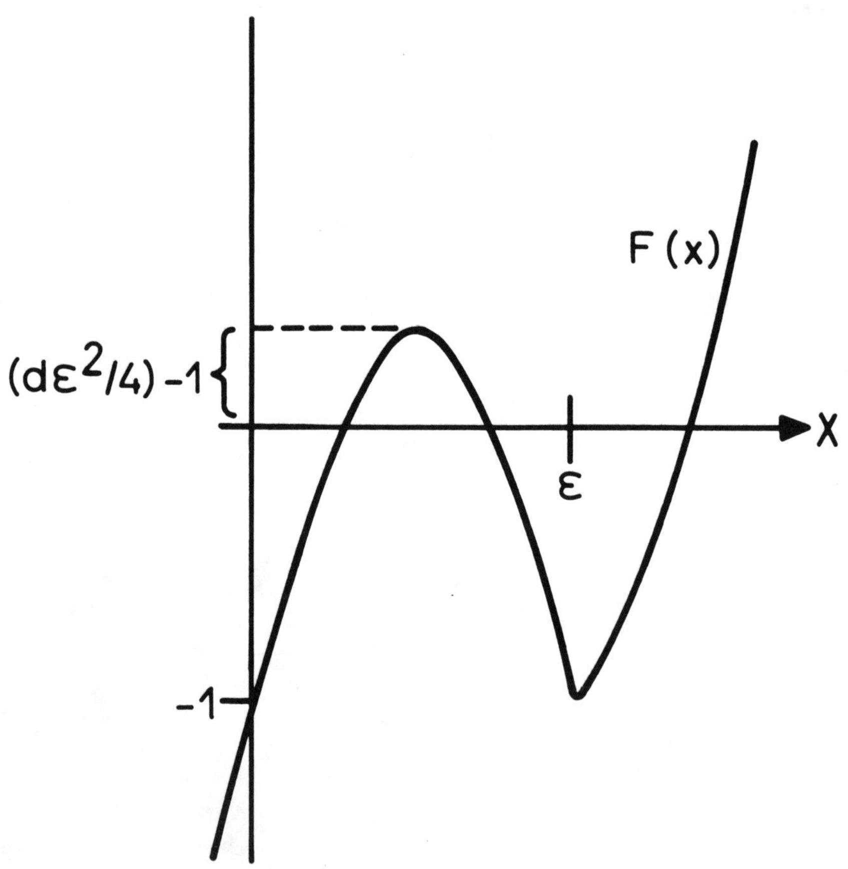

Fig. 8 Sketch of function $F(x) = -1 + d\,|x-\varepsilon|\,x$. If $d\varepsilon^2 > 4$ there are three roots, if $d\varepsilon^2 < 4$ there is one root, to the equation $F(x) = 0$. See equation (28).

nine points. Of the nine steady states four states are stable and five unstable. Two of the stable steady states are symmetric, with $S_3=S_1$, and $q_3=q_1$. One of these has dominating thermal torque and positive flows (q_1, $q_3<0$), the other dominant haline torque and negative flow (q_1, $q_3<0$). We can look at these solutions as a superposition of two of Stommel's two-box solutions, placed back to back. See Figure 5c, d. The two remaining stable steady states are asymmetric, having the appearance of a single pole-to-pole cell. In one half of this single cell the thermal torque dominates, in the other half the haline torque dominates. Again we can look at the solution as a superposition of two cells from the two-box solution, with opposite circulations. See Figure 5e, f. A numerical example is shown in Fig. 9; it shows the family of straight lines determinating the steady states, and the direction field of the trajectories of the transient solutions, in the S_1-S_3-plane. It is finally noted that the previous three-box model, when driven either by fixed temperature and salinities, or by fixed temperature and salinity fluxes, has only one steady state; the calculations of these cases are straightforward but are not shown here.

One may speculate about what happens in case the number of boxes is increased further. If we force the system in corresponding fashion, that is, define an equatorial part which is warm and receiving salinity, and two polar parts which are cold and losing salinity, a multi-box model is believed to give the same basic nine solutions, without appearance of added smaller loops. However, these cases have not yet been explored in theoretical detail. In the continuous model studies, the three-dimensional model by Bryan (1986) and some simplified models studied by this author, only the symmetric two-cell and the asymmetric one-cell solutions were found; however, the study of these models as functions of the model parameters is not completed. If lateral boundaries are removed, and the system is forced in a spatially periodic way, with alternating "equatorial regions" and "polar regions", one may possibly generate cells with much larger horizontal scale, by-passing also the polar regions. In contrast to the standard forced convection problem the corresponding thermohaline problem may, in case of periodic forcing, show an infinity of "subharmonic" steady states.

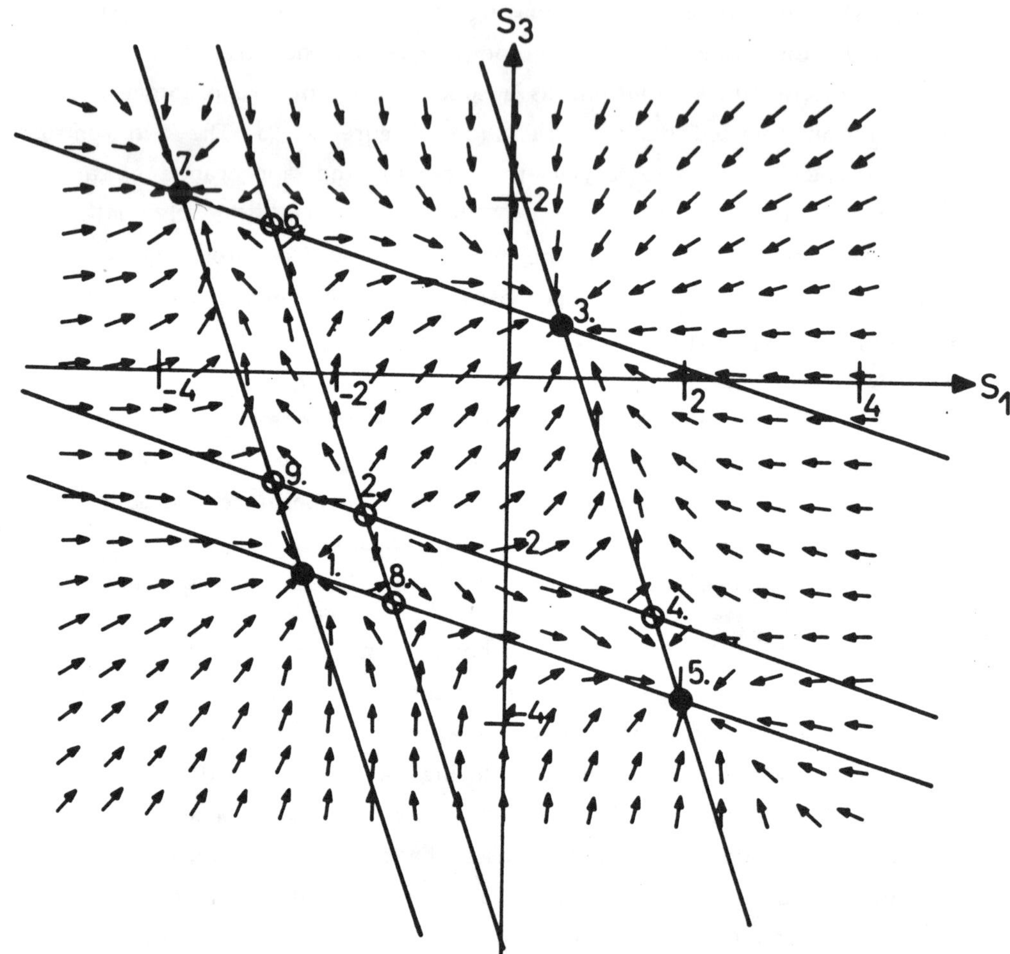

Fig. 9 A numerical example showing the steady states, and the trajectory direction field in the S_1-S_3-plane, for the three-box model decribed by equations (11, 12). There are nine steady states, four of which are stable (numbered 1, 3, 5, 7) and five unstable (numbered 2, 4 , 6, 8, 9). These can be looked on as double Stommel-type solutions, in a back-to-back position. Parameter values m = 3/4, d = 1, ε = 3. See the text for details.

It seems feasible to explore the "Rooth-Bryan flip", and similar phenomena in the thermohaline ocean circulation, by use of two-dimensional models which describe temperature, salinity, and flow fields as functions of a meridional and vertical coordinate. In complexity such models lie between the simple box models and the numerical three-dimensional models, such as the one used by Bryan (1986). A model of this type has actually been explored by the author, in cooperation with J. Willebrand and J. Marotzke at the Institut für Meereskunde in Kiel, with promising results. It has been possible to reproduce the qualitative feature of Bryan's calculation, in particular the "flip" and the change in poleward heat flux, with a model which can be run very economically. It has only two free parameters, and can thus be investigated with little difficulties in the entire parameter space of interest. The results from this study will be published elsewhere.

5. THERMAL AND THERMOHALINE OSCILLATORS

In the previous section an example of a thermocline oscillator was shown, carrying out self-sustained oscillations under steady forcings (see Fig. 7). Actually, a number of such oscillators, driven by thermal or thermohaline forcing, exist. There is no place here to discuss these in detail, but the readers may be interested in looking at the "oscillator gallery" put together in Fig. 10. Brief descriptions of the oscillatory mechanism of these specimens are added below; those interested in learning more about the subject should consult the references. It should be noted that several of the oscillators are easy to build, in a "kitchen version", and that some oscillators may be of practical interest as small "engines", for example, for pumping water in isolated regions, using solar radiation as power. These engines are not very efficient, but have the advantage that only fluids and gases, not mechanical parts, are in motion. There are also examples of unwanted oscillators, for example, harmful thermal oscillations have appeared in fluid cooling systems of nuclear power plants; these apparently are an example of the thermal-inertial oscillator decribed below.

Referring to the different oscillators in the gallery by the numbers indicated, we have the following story to tell. Oscillator 1 is one of the simplest existing thermal oscillators: two fluid-filled containers connected by a U-tube, which is heated at the lower end. One sees an oscillatory flow in the tube, and associated changes in fluid levels in the containers. If the flow starts out in one direction, from an initial symmetric state, with both levels equal, the flow will continue, with warm and lighter fluid rising in one branch and colder fluid sinking in the other branch of the U-tube (it is generally not necessary to add a specific cooling device at the top, as cooling is obtained by conduction and evaporation at the open surfaces). An opposing level difference builds up, and after sufficient time the motion must stop. This state is, however, not in long-time balance, as heat conduction tends to bring the system back to a symmetric state. As the temperature difference between the two branches of the U-tube decreases, a flow in the opposite direction eventually starts, due to the level difference, and then proceeds in a reflected fashion. The period for the oscillations depends on the geometry of the system and the effectiveness of heat source, the thermal conductivity, etc. A certain minimum value is required for the ratio of area of the open surfaces to the tube cross section area. Reference: Welander (1957), an error by a factor of 2 in the stability criterion should be noted. Holloway (priv. comm.) has initiated some accurate measurements of oscillations.

Oscillator 2 is somewhat similar to the previous one, but is commonly driven by a freshwater-salt water difference, rather than by heating and cooling. An open syringe with an attached hypodermic needle is filled with salt water, and placed in a freshwater container as shown in the figure. Salt water will spurt out periodically from the tip of the needle; in between, freshwater is sucked up into the syringe. The phenomenon is easily seen if the salt water is dyed. (The experiment also provides a nice demonstration of small vortex rings). If one starts with the inside and outside levels equal, salt water obviously flows out of the needle. When the level of the salt water has come low enough the flow must stop, but due

The "Oscillator Gallery"

The following oscillators are shown:

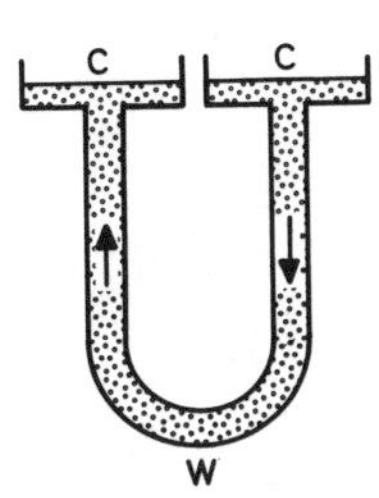

1) U-tube oscillator

2) Martin's salt oscillator

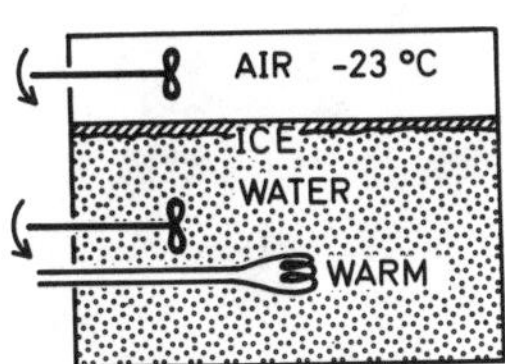

3) heat-ice oscillator

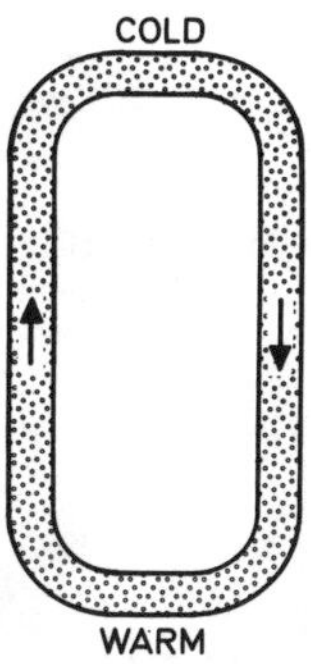

4) thermal-inertial oscillator (Howard-Malkus version)

5) water wheel analogy of thermal-inertial oscillator,

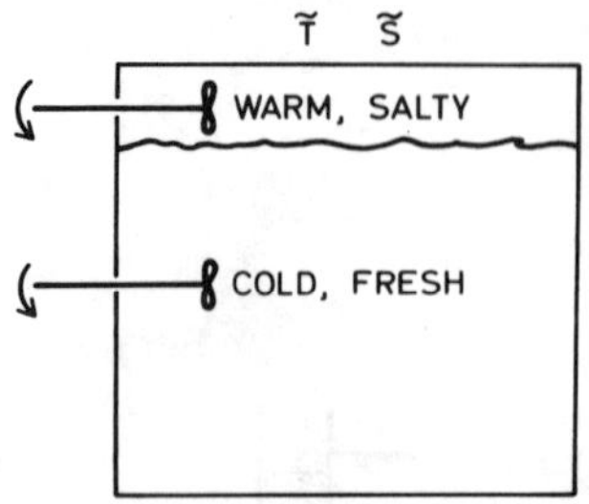

6) two-layer thermohaline oscillator

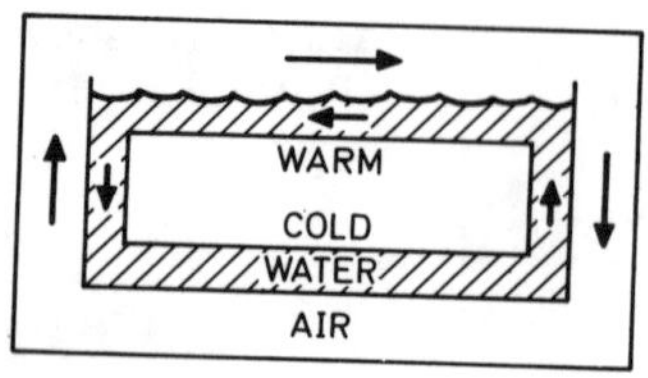

7) air-water oscillator

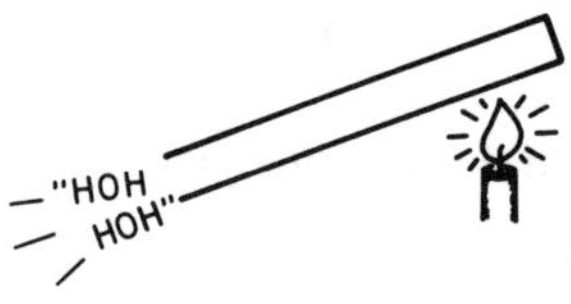

8) glass blower's organ

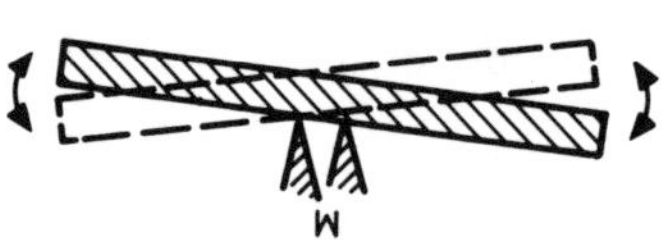

9) Trevelyan rocker

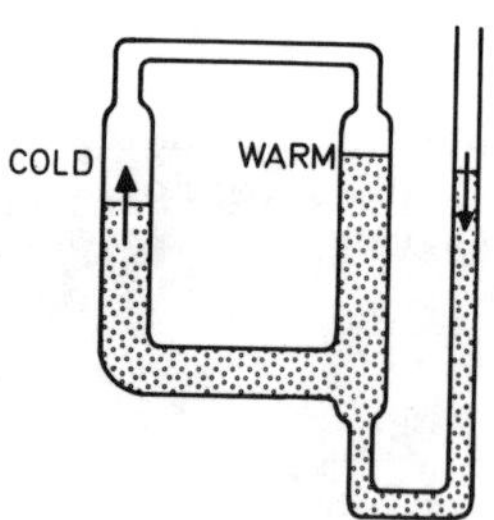

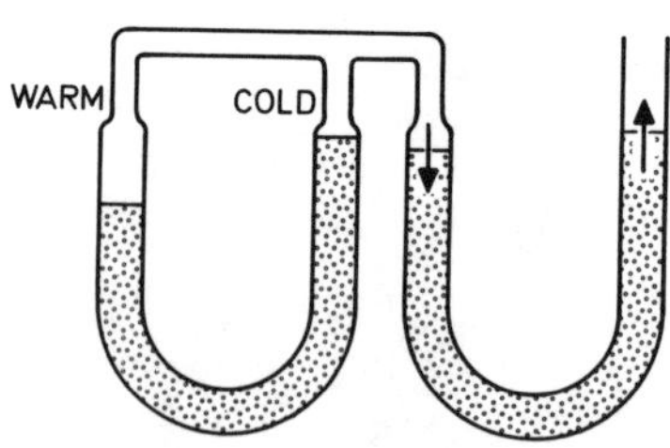

10-11) fluidyne engines

to diffusive effects this is not a perfect equilibrium. After a small amount of salt has diffused from the needle tip into the freshwater, a small upward buoyancy is present, and as soon as upward flow starts this buoyancy amplifies, by the freshwater entering the needle. The level of the salt water rises until a new balance has been reached. The needle is now filled completely with freshwater. A reversal will be caused by some salt diffusing down from the syringe, we get downward flow and a continuing oscillation. The oscillation will eventually die out, when the salt water gets sufficiently diluted; since no new energy is added, such a result is obvious. It is noted that natural oscillators of this type possibly could develop under the Arctic ice. Very cold brine sometimes seeps from the ice into the water, making the sea water freeze around the salt water jet, forming an ice tube which can reach a length of a few meters. References: Martin (1970), Dayton and Martin (1971).

Oscillator 3 is a heat-ice oscillator. Water in an insulated container is in contact at its upper free surface with air kept very cold; inside the water is placed a heat source. The water surface will initially freeze over, adding an insulating layer to the water (heat conduction and evaporation is reduced). The heat source then raises the water temperature enough to melt the ice. Increased heat losses at the water surface adds to the cooling, and the freezing starts again, etc. In one experiment carried out by the author a vessel with 50 l of water was kept in contact with air kept at -23°C, and heated internally by a 100 W electric element; the resulting (somewhat irregular) oscillations had a period of about two days. There have been reports of periodic freezing of lakes on Iceland which possibly could be an example of this oscillator; the heating would then come from a warm, volcanic ground. References: Welander (1977a, 1977b).

Oscillator 4 is a thermal-inertial oscillator. It is the purely thermal version of the thermohaline loop model discussed in the previous section, but now with inertia included. The loop is heated at the lower end and cooled at the upper end, according to the Rayleigh law. It can run steadily in either direction, but such a solution may not be stable, and growing pulsations will then appear, which eventually can be strong enough to reverse the flow direction. The final state may be one of single-periodic oscillations, or more complex multi-periodic or even chaotic variations. To understand the growth of an initial small pulsation in motion, assume that a portion of the fluid in the loop is cooled by an extra amount, adding more weight at this place. This portion of heavier fluid acts as a pendulum when it moves around the loop: when it is at the bottom the loop speeds up, when it is at the top the loop slows down. Therefore the colder portion of the fluid runs faster than normal through the heat source, slower than normal through the heat sink, and it gets further cooled, increasing the pendulum effect, etc. A corresponding reasoning can be made for a portion of the fluid which is heated by an extra amount. This oscillation was firstly predicted for a rectangular loop, but the theory becomes particularly simple for a circular loop, which can be descibed by three ordinary differential equations of Lorenz type. References: Welander (1967), Malkus (1972).

Oscillator 5 is a water wheel version of oscillator 4, which is easy to build. The loop is replaced by a series of plastic cups, placed along the rim of a circular plate. The plate can rotate freely around an axis which is tilted at a suitable angle to the vertical. The cups are filled with water from a continually running faucet, at the top; the water drains through small holes in the bottom of the cups. With this model one can clearly see the "pendulum effect" which leads to the pulsations: if one cup has an extra amount of water it will pass at lower speed by the faucet, and thus receive even more water. This model was designed by L. Howard and W. Malkus at the Mathematics Department, M.I.T., but is not described in print, to this author's knowledge.

Oscillator 6 is a two-layer thermohaline oscillator driven by "mixed boundary conditions", as described in a previous section. The upper layer is heated and also receives a certain salinity flux; a deep lower layer is kept fresh and cold. The heating, acting faster, makes the upper layer become warm and light; the salinity increases only gradually. When the salinity has increased enough the situation becomes statically unstable, overturning occurs, and the process starts over again, with a new cold and fresh upper layer. Reference: Welander (1982).

Oscillator 7 is an air-water one, driven by a stabilizing heat flux (heating at the top, cooling at the bottom). One may think that such an arrangement would prevent growing oscillations, but there is no basic physical law preventing thermal energy to be changed to mechanical energy because the heating is from the top! If we assume that motion in the water loop starts upward in the left branch, downward in the right branch, the left side becomes warm and the right cold. The temperature difference is felt by the surrounding air loop, which then is set in motion in the opposite direction. The air affects the water mechanically, via a stress coupling at the interface (at the top of the figure). The stress on the water tends to oppose the original motion; however, because of the time needed to diffuse heat from the water loop to the air loop there is a certain lag. This lag causes the net work by the stress to be positive over an oscillation cycle, and growing oscillations are seen. The same effect can be seen for a pendulum, when it is subjected to an external restoring force which is delayed in phase. Reference: Welander (1981).

These air-water oscillations may represent, in an idealized fashion, some processes which actually take place in the mutually coupled atmosphere-ocean system. The oceans are storing large amounts of heat, and these can be moved to new positions with relatively little mechanical work (density variations in the ocean are small). Heat that the oceans give to the atmosphere can be relatively effectively transformed to mechanical energy, such as winds (density variations caused by heating are large). Only a small amount of this mechanical energy is recovered by the ocean in the form of wind stress work, but this may be enough to effectively move around the heat sources and sinks, creating an amplifying feedback situation. If the phases are right, the result may be an instability, and growing variations. In the real oceans and atmosphere one cannot expect to see any simple oscillation, if the feedback actually causes an instability, but it may be possible to identify the phenomenon statistically, from studies of temperature records, for example.

Oscillators 8 and 9 are two classics: the first is the "glass blower's organ", the second the "Trevelyan rocker". The glass blower's organ appears when a tube with one closed end is heated to high temperature; the heating produces a standing sound wave in the tube, like in an open organ pipe. The Trevelyan rocker is obtained by placing a hot metal bar on two lead wedges; the bar performs a rocking motion, caused by alternating thermal expansions and contractions of the wedges. These two oscillators are further described in a review article by Walker (1985). The same article also describes some air-water oscillators of a new type called fluidyne engines; two of these are depicted in 10 and 11. There is no place to describe these further, but the imaginative readers may figure out for themselves how they function.

REFERENCES

Broecker, W. and T.-H. Peng, 1982: Tracers in the Sea. Lamont-Doherty Geological Observatory, Columbia University, 690 pp. See Fig. 7-5.

Bryan, F., 1986: Maintenance and Variability of the Thermohaline Circulation. Ph.D.thesis, Princeton University, 254 pp.

Bryan, K., 1963: A numerical investigation of a nonlinear model of a wind-driven ocean. J. Atmos. Science, 20, 594-606.

Bryan, K., S. Manabe and R.L. Pacanowski, 1975: A global ocean-atmosphere climate model. Part II. The oceanic circulation. J. Phys. Oceanogr., 5, 30-46.

Bryan, K. and L.J. Lewis, 1979: A water mass model of the world ocean. J. Geophys. Res., 84, 2503-2517.

Dayton, P.K. and S. Martin, 1971: Observations of ice stalactites in McMurdo Sound, Antarctica. J. Geophys. Res., 76, 1595.

Ekman, V.W., 1905: On the influence of the earth's rotation on ocean currents. Arkiv för Matematik, Astronomi och Fysik, 2(11), 52 pp. See Tafel 3-4.

Ekman, V.W., 1923: Über Horizontalzirkulation bei winderzeugten Meeresströmungen. Arkiv för Matematik, Astronomi och Fysik, 17(26), 74 pp.

Ekman, V.W., 1924: Utkast till en enhettig harsströmsteori (Outline of a unified ocean current theory), private printing, Lund. A translated version published in Tellus (1985), 37a, 379-392.

Gargett, A., 1984: Vertical diffusivity in the ocean interior. J. Mar. Res., 42, 359-393.

Gregg, M., 1985: Turbulent Mixing in the Stratified Ocean: A Review. In preparation.

Hasselmann, K., 1982: An ocean model for climate variability studies. Progress in Oceanogr., Vol. 11, 69-92. Pergamon Press.

Luyten, J.H., J. Pedlosky and H. Stommel, 1983: The ventilated thermocline. J. Phys. Oceanogr., 13, 292-309.

Malkus, W.R., 1972: Non-periodic convection at high and low Prandtl number. Memoires Societe Royale des Sciences de Liege, 6e serie, tome IV, 125-128.

Martin, S., 1970: A Hydrodynamic Curiosity: the Salt Oscillator. Geophys. Fluid Dyn., 1, 143-160.

Meehl, G., Washington and Semtner, 1982: Experiments with a global ocean model driven by observed atmospheric forcing. J. Phys. Oceanogr., 12, 301-312.

Munk, W., 1966: Abyssal recipes. Deep-Sea Res., 13, 707-730.

National Science Foundation, 1980: GEOSECS Atlantic Expedition, Vol. 2, Sections and Profiles. U.S. Govt. Printing Office.

Rhines, P. and W. Young, 1982a: A theory of the wind-driven circulation. I. Mid-ocean gyres. J. Mar. Res., 40 (suppl.), 559-596.

Rhines, P. and W. Young, 1982b: Potential vorticity homogenization in planetary gyres. J. Fluid Mech., 122, 347-368.

Robinson, A.R. and H. Stommel, 1959: The oceanic thermocline and the associated thermohaline circulation. Tellus, 11(3), 295-308.

Rooth, C., 1982: Hydrology and ocean circulation. Progr. Oceanogr., 11, 131-149.

Sarkisyan, A.S., 1962: On the dynamics of wind current appearence in the baroclinic ocean. Oceanology, 2(3), 395-409.

Spiegel, E. and G. Veronis, 1960: On the Boussinesq approximation for a compressible fluid. Astrophys. J., 131, 442-447.

Stommel, H., 1948: The westward intensification of wind-driven ocean currents. Trans. Am. Geophys. Union, 29, 202-206.

Stommel, H., 1961: Thermohaline convection with two stable regimes of flow. Tellus, 13, 224-230.

Stommel, H., 1965: The Gulf Stream: A Physical and Dynamical Description, 2nd ed., University of California and Cambridge University Presses, 202 pp.

Sverdrup, H., 1947: Wind-driven currents in a baroclinic ocean; with application to the equatorial currents of the eastern Pacific. Proceedings of the National Academy of Sciences of the U.S.A., 33, 318-326.

Turner, J.S., 1973: Buoyancy Effects in Fluids. Cambridge University Press, 367 pp.

Veronis, G., 1969: On theoretical models of the thermocline circulation. Deep-Sea Res., 16 (suppl.) 301-323.

Walker, J., 1985: Scientific American, 252(4), 140-144.

Welander, P., 1957: Note on the self-sustained oscillations of a simple thermal system. Tellus, 9(3), 419-420.

Welander, P., 1959: An advective model of the ocean thermocline. Tellus, 11(3), 309-318.

Welander, P., 1967: On the oscillatory instability of a differentially heated fluid loop. J. Fluid Mech., 29(1), 17-30.

Welander, P., 1971: The thermocline problem. Philosophical Transactions of the Royal Society of London, Series A, 270, 415-421.

Welander, P., 1971a: Some exact solutions to the equations describing an ideal fluid thermocline. J. Mar. Res., 29(2), 60-68.

Welander, P., 1977b: Observations of oscillatory ice states in the simple convection experiment. J. Geophys. Res., 82(18), 2591-2592.

Welander, P., 1981: Oscillations in a simple air-water system driven by a stabilizing heat flux. Dyn. Atmos. Oceans, 5, 269-280.

Welander, P., 1982: A simple heat-salt oscillator. Dyn. Atmos. Oceans, 6, 233-242.

DIAGNOSTIC MODELS OF OCEAN CIRCULATION

DIRK J. OLBERS

Alfred-Wegener-Institut
für Polar- und Meeresforschung
Bremerhaven, F.R.G.

1. INTRODUCTION

The basic idea of diagnostic modelling of the ocean circulation is to combine dynamical arguments with the information in data sets of the field variables or data of the forcing functions. The data sources available for this purpose mainly consist of hydrographic data (including all tracers), direct current measurements, sea surface elevation (from satellite altimetry) as well as windstress and the surface fluxes of heat and fresh water. The objectives of the diagnostic approach include the determination of circulation parameters (in particular the three-dimensional velocity field) and the testing of dynamical concepts (especially the role of mixing in the budgets of heat, salt and potential vorticity).

At present, the main body of knowledge about the general circulation of the oceans has been drawn from observations of temperature and salinity. These quantities are relatively easy to measure and in contrast to velocity observations, the climatological signal in the T,S fields is less contaminated by energetic smaller scale motions induced by eddies and waves. Inference of flow characteristics from temperature and salinity data has gone along two almost distinct paths: the water mass analysis techniques with the core layer method (Wüst 1935) and the isentropic analysis (Montgomery 1938, Parr 1938) as most prominent representatives, and the methods residing on the dynamical balance of the large-scale flow, such as the dynamic method (Helland-Hansen and Nansen 1909) and the recent extensions which will partly be summarized in this paper.

J. Willebrand and D. L. T. Anderson (eds.), Large-Scale Transport Processes in Oceans and Atmosphere, 201–223.

2. WATER MASS AND ISOPYCNAL ANALYSIS

The water mass analysis attempts to determine the broad scale movement of the water bodies by identifying the location of their sources and sinks which roughly define a pathway along which the water has to move somehow and gradually lose its characteristics by mixing with surrounding water masses.

An example is Wüst's (1935) core layer approach to the spreading of the Subantarctic Intermediate Water in the Atlantic Ocean (Fig. 1a). The water mass is formed at the Polar Front and introduces a low salinity signal below the warmer water of the upper ocean, with remnants still visible north of the equator. A simple mixing analysis leads to the spreading characteristics of the water mass along its core layer (the layer of minimum salinity) shown in Fig. 1b.

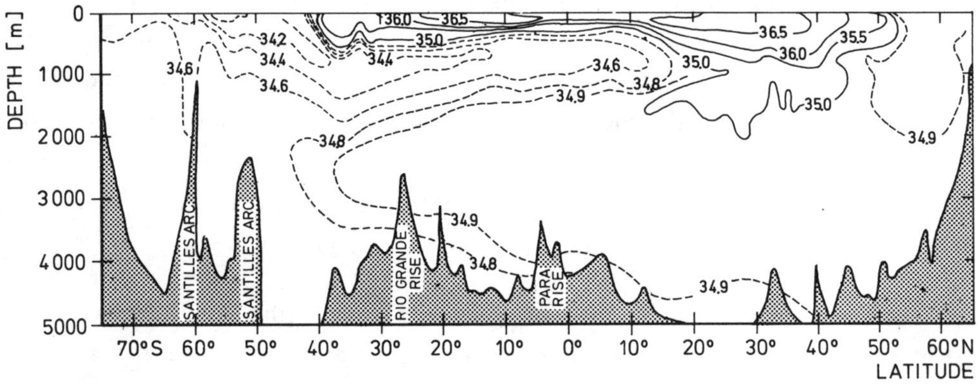

Fig. 1a : Meridional section of the salinity distribution in the western Atlantic Ocean.

The isopycnal analysis is elucidated by the map of potential vorticity $f(\sigma_\theta)_z$ between the isopycnals σ_θ = 27.3 and 27.6 in the North Atlantic

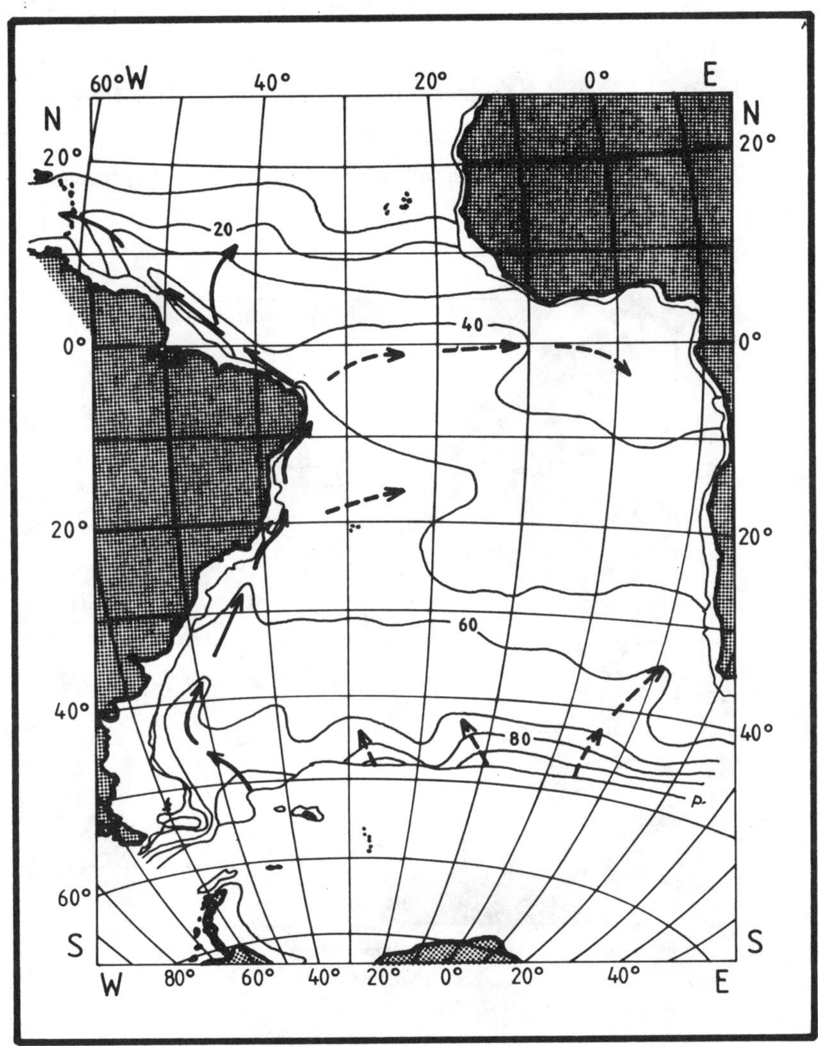

Fig. 1b : Spreading of the Subantarctic Intermediate Water in the core layer (intermediate salinity minimum), represented by the percentage of the Subantarctic water mass component (Wüst 1935).

(Fig. 2, from McDowell at al., 1982). If this quantity were conservative (which basically requires low mixing and neglection of cabbeling) the contours in Fig. 2 would be streamlines of the flow. The flow pattern shows the motion of water from the Irminger Sea to the subtropical Atlantic as well as the extension of Mediterranean water westward across the Atlantic.

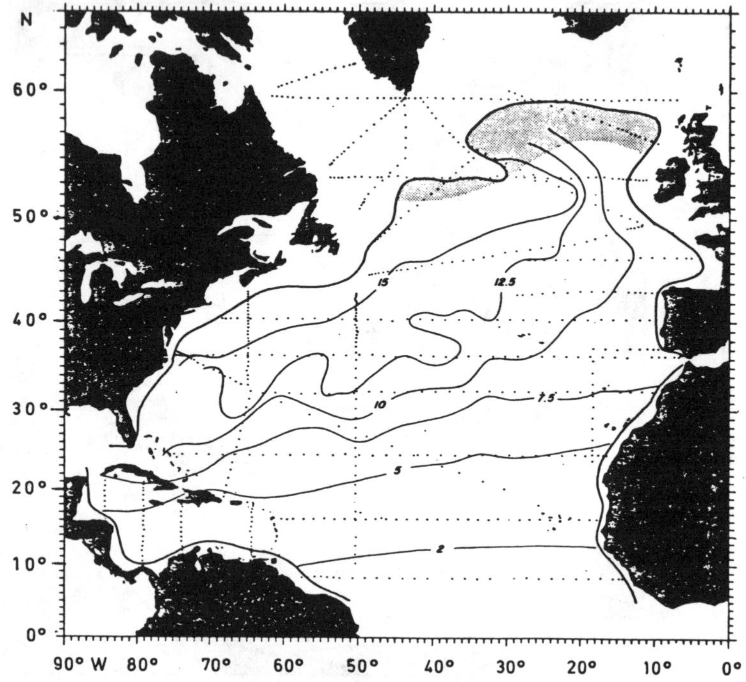

Fig. 2 : Potential vorticity between the σ_θ = 27.3 and 27.6 surfaces. The contours stretch from northeast to southwest, allowing easy communication of fluid between high latitude and the deep subtropical Atlantic (McDowell et al. 1982).

The power of these methods resides in the fact that large-scale connections in the flow field can be inferred directly from the observed large-scale pattern of the properties. However, since local balances are not considered it is not attempted to understand how the fluid in detail establishes these connections. The water mass analysis therefore uses the terminus spreading and avoids the termini advection and mixing or diffusion which have a designated physical meaning in the local and large-scale balance of a fluid property.

3. THE DYNAMIC METHOD

The basic dynamical constraint of the large-scale flow in the ocean is the geostrophic balance between the Coriolis and pressure forces

$$f\underline{u} = \underline{k} \times (\underline{\nabla}\phi)_p \tag{3.1}$$

the latter expressed here in terms of the gradient of the geopotential Φ on a pressure surface. However, hydrographic data only determine the baroclinic part of the pressure arising from density stratification, whereas the barotropic part associated with the inclination of the sea surface remains unknown. In terms of the geopotential

$$\Phi\,(\lambda,\phi,p) = \Phi_o\,(\lambda,\phi) - \int_0^p \frac{dp'}{\rho(T,S,p')} \tag{3.2}$$

the geopotential anomaly (the second term) associated with the oceanic mass field can be computed from hydrographic data whereas the geopotential Φ_o of the surface is unknown and essentially unobservable by standard oceanographic instrumentation. Consequently, the classical dynamical method heavily resides on subjective assumptions about the local barotropic pressure force or, equivalently, the level of no motion. A discussion of early speculations about the level of no motion can be found in Reid (1981). Defant (1941) placed the level of no motion at the depth of minimum vertical shear. Fig. 3 shows the resulting circulation at 2000 m obtained from the METEOR data. This picture surprisingly agrees in many respects with modern concepts of the mid-depth circulation in the Atlantic Ocean (see Reid, 1981 and section 5 for a discussion).

Notice that the geopotential anomaly determines the relative horizontal velocity $\underline{u}_1 - \underline{u}_2$ between two pressure surfaces p_1 and p_2.

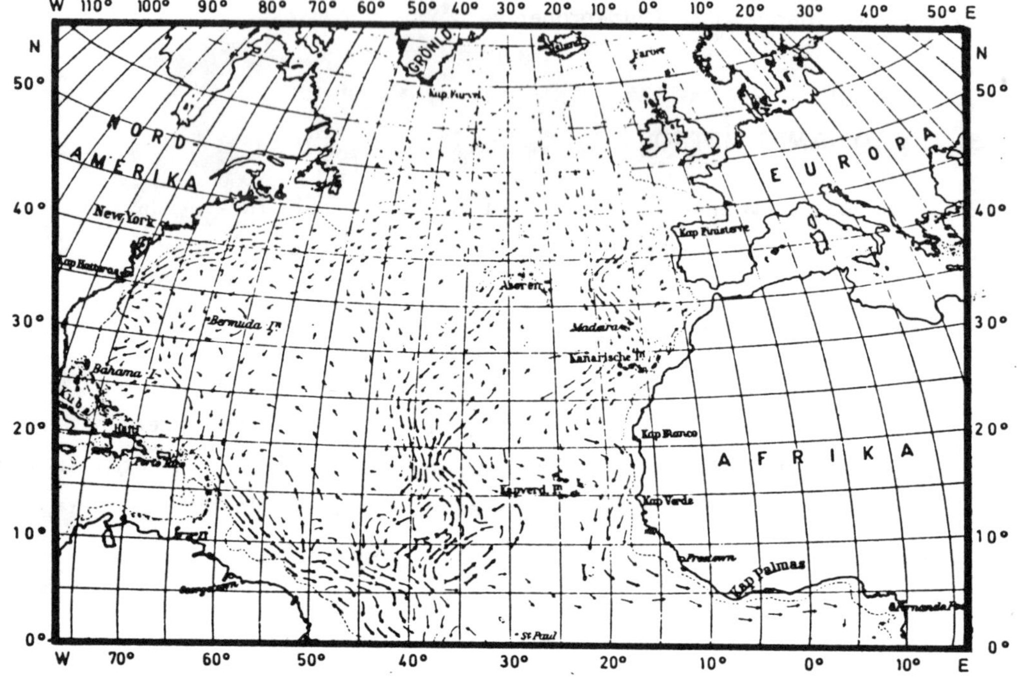

Fig. 3 : The flow field at 2000 m depth, obtained from the dynamic method (Defant 1941).

Generally these are expressed in terms of the dynamic topography $\Phi_1 - \Phi_2$ or the geopotential height $(\Phi_1 - \Phi_2)/g$ between the surfaces p_1 and p_2.

The main problem arising from subjective specifications of reference levels refers to the determination of the deep velocity and the inherent massive uncertainties in the associated values of volume and heat transport. A further questionable issue in the use of the dynamic method is the lack of synoptic data in space and time. The spatial problem is examplified by the various attempts to estimate the transport through the Drake Passage from hydrographic data using current mesurements to provide the reference velocities. The estimates range from 15 Sv westwards (Foster,

1972) to 234 Sv eastwards (Reid and Nowlin, 1971) and seem now to settle at about 130 Sv (Whitworth et al., 1982). The reason is the strong filamentation of Circumpolar Current in the passage displayed in Fig. 4. In such circumstances the loose interpretation of geopotential height (Fig. 5, from Gordon et al., 1978) as absolute flow is quite misleading.

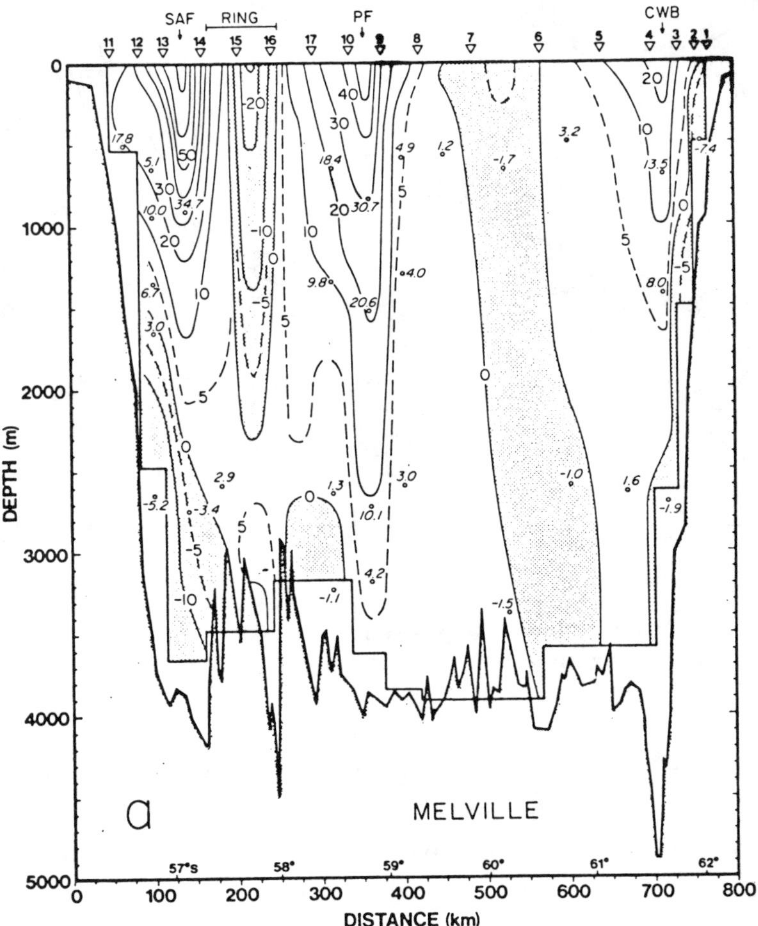

Fig. 4 : Vertical sections of net geostrophic speed (cm s^{-1}) for DRAKE 79 cruise aboard Melville. The values shown for direct speed measurements (circles) are averages for that period of time that produces best agreement with baroclinic shears SAF = Subantarctic Front, PF = Polar Front, CWB = Continental Water Boundary (Whitworth III et al. 1982).

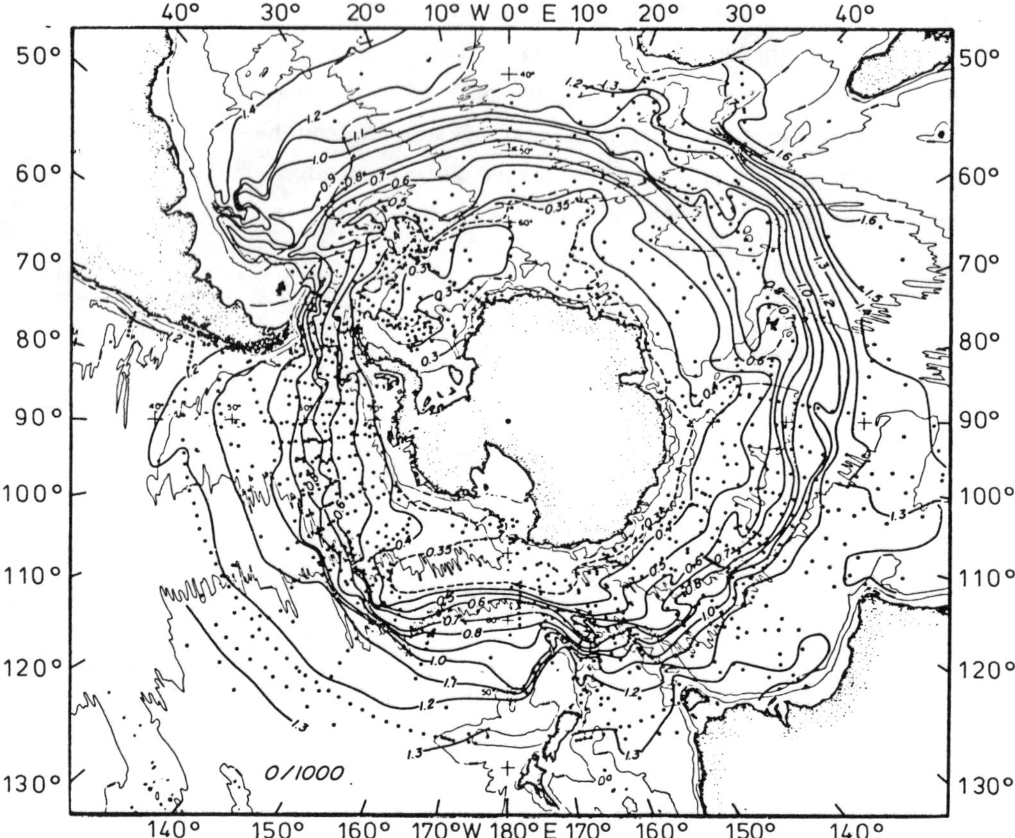

Fig. 5 : Anomaly of geopotential height of the sea surface relative to the 1000-dbar level, expressed in dynamic meters. Stations used in preparing this chart are shown (Gordon et al 1978).

The use of non-synoptic hydrographic sections for diagnostic inverse modelling has frequently been questioned (e.g. Wunsch and Grant, 1982) because in general they do not give an instantaneous view of the oceanic mass field. The resulting model circulation cannot be expected to represent an instantaneous state, nor even a possible state of the real oceanic circulation. The situation may even be rated worse when considering the climatological average of essentially all available hydrographic data as produced by Levitus (1982) which , because of data sampling problems both in space and time, this atlas displays by no means a real climatological average. However, one can argue that a mean in some restricted sense may be physically more meaningful than a composite of non-synoptic sections. In any case, in view of the present severe data restriction for studying ocean circulation both approaches can equally be justified.

4. THE INVERSE METHOD

With the use of the geostrophic balance, or more precisely the thermal wind equations which relate the vertical shear of the horizontal current to the local density gradient, not all information contained in the temperature and salinity observations has been exploited. Each property is subject to advection by the local absolute velocity which constrains the velocity component along the local property gradient. In case of an adiabatic flow, for instance, the absolute velocity vector must lie in the isopycnal surface (or isohaline surfaces and surfaces of constant potential temperature). A water body which is bounded by such surfaces and vertical sections on the sides is thus only affected by the horizontal inflow into these sides. The local inflow of volume can be expressed in terms of the local relative velocity $\underline{u}'$ (which is determined by the known baroclinic pressure gradient) and the unknown local reference velocity $\hat{\underline{u}}$, i.e. the velocity at a specified reference level. Considering larger bodies of water enclosed by hydrographic sections (or even partly by coastlines) one obtains as generalisation of Knudsen's hydrographic theorem (e.g. Proudman 1953) an underdetermined set of equations for the reference velocities $\hat{\underline{u}}$ between the stations

$$\oint ds\, \hat{\underline{u}}\,(s) \cdot \underline{n}\,[h_2(s) - h_1(s)] = - \oint ds \int_{-h_2(s)}^{h_1(s)} dz\, \underline{u}'(s,z) \cdot \underline{n} \tag{4.1}$$

where $h_1(s)$ and $h_2(s)$ are depth of the selected isopycnals, $\underline{n}$ is the unit vector which is localy normal to the section, and s is the coordinate along the closed section loop.

Out of the infinity of solutions one may choose the one which suits a subjective judgement of the large-scale circulation, such as smoothness or minimum or maximum volume transport through the region considered. The subjective choice of a (local) level of no motion is thus replaced by (still subjective) assumptions about large-scale patterns of the flow. However, conservation of properties derived from temperature and salinity is guaranteed. In essence this is the inverse method investigated by Carl Wunsch in various papers (Wunsch 1977, 1978, Wunsch and Grant 1982).

In practice the method is applied to several layers simultaneously. The resulting circulation partly supports the broad, basin-wide recirculation envisioned in the classical description of the wind-driven circulation (see e.g. Reid (1981) for a review). A strange aspect of the inverse models is a frequently break-down of the circulation into a cellular structure of smaller scale. This is the outstanding feature mediated in the sections of horizontal velocities examplified by the 36°N section shown in Fig. 6 (from Wunsch and Grant, 1982): in vast regions the ocean seems to move almost barotropically in alternating columns with horizontal scales of order thousand kilometers. As suggested by Luyten and Stommel (1982) this behaviour of the inverse solution may partly be due to a heavy weighting of deeper layers thus emphasizing topographically induced details, and partly it may arise from small-scale transient irregularities in the data.

5. THE β-SPIRAL METHOD

An alternative scheme for the determination of the reference velocities has been put forward by Stommel and Schott (1977). Likewise, it is based on the thermal wind relations and tracer conservation, but in contrast to Wunsch's method it emphasizes more the local aspect of these balances. The only large-scale element in the data handling arises from the necessity to smooth away eddies in the hydrographic fields in order to obtain reliable estimates of gradients. The method involves a linearized vorticity balance which, for planetary scale motions, relates the vortex stretching to the advection of planetary vorticity.

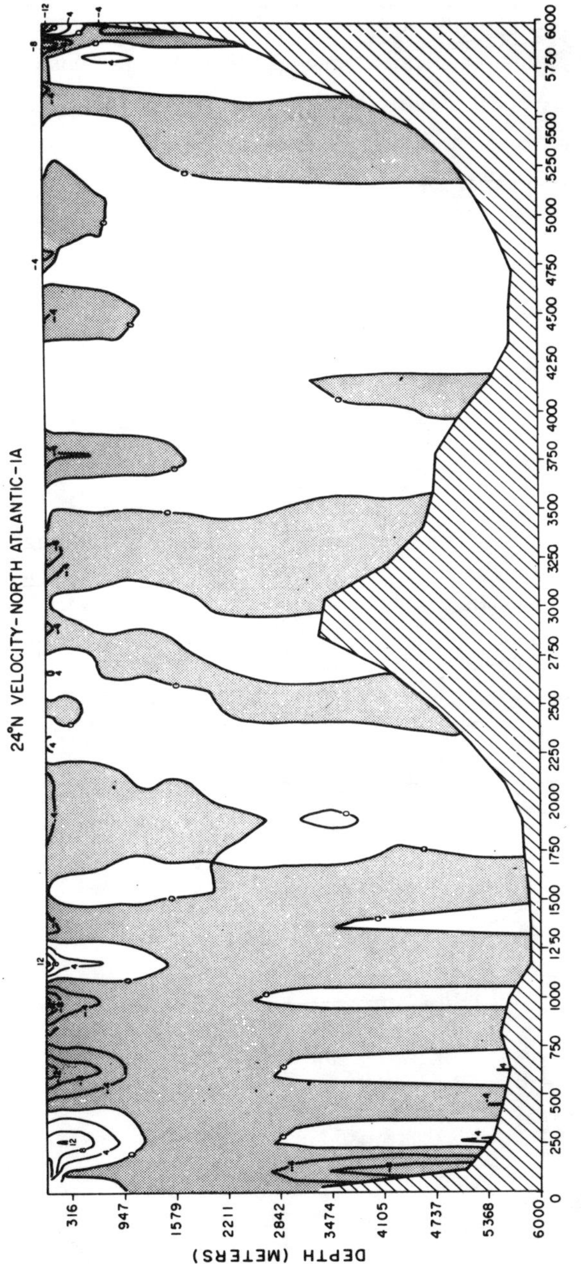

Fig. 6a: Velocity sections across 24°N and 36°N calculated for the North Atlantic-1A-model. Initial reference level was at 2000 decibars and a remnant of that initial zero line is visible in the western part of the section (Shaded areas flow southward) (Wunsch and Grant 1982).

VELOCITY 36° N NORTH ATLANTIC - IA
DEPTH (METERS)
0 316 947 1579 2211 2842 3474 4105 4737 5368 6000
DISTANCE (KILOMETERS)
0 250 500 750 1000 1250 1500 1750 2000 2250 2500 2750 3000 3250 3500 3750 4000 4250 4500 4750 5000 5250 5500 5750 6000 6250
180 -90
30

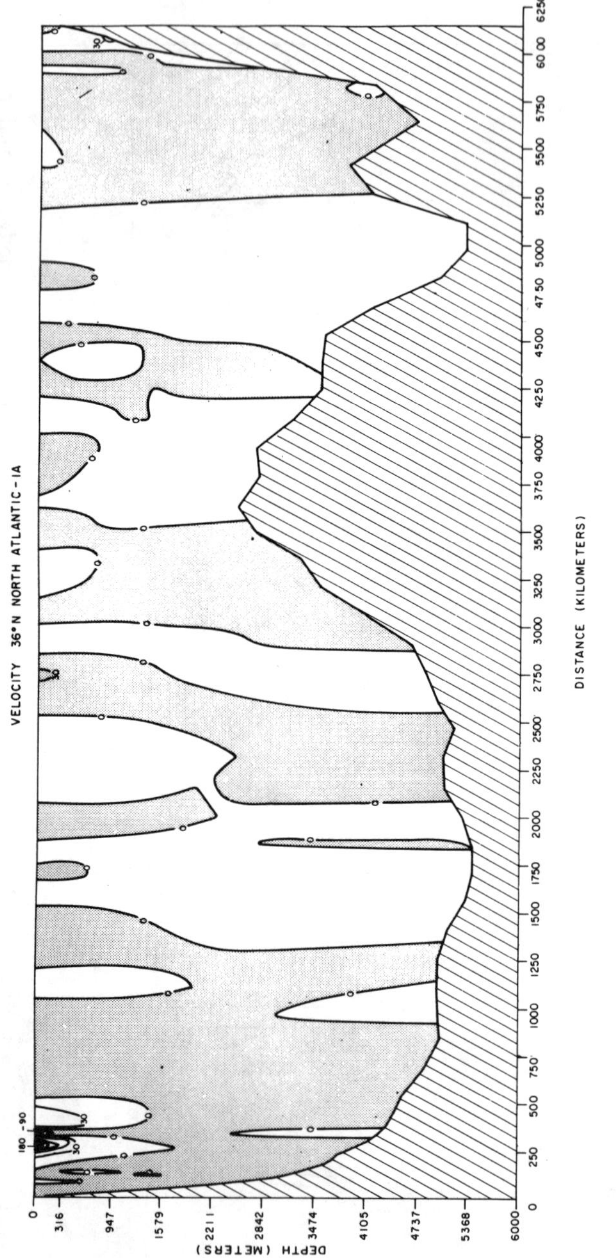

Fig. 6b

The essence of the method can again readily be explained for an adiabatic flow. Since density and potential vorticity are conserved by the system the intersections of isopycnals with surfaces of constant potential vorticity (the geostrophic contours) must be streamlines and thus the direction of the velocity vector is known at each level. So, when considering the horizontal components at two levels it appears obvious that their absolute values are locked by the thermal wind relations since these determine the relative velocity vector in terms of the known density gradients. Apparently the method fails if isopycnals and surfaces of constant potential vorticity coincide so that velocity directions cannot be identified. Another case of failure arises when the absolute horizontal velocity vector does not turn with depth and so incompatability with the relative velocity profiles can occur. A more subtle analysis of the dynamics however reveals (see e.g. Bryden 1980) that the planetary vorticity balance constrains the horizontal vorticity to spiral with depth. The rate of change of the direction ϕ = arctan $^v/u$ of the horizontal velocity vector $\underline{u}$ = (u, v) with depth can be expressed as

$$\frac{d\phi}{dz} = \frac{v_z u - u_z v}{[\underline{u}]^2} = - \frac{g}{f[\underline{u}]^2} \underline{u}\cdot\underline{\nabla}\rho = \frac{g\rho_z}{f[\underline{u}]^2}\left[w_o + \frac{\beta}{f}\int_{z_o}^{z} dz' v \right] \tag{5.1}$$

The second identity follows from the thermal wind relation and the third from the planetary vorticity equation. In fact, the presence of the gradient β of the planetary vorticity guarantees a spiraling hodograph giving the name β-spiral to Stommels's concept.

The adiabatic form of the β-spiral constraint on the reference velocities is already contained in the last identity of (5.1). Expressing the horizontal velocity $\underline{u}$ in term of the reference velocity $\underline{u}_o$ and the relative velocity $\underline{u}'$ known from the thermal wind relations one obtains

$$(5.2) \qquad \underline{u}_0 \cdot \underline{\nabla}\rho + \left[w_0 + \frac{\beta}{f}(z-z_0)\, v_0 \right] \rho_z = -\, \underline{u}' \cdot \underline{\nabla}\rho - \rho_z \frac{\beta}{f} \int_{z_0}^{z} dz'\, v'$$

which constrains the component of $(\underline{u}_0, w_0)$ which is normal to the vector $(\underline{\nabla}\rho + (z-z_0)\frac{\nabla f}{f}, \rho_z)$. In principal the three unknowns $(\underline{u}_0, w_0)$ can be determined by considering (5.2) at three different levels. In practice, when applied to real, noisy hydrographic data the reference velocity vector is determined from a least squares principle, i.e. by considering the spiral between many different levels. The method can easily be extended to diabatic conditions. Schott and Zantopp (1980) have included vertical diffusion of density, and Olbers et al. (1985) describe the attempt to additionally determine horizontal diffusion coefficients of the temperature and salinity and vertical diffusion coefficients of vorticity (or, equivalently, horizontal diffusion coefficients of potential vorticity). Homogeneity of potential vorticity on isopycnals has been observed in a wide layer beneath the wind-driven gyres in the ocean (McDowell et al 1982, Holland et al 1984). Inclusion of data from this layer has very likely caused the break-down of the β-spiral method in some earlier applications (e.g. Schott and Stommel 1978, Behringer 1979) where no unique reference velocities could be identified. The basic problem is known since the early work in the thermocline problem: Needler (1972) has shown that the thermocline equations (without proper boundary conditions) contain an arbitrary barotropic part if potential vorticity is a function of density. Fortunately, however, this relation is broken in the real ocean at deeper levels (say, roughly below 800m in the North Atlantic) and the β-spiral can be used to determined the barotropic velocity.

We present the performance of the β-spiral approach by some results from Olbers et al. (1985) who applied the method to the North Atlantic part of Levitus' (1982) climatological atlas. The North Atlantic is the area of most other inverse modelling activities (because of large data density) which gives the possibility of comparison. There are various computations of absolute β-spiral profiles (e.g. Schott and Stommel 1978, further reference see section 3), and patterns of absolute circulation have been obtained by Wunsch and Grant (1982) on the basis of Wunsch's inverse method and a selection of individual hydrographic sections. A noteworthy early step towards the circulation in the Atlantic is Defant's (1941) attempt to compute maps of absolute geostrophic velocities (see Fig. 3). We should further mention the diagnostic circulation models of Holland and Hirschman (1972) and Mellor et al. (1982) as well as Worthington's (1976) circulation scheme. The results differ significantly in many details, even in such important questions as the vertical structure of the Gulf Stream: Worthington's scheme requires the Gulf Stream to reach the bottom whereas most other work reveals an intermediate level of no motion. Disagreement also exists with respect to the recirculation pattern as well as the extension of the Gulf Stream into the North Atlantic current.

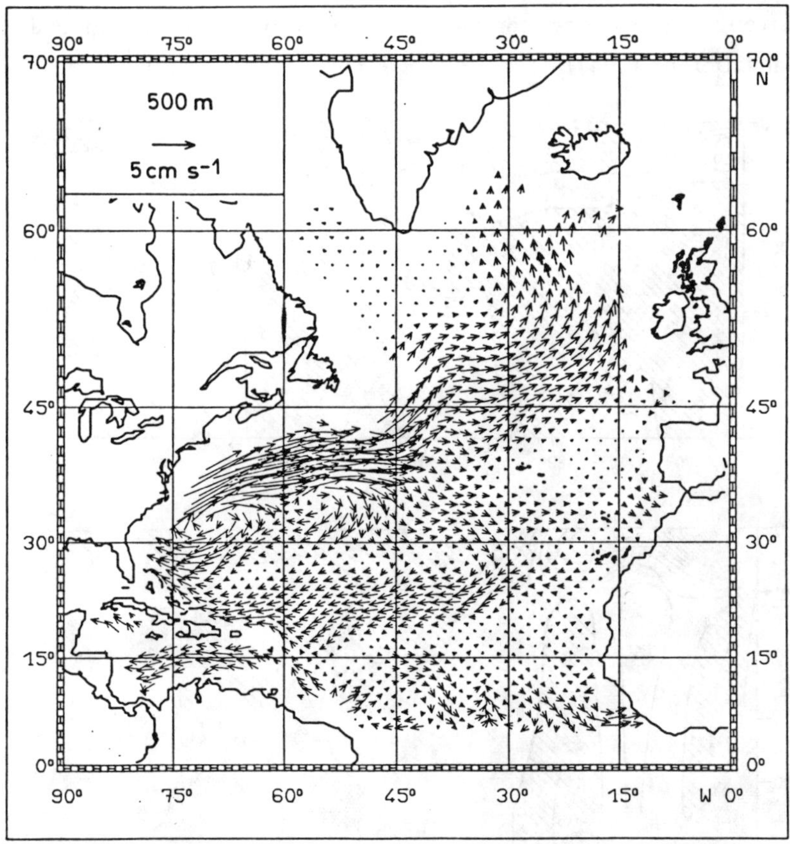

Fig. 7a : The absolute velocity field at 500 m and 2000 m inferred from the Levitus atlas by a β-spiral technique (Olbers et al., 1985).

cells. The results displayed in the maps of Fig. 7 and the sections of Fig. 8 reveal less complex flow patterns than those obtained by Wunsch and Grant (1982). The upper layers are dominated by a broad Gulf Stream which splits at about 40°N and 40°W in a North Atlantic current and the subtropical gyre recirculation which includes the Azores current and the North Equatorial Current as well as a tight recirculation cell in the western region. In the lower layers the outstanding feature is the deep western boundary current fed by overflow water from the Norwegian Sea which penetrates through the Gibbs fracture zone and partly circulates around the Labrador Sea. The vertical sections also clearly demonstrate the broad horizontal circulation which shows one or two current reversals and some indications of almost horizontal levels of no motion. The overall appearance of this circulation bears strong resemblance to the classical flow pattern deduced by Wüst (1935) and Defant (1941).

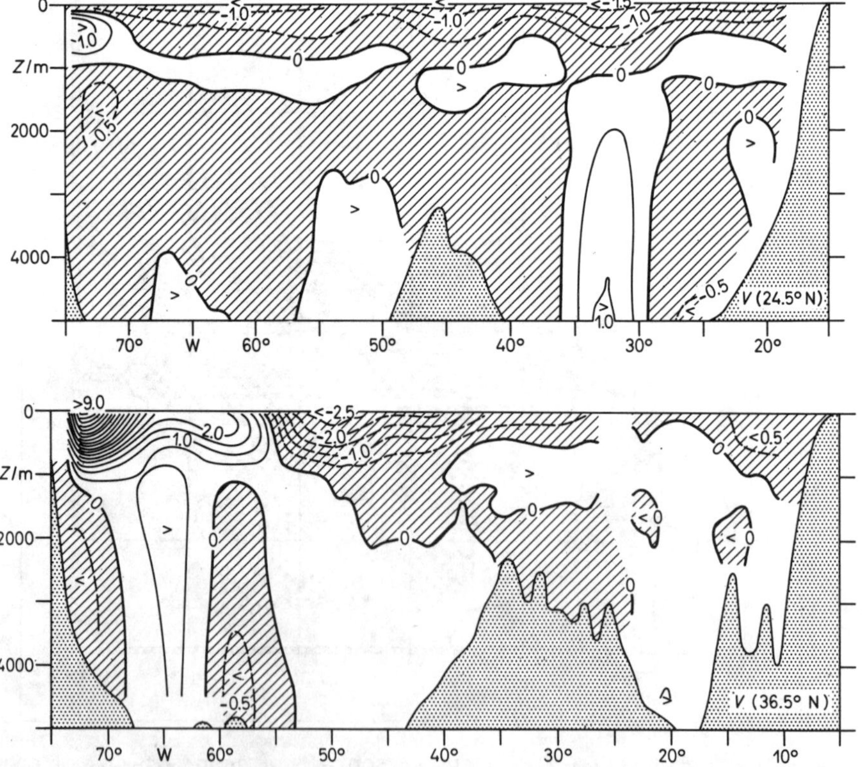

Fig. 8 : Section of the meridional velocity at 24.5°N and 36.5°N of the β-spiral calculation (Olbers et al., 1985).

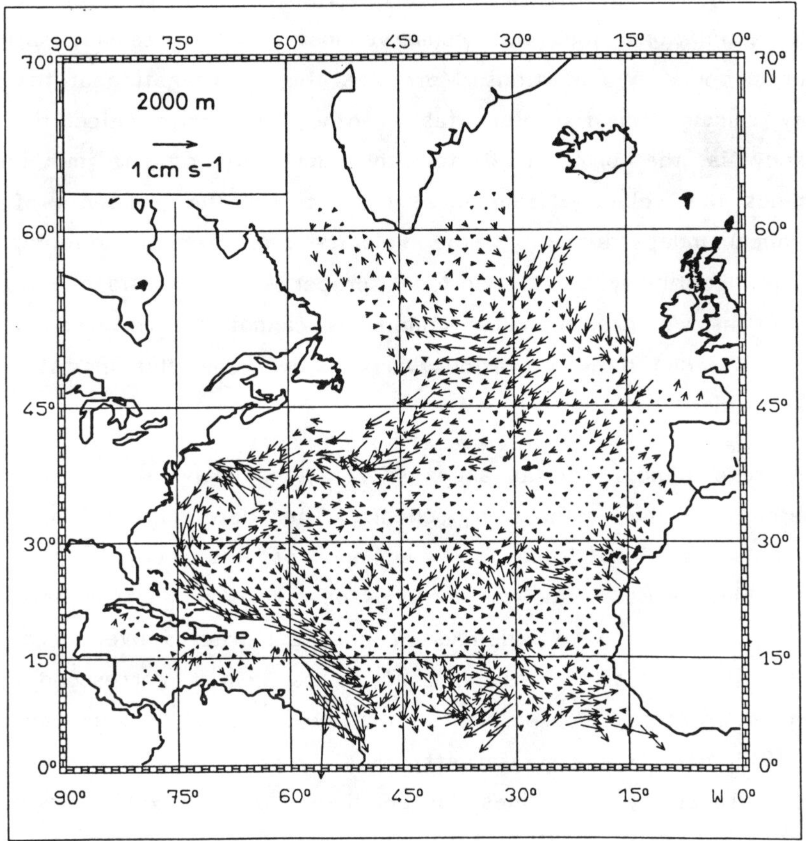

Fig. 7b : Same as fig. 7a, but 2000 m level

The circulation associated with the climatological average not only differs in details from the pictures gained from individual hydrographic sections, but the overall patterns already have a drastically different appearance. The smoothing technique definitely takes away more than the instantaneous eddy transients in the original data. Features which may be rated as long-term mean signal in individual sections appear in the climatological average in a rather blurred shape with considerably reduced gradients. This particularly concerns strong currents which appear in each section as sharp fronts but may have a seasonal or lower frequency oscillation. The circulation gained from these data is a rather simple broad-brush picture dominated by broad currents and basin-wide circulation

We should emphasize here that the velocity field resulting from local β-spiral estimations does not conserve mass and does not satisfy any meaningful boundary conditions. Moreover, the determination of the vertical velocity remains unsatisfactory (as in other diagnostic calculations). This deficiency is the price paid for the simplicity of the method which determines the velocity from local properties (the gradients of density and tracer) independently of the velocity structure at adjoining points. Obviously, continuity is a non-local property and consideration of this constraint as well as boundary conditions cannot be achieved in a local analysis. One can think of various ways to overcome this limitation of the β-spiral method.

A second point of advance is a rational way of including mixing parameters in the estimation scheme. The determination of diffusion coefficients for heat and salt is one of the central problems of oceanographic research and the methods developed range from the study of small-scale turbulence to balances of chemical tracers over ocean basins. The estimates for the diffusion coefficients span a corresponding range, e.g. microstructure measurements indicate vertical mixing coefficients below 10^{-6} m^2/s (see e.g. Gargett, 1984) whereas large-scale balances of chemical tracers yield values larger than Munk's (1966) abyssal recipe 10^{-4} m^2/s (Broecker and Peng 1982). The role of vorticity mixing and mixing of potential vorticity has received new attention with the introduction of the theories of Rhines and Young (1982) as well as Luyten et al. (1983) which attempt to explain the vertical structure of the wind-driven circulation on the basis of the potential vorticity balance. The theories are based upon low or vanishing potential vorticity mixing , respectively. All relevant diffusion coefficients can principally be determined within the β-spiral framework: the coefficients describing diapycnal and isopycnal mixing of heat and salt as well as the coefficient describing vertical diffusion of vorticity, or equivalently, horizontal diffusion of potential vorticity. Results may become rather noisy since second order derivatives of the data are involved (the reference velocities are obtained from first order derivatives in our scheme). Moreover, with increasing number of parameters to be resolved their variance will generally also increase.

Some definite statements about the resulting diffusion coefficients can be made. Within the subtropical gyre and, more generally, outside the regimes of stronger currents the diapycnal and isopycnal diffusivities (as well as their standard deviations) are considerable smaller than the classical values, e.g. those used in numerical models of the ocean circulation. The diapycnal diffusivity in the subtropical gyre is less than 10^{-5} m^2/s and the isopycnal is about 10^2 m^2/s. Towards the surrounding strong current region these values increase roughly by an order of magnitude. The pattern thus resembles maps of eddy activity. As an example Fig. 9 shows the estimates of the vertical diffusivity for the depth interval 800 m to 2000 m. We have to conclude that mixing of heat and salt in the ocean is a relatively weak process that seems to be overemphasized

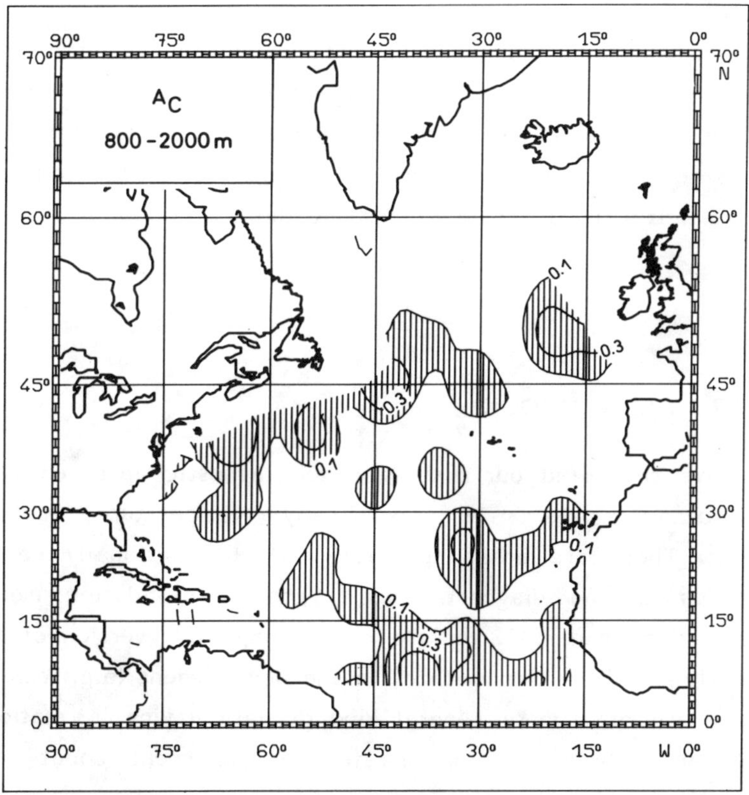

Fig. 9 : Estimate of the diapycnal diffusion coefficient for the depth range 800 m to 2000 m. Units are cm^2/s (Olbers et al., 1985).

in numerical models. This conclusion becomes even more stringent when realizing that our values are likely to be overestimated due to the climatological averaging of the hydrographic data.

Vertical diffusitives of vorticity are of order 10^{-1} m^2/s in the regimes of strong currents and an order of magnitude smaller in quiet regions. If associated with the vertical tranfer of horizontal momentum, diffusivities of order 10^{-1} m^2/s appear rather large. However, as pointed out by P. Rhines (pers. communication) these large values do not succeed in carrying all the momentum downward because the momentum transfer is almost balanced by adjustment of the density field: when the horizontal momentum profile of a board current begins to deepen the readjustment of the density field is accompained by horizontal geostrophic circulation which nearly compensates the acceleration induced by the momentum transfer.

The corresponding diffusivities of potential vorticity are $(^N/f)^2$ times those of vorticity. They thus range between 10 m^2/s and 10^3 m^2/s. These values agree favourably with the range of estimates (from Lagrangian diffusivities of fluid particles which should equal the diffusivity of potential vorticity) discussed by Rhines and Holland (1979).

6. SUMMARY AND OUTLOOK

We have restricted our discussion to diagnostic approaches based on interior oceanic data, i.e. specifically temperature and salinity observations. There are other approaches which additionally consider the forcing fields as the diagnostic ocean general circulation models (e.g. Holland and Hirschman, 1972) or the robust diagnostic version of Sarmiento and Bryan (1982). There are methods in the development (e.g. Wunsch, 1984) which make use of a more general kind of data having a relation to the circulation and which can incorporate any statement about circulation properties that can be formulated as a mathematical constraint (in particular bounds on circulation parameters). In a way these methods can be viewed as a logical combination of the dynamical constrains and the Wüstian type constaints.

This is contribution No. 10 of the Alfred-Wegener-Institute for Polar- and Marine Research

REFERENCES

Behringer, D. W., 1979. On computing the absolute geostrophic velocity spiral. Journal of Marine Research 37: 459-470

Broecker, W. S. and T.-H. Peng, 1982. Tracers in the Sea. Lamont-Doherty Geophysical Observatory, Columbia University, NY.

Bryden, H.L., 1980. Geostrophic vorticity balance in midocean. Journal of Geophysical Research 85: 2825-2828.

Defant, A., 1941. Quantitative Untersuchungen zu Statik und Dynamik des Atlantischen Ozeans. Die absolute Topographie des physikalischen Meeresniveaus und der Druckflächen sowie die Wasserbewegungen in Raum des Atlantischen Ozeans. In: Wissenschafliche Ergebnisse der Deutschen Atlantischen Expedition auf dem Forschungs- und Vermessungsschiff 'Meteor' 1925-27, 6:2nd Part,1,191-260.

Foster, L.A., 1972: Current measurements in theDrake Passage. M.S.thesis, Dalhousie University, 61pp.

Gargett, A. E., 1984. Vertical eddy diffusivity in the ocean interior. Journal of Marine Research 42:359-393

Gordon, A.L., E. Molinelli and T. Baker, 1978: Large-Scale Relative Dynamic Topography of the Southern Ocean, Journal of Geophys. Res., Vol. 83, No.C6.

Helland-Hansen,B., and F. Nansen, 1909. The Norwegian Sea. Its physical oceanography based upon the Norwegian researches 1900-1904. Rep. Norw. Fish. Mar. Invest. 2:1-390 (and suppl.).

Holland, W. R. and A. D. Hirschman, 1972. A Numerical Calculation of the circulation in the North Atlantic Ocean. Journal of Physical Oceanography 2:336-354

Holland, W. R., T. Keffer, and P. B. Rhines, 1984. Dynamics of the oceanic general circulation: the potential vorticity field. Nature 308:698-705.

Levitus, S., 1982. Climatological Atlas of the World Ocean. NOAA Technical Paper 3:173 pp.

Luyten, J. R., and H. Stommel, 1982. Recirculation revisited. Journal of Marine Resarch 40 (Suppl.):407-426

Luyten, J. R., J. Pedlosky, and H. Stommel. The ventilated thermocline. Journal of Physical Oceanography 13:292-309.

McDowell S., P. Rhines, and T. Keffer, 1982. North Atlantic potential vorticity and its relation to the general circulation. Journal of Physical Oceanography 12:1417-1436.

Mellor, G. L., C. R. Mechoso, E. Keto, 1982. A diagnostic calculation of the Atlantic Ocean. Deep-Sea Research 29:1171-1192.

Montgomery, R. B., 1938. Circulation in upper layers of southern North Atlantic deducted with use of isentropic analysis. Paper in Physical Oceanography and Meteorology 6:2,55 pp.

Munk, W. H., 1966. Abyssal recipes. Deep-Sea Research 13:707-730

Needler, G. T., 1972. Thermocline models with arbitrary barotropic flow. Deep-Sea Research 18:895-903.

Olbers, D.J., J. Willebrand and M. Wenzel, 1985: The inference of North Atlantic circulation parameters from climatological hydrographic data. R. Geophys. 23:4, 313-356

Parr, A. E., 1938. On the validity of the dynamic topographic method for the determination of ocean current trajectories. Journal of Marine Research 1:119-132.

Proudman, J., 1953. Dynamical Oceanography. Methuen, 409.

Reid, J. L., 1981. On the Mid-depth circulation of the World Ocean. In: Evolution in Physical Oceanography, B. Warren and C. Wunsch, Eds., MIT Press., 70-111.

Reid, J.L. and W.D. Nowlin Jr., 1971: Transport of water through the Drake Passage. Deep-Sea Res., 18, 51-64.

Rhines, P.B. and W.R. Holland, 1979: A theoretical discussion of eddy-induced circulation. Dyn. Atmos. Oceans 3, 285-325.

Rhines, P. B., and W. R. Young, 1982. Homogenization of potential vorticity in planetary gyres. Journal of Fluid Mechanics 122:347-367.

Sarmiento, J.L. and K. Bryan, 1982: An Ocean transport model for the North Atlantic. J. Geophys. Res. 87, 394-408.

Schott, F. and H. Stommel, 1978. Beta spirals and absolute velocities in different oceans. Deep-Sea Research 25:961-1010.

Schott, F. and R. Zantopp, 1980. On the effect of vertical mixing on the determination of the absolute currents by the beta spiral method. Deep-Sea Research 27A:173-180.

Stommel, H. and F. Schott, 1977. The beta spiral and the determination of the absolute velocity field from hydrographic station data. Deep-Sea Research 24:325-329.

Whitworth III, T., W.D. Nowlin Jr. and S.J. Worley, 1982: The Net Transport of the Antarctic Circumpolar Current through Drake Passage. J. of Geophys. Oceanogr., 12, 960-971.

Worthington, L. V., 1976. On the North Atlantic Circulation. Johns Hopkins Oceanographic Studies 6:110pp, Baltimore.

Wunsch, C., 1977. Determining the general circulation of the oceans: a preliminary discussion. Science 196:871-875.

Wunsch, C., 1978. The general circulation of the North Atlantic west of 50°W determined from inverse methods. Review of Geophysics and Space Physics 16:583-620.

Wunsch, C., 1984. An Eclectic Atlantic Ocean Circulation Model. Part I: The Meridional Flux of Heat. Journal of Physical Oceanography 14:1712-1733.

Wunsch, C. and B. Grant, 1982. Towards the general circulation of the North Atlantic Ocean. Progress in Oceanography 11:1-59.

Wüst, G., 1935. Schichtung und Zirkulation des Atlantischen Ozeans. Das Bodenwasser und die Stratosphäre. Wissenschaftliche Ergebnisse der Deutschen Atlantik Expedition 'Meteor' 1925-1927, 6:288pp, Berlin.

WIND DRIVEN OCEAN CIRCULATION THEORY - STEADY FREE FLOW

JOHN C. MARSHALL

Imperial College of Science and Technology
Department of Physics
London, U.K.

ABSTRACT

Classical homogeneous ocean circulation theory is reviewed from the perspective provided by the quasi-conservation of absolute vorticity q following the motion. Particular emphasis is placed on steady free flows in which q is constant along a streamline of the flow ψ. The importance of weak forcing and dissipation in removing the indeterminacy of inviscid theory is stressed. Ways of extending the theory to baroclinic flow are suggested.

1. INTRODUCTION

Here we study the dynamics of a homogeneous layer of water on a ß-plane driven by an imposed wind-stress curl and frictionally retarded. The motion is two dimensional and horizontally non-divergent. The model is extremely crude with no attempt to descibe the vertical structure of the ocean current. Nevertheless, it is instructive since it includes the essential ingredients of all ocean models: a vorticity source provided by the wind-stress curl, a representation of the redistribution of vorticity by ocean currents, and a vorticity sink. It is the simplest dynamical model of ocean circulation, the standard or reference.

There are several excellent reviews of homogeneous ocean circulation theory in text books (see for example Robinson, 1963; Stommel, 1965; Pedlosky, 1979; Veronis, 1981 and Gill, 1982). For a comprehensive account the reader should consult these standard texts. In this lecture however I

J. Willebrand and D. L. T. Anderson (eds.), Large-Scale Transport Processes in Oceans and Atmosphere, 225–245.

shall attempt to present the material from a slightly different perspective, keeping in mind at all times the quasi-conserved "potential vorticity" of the model - the absolute vorticity q. I shall pay particular attention to steady inertial theory. It is hoped that this review will provide a useful starting point from which to contemplate more detailed models and oceanographic observations.

2. FORMULATION

In standard notation

x momentum $$\frac{Du}{Dt} - fv + \frac{1}{\rho}\frac{\partial p}{\partial x} = F_x \tag{1}$$

y momentum $$\frac{Dv}{Dt} + fu + \frac{1}{\rho}\frac{\partial p}{\partial y} = F_y \tag{2}$$

continuity $$\frac{\partial u}{\partial x} + \frac{\partial v}{\partial y} = 0 \tag{3}$$

where x is east, y is north and t is time

$\underline{v} = (u, v)$ is the velocity

p is pressure

ρ is a constant density

$f = f_o + \beta_o y$ is the coriolis parameter

$\frac{D}{Dt} = \frac{\partial}{\partial t} + \underline{v}\cdot\nabla$ is the substantial derivative

and $F = (F_x, F_y)$ are the momentum sources and sinks due to tangential stresses at the ocean surface associated with prevailing wind systems.

(a) The Vorticity equation

Eqs (1), (2) and (3) are three partial differential equations in three unknowns u, v and p. This apparently complicated problem can be greatly simplified. The method, which can be applied to many other problems, is very important. We begin by forming the vorticity equation by eliminating the pressure terms between Eq (1) and (2): $\frac{\partial}{\partial x}(2) - \frac{\partial}{\partial y}(1)$ to give (using Eq (3)) the vorticity equation

$$\frac{D}{Dt}\left(\frac{\partial v}{\partial x} - \frac{\partial u}{\partial y} + f\right) = G \tag{4}$$

where $\frac{\partial v}{\partial x} - \frac{\partial u}{\partial y} + f$ is the absolute vorticity, conserved but for vorticity sources and sinks $G = \frac{\partial F_y}{\partial x} - \frac{\partial F_x}{\partial y}$.

Now we have two equations ((3) and (4)) in two unknowns (u and v). However because the flow is horizontally non-divergent (Eq(3)) a streamfunction can be defined

$$v = \frac{\partial \psi}{\partial x} \quad ; \quad u = -\frac{\partial \psi}{\partial y} \tag{5}$$

enabling Eq (4) to be written thus:

$$\left(\frac{\partial}{\partial t} - \frac{\partial \psi}{\partial y}\frac{\partial}{\partial x} + \frac{\partial \psi}{\partial x}\frac{\partial}{\partial y}\right)(\nabla^2 \psi + f) = G$$

which is one equation in one unknown. Finally defining the absolute vorticity q

$$q = \nabla^2 \psi + f \tag{6}$$

we can write the above in conservative form

$$\frac{\partial q}{\partial t} + J(\psi, q) = G \tag{7}$$

where J is the Jacobian of ψ and q

Thus, given an initial vorticity distribution and the forcing as a function of space and time, Eq (7) can be integrated forward to predict for the new vorticity at any later time. Eq (6) is inverted to give the streamfunction and hence, through (5) the horizontal velocity. Should we be interested in the pressure field it can be calculated from the momentum equations (1) and (2). Although there are certain computational advantages in expressing the dynamics in the single prognostic equation (7) for a single unknown, the main advantage is that it enables us to gain physical insight and so use our intuition. This is because q is conserved like any other fixed feature of a particle of fluid in the absence of sources and sinks ($G = 0$). Eq (7) makes it clear that ocean circulation theory is really vorticity transport theory (for an excellent discussion of the wind-driven ocean circulation viewed from this perspective see Stewart, 1964). The ocean moves ($\underline{v} \cdot \nabla q \neq 0$) because locally there is an imbalance in the vorticity sources and sinks, i.e. $G \neq 0$ locally.

Eq (7) is the appropriate starting point for our discussion: it is not as restrictive as it might seem because in baroclinic fluids the dynamics can again be expressed in conservative form provided we reinterpret ψ and q. So, for example, ψ could be interpreted as the streamfunction for the geostrophic flow in an isentropic layer and q as the isentropic potential vorticity $(\nabla^2\psi+f)/h$ where h is the layer depth. Eq (7) suggests immediately that there should be an interest in the geometry of the q contours, for they are the reference contours along which steady unforced flow moves.

(b) Source and Sinks of Vorticity

We write $G = S_o - S_i$

where S_o is a potential vorticity source and

S_i is a potential vorticity sink.

In wind-driven ocean circulation theory S_o is the wind-stress curl imposed at the upper surface which will be chosen to have the simple form

$$S_o = -\frac{1}{\rho H}\frac{\partial \tau_x}{\partial y} = -\frac{\pi\,\tau}{\rho H L}\sin\left(\frac{\pi y}{L}\right) \qquad \text{(see Fig. 1)} \tag{8a}$$

Depending on one's taste and prejudices the vorticity sink S_i is variously parametrized as a bottom Ekman layer ($\varepsilon\,\nabla^2\psi$) or through a "lateral friction" term ($-\kappa\,\nabla^2 q$) representing the lateral tranfer of momentum to a coast in turbulent stresses

$$S_i = \varepsilon \nabla^2 \psi \tag{8b}$$

$$S_i = -\kappa \nabla^2 q \tag{8c}$$

The most apropriate parametric representation of these sub-grid scale processes is not known, although solutions to Eq (7) are sensitive to the form adopted for S_i, and the boundary conditions chosen at the lateral walls.

c) Boundary Conditions

Solutions to Eq (7) are sought in a closed domain through which there can be no flow (see Fig. 1), i.e. ψ = constant, chosen to be zero on the boundary

However, in addition the vanishing normal flow, other boundary conditions may be required depending on the form of S_i. In particular if $S_i = -\kappa \nabla^2 q$, a no-slip boundary condition would seem appropriate. However, if $S_i = \varepsilon \nabla^2 \psi$ no additional boundary condition is required.

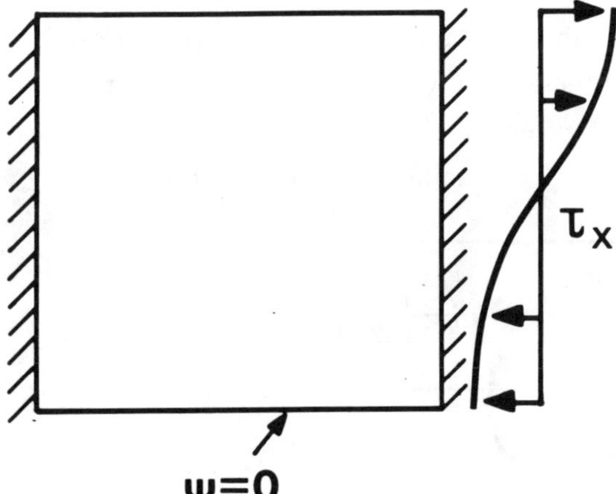

Fig. 1 The geometry of the basin in relation to the wind-stress forcing τ. The boundary is a streamline for the flow.

3. STEADY-STATE SOLUTIONS

(a) Linear

In the interior of the ocean the Rossby number is small

$$R_o = \frac{V}{fL} \sim \frac{\zeta}{f} << 1$$

but also

$$\frac{\nabla\zeta}{\beta} << 1 \qquad \text{where } \zeta = \nabla^2 \psi$$

i.e variations in f are very important. In other words, in the barotropic model the interior q geometry is dominated by variations in f. In linear models $q = \beta y$ everywhere and so the q contours are blocked by meridional walls (see Fig. 2). This is the most fundamental difference between meteorological and oceanographic flows. In the ocean the basin geometry is incompatible with that of the q contours: free flow along q contours is halted by their intersection with the coasts. How then does the barotropic ocean circulate in gyres despite the tendency of contours to be coincident with latitude contours beginning and ending at coasts?

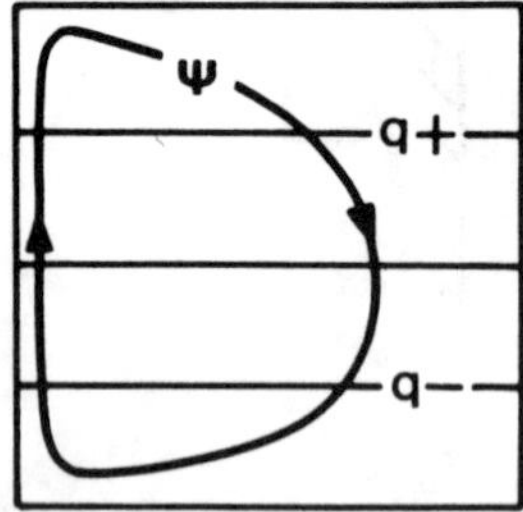

Fig. 2 The q geometry of the linear model, $q = \beta y$. The interior flow is driven across the q contours by the wind-stress curl, Eq (9). It returns in the western boundary current where frictional effects remove vorticity imparted in the interior, Eq (10).

In linear models the flow is driven across the q contours by the wind-stress curl

$$\frac{Df}{Dt} = S_o \tag{9}$$

expressing a balance between the rate of change of planetary vorticity and the rate of vorticity input by the wind-stress curl. The flow returns in the western boundary current where dissipative processes remove the vorticity imparted to the fluid in the interior by the wind-stress curl.

$$\frac{Df}{Dt} = - S_i \tag{10}$$

The detailed structure of the boundary current depends on the form chosen for S_i. It is important to realize, however, that assumptions regarding the laws of turbulent stress are a matter of detail affecting only the precise width and structure of the return current.

This is a familiar cycle of vorticity input over the interior of the gyre driving steady flow across the q contours, and dissipation of vorticity in frictional western boundary layers (Stommel, 1947; Munk 1950).

The intensity of the circulation, set by the interior vorticity balance Eq (9), is (using Eq. 8a)

$$U_s = \frac{\pi\tau}{\rho\beta HL} \tag{11}$$

(b) Nonlinear

The most serious limitation of linear barotropic models is that they assume $q = \beta y$ everywhere - significant features of the ocean circulation cannot be modelled if non-linearities are neglected. Recent mapping of the potential vorticity field, q, of the world's oceans (see, for example, McDowell et al., 1982; Keffer, 1985) show that there is strong non-linear control of the q contours. Rather than the q contours being coincident with latitude circles, as one would expect if the flow were weak, the q contours are strongly perturbed by the motion field particularly in the "bowl" of the wind-driven circulation in the upper kilometer or so.

Fofonoff (1954) considered some of the constraints that must be satisfied by non-linear flow patterns in a homogeneous ocean by studying steady free circulation

$$\frac{\partial}{\partial t} = 0 \; ; \; S_o = S_i = 0 \text{ in equation (7).}$$

$$\therefore J(\psi, q) = 0$$

$$\text{and so} \quad q = q(\psi) \tag{12}$$

i.e. q is constant along a streamline ψ. This is a very stringent constraint. Thus

$$\nabla^2 \psi + \beta y = q(\psi) \tag{13}$$

We can see from Eq (13) that unless the relative vorticity becomes comparable with βy, ψ will be a function of y only, and the flow will be along latitude circles. Near meridional boundaries where flow is across latitude circles, the relative vorticity must be large enough to "close-off" the q contours - this is schematically shown in Fig. 3. This closing off of the q contours isolates the flow from the influence of the lateral boundaries.

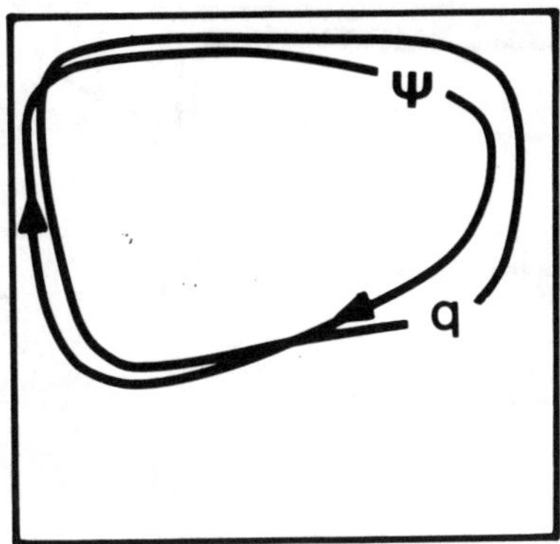

Fig. 3 Schematic diagram showing the close coincidence of ψ and q possible in a strongly non-linear flow.

In order to proceed further, Fofonoff (1954) adopted a linear q/ψ relationship and set

$$\frac{dq}{d\psi} = -\frac{\beta}{U_I} > 0 \tag{14}$$

(with $U_I < 0$)
Eq (13) thus becomes

$$\nabla^2 \psi + \beta y = \frac{dq}{d\psi}\psi = -\frac{\beta}{U_I}\psi \tag{15}$$

Eq (15) can easily be solved for ψ and is sketched in Fig. 4. The solution has distinctive features due to the variation in f. There is a slow interior westward drift,

$$\psi_I = -U_I\, y$$

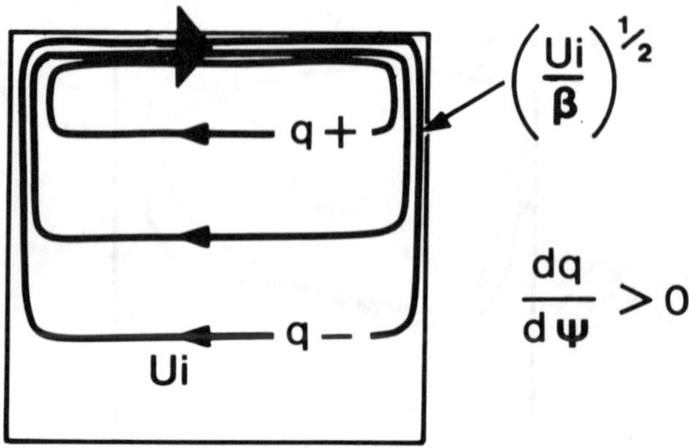

Fig. 4 The Fofonoff solution, Eq (15) with $dq/d\psi > 0$ and interior westward flow. The return flow to the east occurs in a boundary current (here arbitrarily chosen along the northern wall).

and the eastward current is swift and narrow, the scale being given by $(|U_I|/\beta)^{\frac{1}{2}}$.

Although highly idealized, Fofonoff's is one of the survivors of early inertial theory. As shown in the numerical integrations of Veronis (1966) (see Fig. 5) strongly non-linear versions of Stommel's model resemble Fofonoff gyres with closed q contours and almost east/west symmetry (provided, as discussed in Blandford (1971), slippery boundary conditions are employed with a bottom friction parameterization). The high resolution quasi-geostrophic model of Holland (1984) exhibits highly inertial mean flows (even in the uppermost layer) with broad westward return flows and a narrow swift eastward interior jet. The recirculation south of the Gulf Stream is probably strongly inertial in character, some aspects of which may be captured by generalization of Fofonoff's solution to baroclinic flow.

The difficulty with purely inertial theory Eq (12) is that the intensity of the circulation is arbitrarily set. However, Niiler (1966) argued that although local effects of forcing and dissipation may be small, they must be important in the overall equilibrium of the gyre. So integrating equation (7) over a closed streamline

$$\int S_o \, dxdy = \int S_i \, dxdy$$

For the S_o given by Eq (8a) and the S_i given by Eq (8b)

$$\underbrace{\frac{1}{\rho H} \oint_{\psi} \underline{\tau} \cdot \underline{dl}}_{\text{dominated by the interior}} = \underbrace{\varepsilon \oint_{\psi} \underline{u} \cdot \underline{dl}}_{\text{dominated by the boundary currents}} \tag{16}$$

Niiler (1966) used Eq (16) to set the intensity of the inertial circulation to give a boundary current velocity

$$U_{BC} \sim \frac{\tau}{\rho H \varepsilon}$$

and hence by mass continuity an interior velocity

$$L\, U_I \sim L_{BC}\, U_{BC} \sim \left(\frac{|U_I|}{\beta}\right)^{1/2} \frac{\tau}{\rho H \varepsilon}$$

$$\text{or } U_I \sim \frac{1}{\beta} \left(\frac{\tau}{\rho \varepsilon L H}\right)^2 \tag{17}$$

This should be contrasted with the Sverdrup velocity scale given by linear theory U_s, Eq (11).

Inserting values appropriate to the model integration of Fig. 5 (see figure legend), the ratio

$$\frac{U_S}{U_I} = \frac{\rho\, L\, H\, \varepsilon^2\, \pi}{\tau}$$

is accordingly about 1/3, showing that the gyre has "spun-up" in excess of the Sverdrup velocity scale. This spin-up can be attributed to the lack of lateral frictional boundary layers in the model.

So the integral balance between weak forcing and dissipation Eq (16) sets the intensity of the circulation. However, Niiler did much more than this with Eq (16): he used it to determine (or at least limit the possible choices of) the functional relationship between ψ and q. This idea of invoking weak forcing and dissipation to pick out a unique solution (the purely inviscid theory admits an infinity of solutions) has recently been exploited by Rhines and Young (1982) (in an elegant theory of mid-ocean gyres) and Pierrehumbert and Malguzzi (1984) (in a study of non-linear resonance and atmospheric blocking). The approach is well illustrated in the present context, and so we go on to study the forcing and dissipation processes compatible with the maintenance of a Fofonoff gyre.

(i) Forcing and dissipation balances and the $q(\psi)$ relationship

Following Niiler (1966) (see also Pierrehumbert and Malguzzi (1984)) we suppose that the circulation can achieve all the transfers necessary to offset potential vorticity sources and sinks by making only small adjustments to purely free inertial flow ψ_o, q_o.

$$\psi = \psi_o + \psi_1$$
$$q = q_o + q_1$$

were ψ_1, q_1, are "small" corrections (formally we should expand about ψ_o, q_o in terms of a small parameter which Niiler chose to be the ratio U_S/U_I given above). At zero order

$$J(\psi_o, q_o) = 0$$
$$\text{and so} \qquad q_o = q_o(\psi_o) \tag{18}$$

In the inviscid problem Eq (18) there are an infinite number of possible choises $q_o = q_o(\psi_o)$, but not all yield a valid solution at next order. In other words, not all solutions are compatible with forcing and dissipation balances at next order.

The equation for the correction ψ_1, q_1 is

$$J(\psi_o, q_1) + J(\psi_1, q_o) = S_o - S_i$$

and so a necessary condition for the existence of the ψ_1 field is that potential vorticity sources should balance sinks when integrated over a closed ψ_o contour.

$$\int S_o \;\; dxdy = \int S_i \;\; dxdy \tag{19}$$

As Niiler puts it "the condition on the existence of the 1st order problem yields an integral constraint on the zero order problem". It is important to note that sources and sinks need not balance locally but only in an integral sense. Eq (19) is just the balance Eq (16) taking S_o as the wind-stress curl and S_i a bottom friction

$$\frac{1}{\rho H} \oint_{\psi_o} \underline{\tau} \cdot \underline{dl} = \varepsilon \oint_{\psi_o} \underline{U}_o \cdot \underline{dl} \tag{20}$$

Given $\underline{\tau}$ it appears that no direct method of solution to Eq (18) and (20) can be developed. Instead, Niiler (1966) adopted the linear $q_o(\psi_o)$ relationship chosen by Fofonoff, Eq (14), solved Eq (18) for ψ_o and then used Eq (20) to determine which wind-stress forcing was compatible with it. The details are not of interest. The important point is that the integral balance Eq (20) can evidently equilibrate an inertial gyre since $\int \underline{\tau} \cdot \underline{dl}$ has the same sign as $\int \underline{u} \cdot \underline{dl}$ (the sense of circulation must refect the sign of vorticity source). In other words, the strongly inertial gyre spun up in Fig. 5 by an anticyclonic wind-stress curl can be equilibrated by bottom friction. Thus our model of circulation yields unambiguous results, both numerically and analytically.

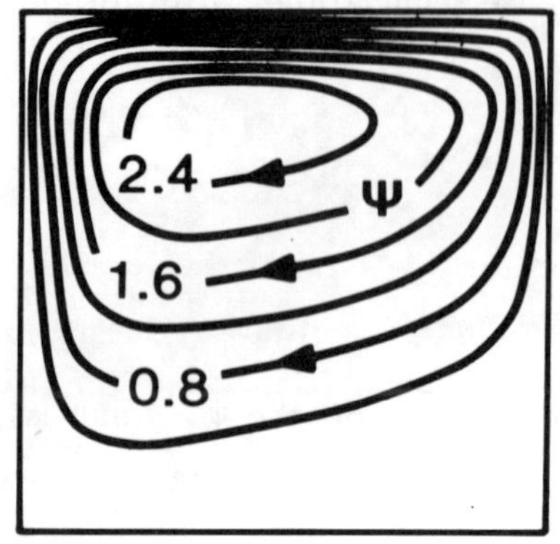

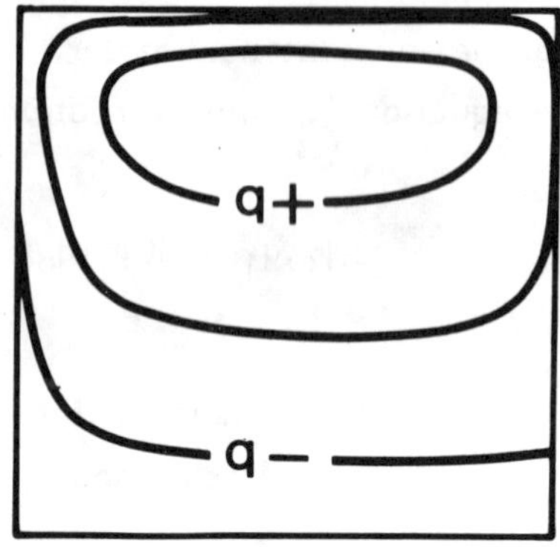

Fig. 5 Numerically obtained steady state solution to

$$J\,(\psi, \nabla^2\psi + \beta y) = -\frac{\pi\tau}{\rho HL}\,\sin\!\left(\frac{\pi y}{L}\right) - \varepsilon\,\nabla^2\,\psi$$

with

$$\tau = 10^{-1}\ \mathrm{N\,m^{-2}},\ \rho = 10^3\ \mathrm{kg\,m^{-3}},\ \beta = 10^{-11}\ \mathrm{s^{-1}\,m^{-1}}$$
$$\varepsilon^{-1} = 10^7\ \mathrm{S},\ H = 10^3\ \mathrm{m},\ L = 10^6\ \mathrm{m}.$$

(a) streamfunction (units of $\pi\tau/\rho\beta H$) **(b)** potential vorticity.

(ii) Lateral Transfer of q

If instead of a bottom friction parametrization $\varepsilon\,\nabla^2\,\psi$, we chose to transfer q laterally $-\nabla\cdot(\kappa\nabla q)$ supposing plausibly that geostrophic eddies transfer potential vorticity systematically down-gradient, how might our results be expected to change? Again the key is the integral constraint Eq (19).

Now, instead of Eq (20) we have

$$\frac{1}{\rho H}\oint_{\psi_o} \underline{\tau}\cdot\underline{d}l = -\oint_{\psi_o} \kappa\nabla q_o\cdot\underline{n}\ dl$$

where $\underline{n}$ is a unit vector normal to ψ_o or, noting that $\nabla q_o = \frac{dq_o}{d\psi_o}\nabla\psi_o$ the above may be written

$$\frac{dq_o}{d\psi_o} = -\frac{1}{\rho H}\,\frac{\oint \underline{\tau}\cdot\underline{d}l}{\oint \kappa\underline{u}_o\cdot\underline{d}l} \qquad (21)$$

Thus the requirement that net forcing must balance net dissipation over a closed streamline, tells us that

$$\frac{dq_0}{d\psi_0} < 0$$

if the sink of vorticity is to be represented by a geostrophic eddy field fluxing potential vorticity down the q gradient. It follows from Eq (15) that inertial boundary layers cannot now be supported (solutions Eq (15) are oscillatory rather than exponential) and so weak lateral transfer of q cannot equilibrate a Fofonoff gyre. Blandford (1971) confirmed this numerically and explicitly demonstrated that the qualitative difference between the Bryan (1963) [lateral friction] and Veronis (1966) [bottom friction] integrations can be attributed to their different choice of frictional parametrizations.

It is important to realize that the result $dq/d\psi < 0$ is a consequence of the hypothesis that geostrophic eddies flux potential vorticity systematically down-gradient. However, there is strong theoretical and observational support for this conjecture. It should not be concluded, however, that this necessarily implies that free circulations of the type discussed here are unlikely to be relevant to actual oceanic flows. The problem lies more in the inappropriateness of the barotropic formulation as a representation of the dynamics of the baroclinic ocean. In illustration we present a simple, but hopefully interesting extension of barotropic inertial theory in the following section.

(iii) A 1½ Layer Inertial Gyre Equilibrated by Lateral Transfer of Potential Vorticity

Instead of the absolute vorticity defined in Eq (6) we adopt an "equivalent barotropic" formulation:

$$q = \nabla^2 \psi + \beta y - \frac{\psi}{L_\rho^2}$$

where $L_\rho = \sqrt{\dfrac{g'h}{f^2}}$ is the Rossby radius, h is the mean depth of the layer and g' is the "reduced" gravity. It is difficult to unambiguously interpret the equivalent barotropic model in terms of the baroclinic ocean. Nevertheless we will try to do so and associate high values of ψ with a raised sea-level and isopycnals bowing downwards into the thermocline, and low values of ψ with a lowered sea-level and isopycnals bowing upwards. The term ψ/L_ρ^2 can then be interpreted as a crude (linearized) representation of the dynamical effects of vortex stretching associated with changes in layer depth. The term is absent in the barotropic formulation. We look for solutions of the form

$$\nabla^2 \psi + \beta y - \frac{\psi}{L_\rho^2} = \frac{dq}{d\psi} \psi \tag{22}$$

where $dq/d\psi$ is assumed constant. From Eq (21) we see that if such a flow is to be equilibrated by lateral transfer of q then

$$\frac{dq}{d\psi} < 0$$

Thus from (22) our gyre will have interior westward flow supporting inertial boundary layers (even though $dq/d\psi < 0$) provided that now only

$$\frac{dq}{d\psi} + \frac{1}{L_\rho^2} > 0.$$

So the requirement for inertial boundary layers is that

$$-\frac{\beta}{U_I} = \frac{dq}{d\psi} + \frac{1}{L_\rho^2} > 0 \tag{23}$$

(cf. Eq (14)).

The solution, sketched in Fig. 6, has some interesting properties. The inclusion of a representation of vortex stretching enables the interior potential vorticity gradient to be reversed ($\partial q/\partial y$ is no longer dominated by β and is reversed in the interior return flow). It is this reversal which enables the gyre to be equilibrated by lateral transfer of q.

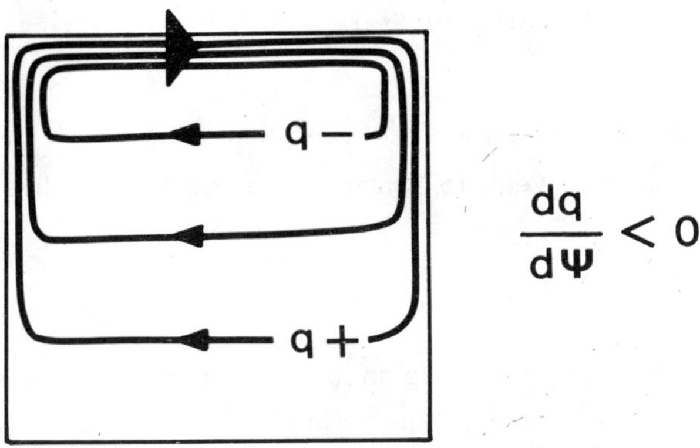

Fig. 6 Inertial flow in a $1\frac{1}{2}$ layer model, Eq (22), with $dq/d\psi < 0$. The thermocline now slopes down towards the northern boundary of the subtropical gyre, more than offsetting the change in planetary vorticity across the westward flowing interior flow.

Although our solution can be no more than suggestive it does have some contact points with reality, and particularly with more detailed ocean models. The upper mean flow of Holland's (1984) eddy resolving model is strongly inertial with eastward flow confined to a narrow jet and broad gentle return flows. Moreover, $dq/d\psi < 0$ here, with reversed q gradients in the interior where vortex stretching reverses the planetary vorticity gradient. Indeed, the N. Atlantic potential vorticity maps of McDowell et al. (1982) show just such a reversal in the uppermost layers of the North Equatorial current (which is thus a prime region for baroclinic instability). Our simple model suggests that $dq/d\psi < 0$ in the near surface layers in order that potential vorticity sources can be balanced by the lateral transfer of q achieved by the geostrophic eddy field. This is certainly the dynamical balance at work in the weakly dissipative, highly non-linear, thermodynamically inactive eddy resolving models. To what extent this limit is an appropriate description of the real ocean must await more detailed diagnostic and modelling studies.

(iv) The Uniform Potential Vorticity State

If the geostrophic eddies are very efficient (κ large) or the potential vorticity sources weak (S_o small), then Eq (21) implies that

$$\frac{dq}{d\psi} \simeq 0$$

i.e. the inertial gyre sits in a region of almost uniform q. Our solution has chosen the uniform q state. This result has been exploited by Rhines and Young (1982) in a theory of mid-ocean gyres ($\nabla^2\psi$ neglected) where the Sverdrup mass transport is distributed in the vertical by hypothesising that q is uniform in layers which are isolated from the effects of surface forcing. Obviously this is a natural end state if there are no interior potential vorticity sources and geostrophic eddies smooth out potential vorticity gradients.

In our $1\frac{1}{2}$ layer inertial theory the uniform q state corresponds to

$$\nabla^2 \psi + \beta y - \frac{\psi}{L_\rho^2} = \text{constant}$$

which gives

$$\psi_I \sim \beta L_\rho^2 y$$

and

$$\psi_{BC} \sim \exp\left(\frac{x}{L_\rho}\right)$$

independent of the eddy tranfer coefficient κ.

In this limit case all three components of the potential vorticity, relative vorticity, stretching and planetary vorticity collude together to make q uniform.

4. CONCLUDING REMARKS

Here we have only had time to discuss some simple steady-state solutions for homogeneous wind-driven ocean circulation. The real value of such studies is that they provide firm reference points from which to interpret more complex models, numerical model simulations and particularly oceanographic observations. These simple models are important because they allow us to isolate and elucidate important aspects of the physics. It is only then that we can have the confidence to move on to more complex models with more detailed physics.

The barotropic formulation although extremely crude can nevertheless capture many of the characteristics of more detailed models. For example, despite the absence of baroclinity, the barotropic model can be used in an eddy-resolving mode (see Marshall, 1984) in which pysically resolved eddies (generated in a barotropically unstable internal jet) play a central and plausible role in the gyre-scale dynamics.

The extension to layered quasi-geostrophic models in straightforward for the govering equations can again be expressed in the form Eq (7): in abiabatic frictionless motion fluid parcels now conserve their quasi-geostrophical potential vorticity in each layer, the layers being coupled in the vertical through vortex stretching. Geostrophy and hydrostatic balance now implies a thermal-wind relation, sloping isentropes and hence vast stores of potential energy available for conversion. Baroclinic instability is thus an ubiquitous feature of the ocean circulation tending to smooth out interior potential vorticity gradients and driving mean flows. Just as the surface wind pattern in the atmosphere is a sensitive indicator of eddy tranfer processes, so the pattern of subsurface, and particularly deep ocean currents, should reflect the transfer of mass and heat by the oceanic eddy field.

ACKNOWLEDGEMENT

I should like to thank George Nurser for many helpful comments.

REFERENCES

Blandford, R.R., 1971: Boundary conditions in homogeneous ocean models. - Deep-Sea Res. 18, 739-751.

Bryan, K., 1963: A numerical investigation of a non-linear model of a wind-driven ocean. - J. Atmos. Sci. 20, 594-606.

Fofonoff, N.P., 1954: Steady flow in a frictionless homogeneous ocean. - J. Mar. Res. 13, 254-262.

Gill, A.E., 1982: Atmosphere-ocean dynamics. - Academic Press, pp. 512-522.

Holland, W.R., T. Keffer and P.B. Rhines, 1984: Dynamics of the oceanic general circulation: the potential vorticity field. - Nature 308, 698-705.

Keffer, T., 1985: The ventilations of the World's oceans: maps of potential vorticity field. J. Phys. Oceanogr. 15, 509-523.

Marshall, J.C. 1984: Eddy-mean-flow interaction in a barotropic ocean model. Quart.J.Roy.Met.Soc. 110, 573-590.

McDowell, S., P. Rhines and T. Keffer, 1982: North Atlantic potential vorticity and its relation to the general circulation. - J. Phys. Oceanogr. , 1417-1436.

Munk, W.H., 1950: On the wind-driven circulation. - J. Met. 1, 79-93.

Niiler, P.P., 1966: On the theory of wind-driven ocean circulation. - Deep-Sea Res. 13, 597-606.

Pedlosky, J., 1979: Geophysical Fluid Dynamics. - Springer-Verlag, pp. 236-313.

Pierrehumbert, R. and P. Malguzzi, 1984: Forced coherent structures and local multiple equilibria in a barotropic atmosphere. - J. Atmos. Sci. 41, 246-257.

Rhines, P. and W. Young, 1982: A theory of wind-driven circulation - I. Mid-ocean gyres. - J. Mar. Res. 40, 559-596.

Robinson, A., 1963: Wind-driven ocean circulation. - Blaisdell Publishing Company.

Stewart, R., 1964: The influence of friction on inertial models of oceanic circulation. In: Studies in Oceanography, Tokyo University, 3-9

Stommel, H., 1948: The westward intensification of wind-driven ocean currents. - Transactions of the American Geophysical Union, 29, 202-206.

Stommel, H., 1965: The Gulf Stream, A physical and dynamical description. Unversity of California Press, 248 pp.

Veronis, G., 1966: Wind-driven circulation. Part II Numerical solutions of the non-linear problem. - Deep-Sea Res. 13, 31-55.

Veronis, G., 1981: Dynamics of large-scale circulation. - In: Evolution of Physical Oceanography, B.A. Warren and C. Wunsch, eds., MIT Press, Cambridge, pp. 140-183.

COUPLED OCEAN-ATMOSPHERE MODELS OF EL NINO AND THE SOUTHERN OSCILLATION

JULIAN P. McCREARY, JR.

Nova University Oceanographic Center
Dania, FL 33004, U.S.A.

ABSTRACT

A coupled ocean-atmosphere model of El Nino and the Southern Oscillation is discussed. The model ocean is a reduced-gravity system that also includes an equation for the temperature of the layer, T. The model atmosphere is a linear, single-baroclinic mode system driven by convection, Q. The wind stress used to drive the model ocean is proportional to the wind velocity produced by the model atmosphere, while Q over the ocean is a function only of T. Some of the solutions also involve land as well as ocean; in that case, Q over land is either a fixed forcing function or it is set to zero.

For a broad range of model parameters, solutions oscillate at the long time scales associated with El Nino and the Southern Oscillation (ENSO). In all cases, the basic process causing the oscillation is the growth of unstable disturbances to finite amplitude and their subsequent slow eastward propagation. The dynamics of the instability are nicely illustrated in the simpler analyses of Lau (1981), Philander et al. (1984) and others.

Solutions compare favorably with observations of ENSO in several respects. For example, for a basin geometry that most resembles the Indian and Pacific Oceans, strong permanent convection exists in the eastern Indian Ocean, and there is a oscillation in the Pacific Ocean with a period of about five years. Associated with this oscillation is a patch of convection that develops in the central and western ocean and propagates into the eastern ocean before dissipating. Solutions do not compare favorably in other ways. A major limitation is that they do not simulate the rapid onset or intermittancy of ENSO.

J. Willebrand and D. L. T. Anderson (eds.), Large-Scale Transport Processes in Oceans and Atmosphere, 247–280.

1. INTRODUCTION

The phenomenon of El Niño and the Southern Oscillation (ENSO) exhibits a time scale of the order of 2-9 years. Atmospheric models suggest that the atmosphere itself does not have such low-frequency variability (Manabe and Hahn, 1981; Lau, 1981; Lau and Oort, 1985). An intriguing hypothesis is that the atmosphere and ocean form a strongly coupled system in the tropics with an inherent periodicity of 2-9 years. Several studies using simple, coupled ocean-atmosphere models have begun to explore this idea. These systems use ocean and atmosphere models that are dynamically considerably less sophisticated than general circulation models. Their advantage is that, due to their simplicity, it is easier to isolate and to understand the various processes at work in them. Their limitation, of course, is that they necessarily involve many simplifying assumptions.

This introduction provides a brief overview of some of these models, and the body of this paper describes one of them (Anderson and McCreary, 1985a,b) in detail. The models generally separate into two types: those that can develop two equilibrium states, and those that generate unstable disturbances. (An exception to this categorization scheme is the model of McCreary, 1983, which has neither property.) Another point of separation is that some solutions oscillate through a full ENSO cycle, whereas others treat only one aspect of the phenomenon.

McCreary (1983) first developed a coupled system that oscillated at the long time scales associated with ENSO. The model ocean consisted of the single baroclinic mode of a two-layer ocean, and sea surface temperature (SST) was either warm or cold according to whether the depth of the model interface, h, was less than or greater than a specified depth. The model atmosphere consisted of two patches of zonal wind stress, τ_w and τ_h, that were assumed to interact with the ocean in a way suggested by the ideas of Bjerknes (1966, 1969): when the eastern ocean was cool the equatorial easterlies, τ_w, were assumed to strengthen in the central ocean, thereby simulating Bjerknes' Walker circulation; when the eastern ocean was warm the extra-equatorial easterlies, τ_h, were assumed to strengthen in the eastern ocean, thereby simulating an enhanced Hadley circulation there. The pattern τ_h provided a restoring force that prevented solutions from ever reaching an equilibrium state. Wind curl associated with τ_h deepened the model interface to form a ridge in a region centered off the equator. This ridge propagated westward slowly as a Rossby wave, reflected from the western boundary as an equatorially trapped Kelvin wave, eventually affecting SST in the eastern ocean. It was the long time it took the ridge to propagate across the ocean basin that governed the period of the oscillation.

Hughes (1979, 1984) first discussed a coupled system relevant to ENSO that possessed two equilibrium states. His model ocean consisted simply of an equatorial pycnocline which tilted in response to zonal wind stress, and all other properties of the tropical ocean were ignored. The model atmosphere was an externally specified equatorial easterly wind stress (a Walker circulation), that responded to the depth of the pycnocline in the eastern ocean. One equilibrium state had a large pycnocline tilt and strong easterlies, and the other had a small tilt and weak easterlies. In the latter study Hughes included the effect of the annual cycle of the trades, and suggested that El Nino events which are initiated during the warm part of the annual cycle tend to be large.

McCreary and Anderson (1984) also considered a coupled system that allowed for the possibility of two equilibrium states. Their model was similar in structure to that of McCreary (1983), differing primarily in that τ_h was replaced by another wind pattern, τ_s, that represented the annual cycle of the Pacific trades. Without τ_s, the system adjusted to one or the other of two equilibrium states: one with τ_s switched on (a "normal" state) and another with it off (an "El Niño" state). With τ_s, no equilibrium state was possible; instead τ_s acted as a "trigger" that continued to flip the system from being near one equilibrium state to being near the other. For realistic choices of model parameters solutions oscillated at ENSO time scales. A nice property of these solutons was that El Niño events were phase-locked to the annual cycle, in a manner consistent with the observations.

Lau (1984), Welander and Long (1985) and Long (1985, private communication) also produced oscillating solutions with coupled systems that allowed for two equilibrium states. Lau presented observational evidence for a bimodal climatic state associated with ENSO, and compared observations to the response of a stochastically forced, nonlinear oscillator. Similarly, Welander and Long "triggered" a flip of their model from one near-equilibrium state to the other with stochastic forcing.

Lau (1982) and Philander et al. (1984) first developed coupled models that produced unstable disturbances relevant to ENSO (Philander and co-workers explicitly pointed out this relevance). The model ocean in both studies was a linear, reduced-gravity model. The model atmosphere was a linear, single-baroclinic-mode model that was forced by a heating function, Q [as in equations (15)], where Q was taken to be directly proportional to the interfacial-depth anomaly, h-h. Lau greatly simplified his system by considering only coupling between equatorial Kelvin waves in the ocean and atmosphere, and found a simple dispersion relation for the coupled waves. One of the roots always had a phase speed very close to that of an atmospheric Kelvin wave. The other was an oceanic Kelvin wave for weak coupling. As coupling increased, the phase speed of the wave slowed to zero and then became positive imaginary, indicating that it was now a stationary, growing instability. Philander and co-workers solved their system numerically without resorting to Lau's simplification, and also found a rapidly growing instability. They discussed the relevance of this unstable growth to ENSO. See section 4a for a further discussion of this type of instability.

Hirst (1985) and Yamagata (1985) have studied coupled systems that are similar to the two described in the preceeding paragraph. Not surprisingly, they also find eastward-propagating instabilities when Q is assumed to be proportional to h-h. In addition, they describe another type of coupled instability that occurs when Q is proportional to the zonal advection of temperature. Gill (1985) produced oscillatory solutions in a model that involved this mechanism; however, he concluded that these solutions were probably not relevant to ENSO, because their period was so short (only being somewhat longer than one year).

The coupled model of Anderson and McCreary (1985a,b), discussed in detail in the body of this text, develops unstable disturbances of the sort discussed by Lau (1981) and Philander et al. (1984). The model ocean is a reduced-gravity model that also has explicit thermodynamics, including both horizontal advection and vertical redistribution of heat. The model atmosphere is a linear, single-baroclinic-mode model, with Q specified to be a function of ocean temperature. This specification of Q prevents instabilities from growing indefinitely; rather, they attain a maximum amplitude and then propagate slowly eastward. Oscillatory solutions exist for a broad range of model parameters, and they compare favorably with observations of ENSO in some respects; however, they fail to simulate adequately the rapid onset and intermittancy of El Nino events.

Recently, Zebiak (1985) has studied a coupled system that is both intriguingly similar and different from the Anderson and McCreary model.(See Cane and Zebiak, 1985, for an overview of this model.) The model ocean is a linear, reduced-gravity model, in which a thin, frictional, surface sublayer is also included, thereby allowing surface currents to be quite large. The model atmosphere differs in that Q is related to convergence in the wind field as well as to ocean temperature. Finally, only anomaly fields are solved for, so that monthly mean values of observed SST and (model generated) surface currents are included in the temperature anomaly equation. Solutions oscillate at time scales associated with ENSO, and an essential process for the existence of these oscillations is again the growth of unstable disturbances of the type discussed by Lau (1981), Philander et al. (1984), and Anderson and McCreary (1985a,b). The solutions also exhibit interesting aperiodicities that are clearly related to the presence of the thin surface sublayer and to the convergence feedback mechanism in the model atmosphere.

2. THE MODEL OCEAN

The model ocean is an extension of the usual reduced-gravity system that allows the temperature of the layer to vary. The equations are

$$
\begin{aligned}
(hu)_t + (huu)_x + (huv)_y - \beta yhv &= -\left(\tfrac{1}{2}\alpha gh^2T\right)_x + \tau^x + \nu_h \nabla^2(hu), \\
(hv)_t + (huv)_x + (hvv)_y + \beta yhu &= -\left(\tfrac{1}{2}\alpha gh^2T\right)_y + \tau^y + \nu_h \nabla^2(hv), \\
h_t + (hu)_x + (hv)_y &= \frac{2\delta}{hT} - \omega + \gamma\left(\frac{T-T^*}{T}\right), \\
T_t + uT_x + vT_y &= \frac{2}{h}\left[-\gamma(T-T^*) - \frac{\delta}{h}\right] + \nu_h \nabla^2 T,
\end{aligned}
\tag{1}
$$

where T is the temperature excess of the surface layer over the deep inert layer, ρ_o, α and c_p are the density, thermal expansion coefficient and specific heat of water, respectively, and other quantities have their usual definitions. The values of α, β and ν_h are fixed throughout to be $.0003°C^{-1}$, $2.28 \times 20^{-11}\ m^{-1}s^{-1}$, and $2 \times 10^3\ m^2s^{-1}$, and the value of c_p is never needed. The dynamical form of these equations is discussed more fully in Anderson (1984), while the derivation of the forcing terms for the mixed layer follows closely that of Hughes (1980).

The surface stress is linearly related to the atmospheric velocity field according to

$$
\tau^x = \rho_a C_D' U, \qquad \tau^y = \rho_a C_D' V, \tag{2}
$$

where U and V are defined in (6), ρ_a is the density of the air taken to be $1.3\ kg\ m^{-3}$, and C_D' is a drag coefficient with the value .008 m/s. This choice for C_D' ensures that the usual drag coefficient, $C_D = C_D'/U$, has a value of .00125 when the wind speed is 5 m/s.

Equations (1) involve three thermodynamic processes: Q_S, δ, and w. Q_S is the heat flux into the layer. In reality, Q_S depends on many atmospheric parameters. Since the objective of this paper is to explore simple coupling mechnisms, the simple parameterization

$$Q_s = - \rho_o c_p \gamma (T-T^*) \tag{3}$$

is used, a form first suggested by Haney (1973). In this model T is always less that T*, so Q_S always acts as a heat source for the layer. Turbulent mixing, $\rho_o \alpha g \delta$, represents the tendency for wind stirring to entrain fluid into the layer from below, thereby deepening the mixed layer and increasing its potential energy. Again for simplicity, we take $\delta = \delta_o/h$. w is a slow upwelling velocity in the deep ocean, presumably driven by a background thermohaline circulation.

The three thermodynamic processes involve four parameters: w, δ_o, γ and T*. The values of w, δ_o and γ are generally 4×10^{-7} m/s, 4×10^{-2} m^3°C s^{-1} and 3×10^{-6} m/s, respectively. T* varies with latitude according to

$$T^* = 4 + (11.33-4)\ (1+\cos 2\pi y/y_N)/2 \ , \tag{4}$$

where y_N = 9000 km. T* has a maximum value of 11.33°C at the equator and decreases to 4°C at a distance of 4500 km from the equator.

The choices of parameters in the preceeding paragraph ensure that various aspects of the ocean reponse are physically realistic. For example, in the absence of dynamical processes (so that u = v = 0) and when T is uniform, the latter two equations of (1) have the steady solution

$$\bar{T} = \frac{\gamma\ T^*}{(\gamma+\omega)} \ , \qquad \bar{h} = \left[\frac{\delta_o(\gamma+\omega)}{\gamma\omega T^*}\right]^{1/2} \tag{5}$$

Thus, the values of T and h at the equator are 10°C and 100 m, respectively. Note that if σ, δ_o or w are set to zero, so that any one of the thermodynamic processes is neglected, the model cannot attain a sensible state of rest.

The model ocean consists of the finite-difference versions of equations (1) evaluated on an Arakawa C-grid, with T values stored at h points. The domain extends from the equator to 4500 km north and may be either cyclic in x or a bounded basin. The horizontal resolution of the grid is 150 km for all the solutions shown here. Kelvin waves along the northern boundary are artificially damped by including a damping term $-\kappa(h-\bar{h})$ in the third of equations (1), where $\bar{h}$ is defined in (5) and κ decreases linearly from a value of $2.5 \times 10^{-7}s^{-1}$ at the boundary to zero 450 km away. The equations are integrated forward in time with a time step of 1/8 day.

3. THE MODEL ATMOSPHERE

The equations of motion of the model atmosphere are

$$\begin{aligned} -\beta yV &= -P_x - rU\ , \\ \beta yU &= -P_y - rV\ , \\ U_x+V_y &= \frac{Q}{c^2} - r\frac{P}{c^2}\ , \end{aligned} \tag{6}$$

where Q is the forcing by latent heat release, c is the speed Kelvin waves have in the undamped system and r is a coefficient of Newtonian cooling. These equations describe the response of the first baroclinic mode of the atmosphere, a model that has already been successfully used in several studies of the tropical wind field (Webster, 1972; Egger, 1977; Gill, 1980; Zebiak, 1982). The two free parameters in this model, r and c, have the values $3 \times 10^{-5}s^{-1}$ and $60\ ms^{-1}$, respectively, the same values used by Gill and by Zebiak. For these choices an equatorial Kelvin wave will be damped on a length scale of $c/\beta \sim 2000$ km, while the lowest-order equatorial Rossby wave will decay on the scale $1/3(c/\beta) \sim 700$ km.

Latent heat release over the ocean is assumed to be related to T according to

$$Q = Q_0 \frac{T - T_c}{\bar{T}(0)-T_c} \theta(T-T_c) , \tag{7}$$

where θ is the Heaviside step function, $\bar{T}(0)$ is the temperature calculated at the equator using (5) and Q_o is an amplitude factor. Thus the strength of the forcing is taken to be proportional to the amount T is above some critical value T_c, and is zero when T is below T_c; unless stated otherwise, the value of T_c is always T(0)-1.5. Q_o is chosen such that the strength of the surface wind stress as calculated from (6) is consistent with the typical observed equatorial values, and usually has the value of .05 m^2s^{-3}.

The response of the atmosphere over land is also found for the solutions in section 3d. Latent heat release over land is parameterized according to

$$Q = Q_1 X(x) \frac{\bar{T} - T_c}{\bar{T}(0)-T_c} \theta(\bar{T}-T_c) , \tag{8}$$

where Q_1 is an amplitude factor usually taken equal to Q_o, and X(x) is a modulating factor defined below.

A number of assumptions are inherent in this model. First, it is assumed that the atmosphere adjustment times are short compared with those of the ocean, and therefore that the atmosphere will be always in equilibrium with any evolving ocean surface temperature field. Second, the main forcing of the tropical atmosphere is considered to be the result of convection which occurs mainly in the middle-troposphere (500-400 mb), thereby exciting a first-mode type of structure (Gill, 1980). Third, the atmospheric response to convection is assumed linear; this assumption is probably valid where convection is weak, but not where convection is strong. Fourth, it is assumed that convection depends only on local SST; this dependence again is unlikely to be strictly true, since convection also depends on surface fluxes and advection of moisture.

Equations (6) are solved numerically using the same technique as Zebiak (1982). Solutions are found by Fourier transforming the equations of motion in x and then solving for an equation in $\tilde{V}$ (the transform of V) alone. This equation is an ordinary differential equation in y, and in finite-difference form it reduces to a tridiagonal matrix equation that can be quickly solved for $\tilde{V}$. The quantities $\tilde{U}$ and $\tilde{P}$ are easily expressed in terms of $\tilde{V}$, and the solution is calculated by finding their discrete, inverse transforms. The domain extends from the equator to 4500 km north, and is cyclic in x with a horizontal resolution of 150 km. The atmosphere is updated only every 5 days, a procedure which is valid because the time scale of the solution is so long.

4. RESULTS

This section contrasts the model response for the six basin configurations shown in figure 1. Note that the first three configurations involve only one ocean basin, whereas the latter three involve two. As we shall see, in each case coupled oscillations develop that involve <u>all</u> the oceanic and atmospheric variables.

Recall that the temperature field is not SST, but rather is the temperature excess of the surface-layer water over the deeper water that is entrained by the layer. A realistic value for the temperature of entrained water in the tropics is 20°C, so that SST is given by T + 20°C. Also recall that convection occurs whenever $T > T_C = T(0) - 1.5 = 8.5°C$, so that plots of T also show convection.

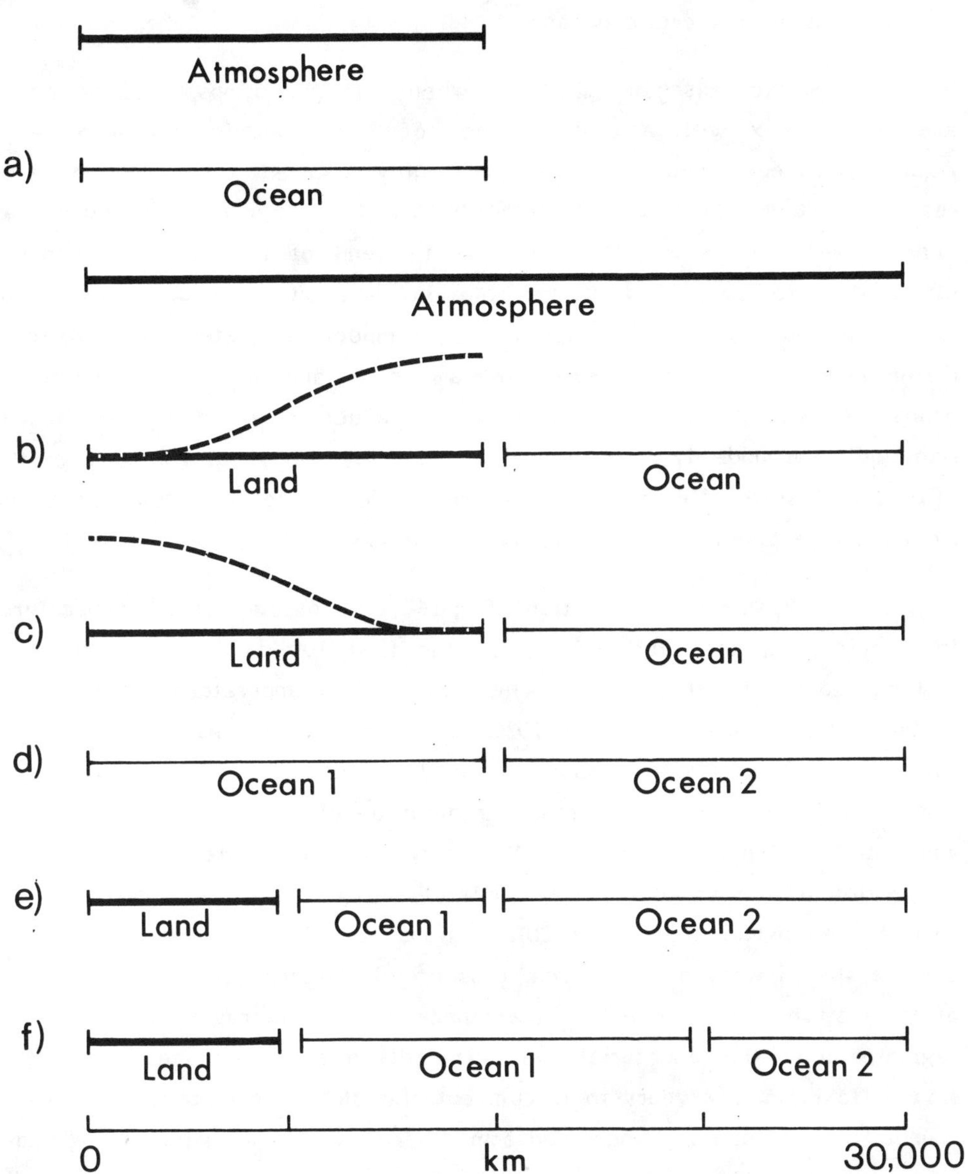

Fig. 1: A schematic diagram showing the location of land (solid lines) and ocean (thin lines) for the six configurations discussed in the text. In all cases the atmosphere has cyclic boundary conditions either with a circumference of 15,000 km or 30,000 km. Configurations a-f correspond to the solutions shown in figures 2-7, respectively. For configurations b and c, convection is fixed over land according to (8) with X(x) given by (9) and (11), respectively, and the thin curves over land illustrate these choices for X(x). There is no land convection for configurations e and f. Over the ocean, convection is always specified as in (7), and so changes as SST evolves.

(a) Solutions with a cyclic ocean

This section discusses solutions when both the atmosphere and ocean are cyclic in x with a circumference of 15,000 km (configuration a in figure 1). In each run, the model is initially assumed to be in a state of rest with values of h and T given by (5) and it is then forced by an imposed wind stress for 100 days. At the end of this time the imposed stress is switched off, and the solution is allowed to develop without further interference. For a wide range of model parameters, a large-scale disturbance subsequently grows to a finite amplitude and begins to propagate slowly eastward. The zonal structure of the disturbances (labelled here mode 1, 2, etc.) is determined by the zonal structure of the initial wind state. The disturbances are stable in the sense that once the model is in a given mode it remains in that mode.

Figure 2 shows time sections for 16 years of equatorial temperature, layer thickness, and wind stress. For the first 100 days, while the ocean is being forced by the imposed wind stress, the temperature remains quite uniform at the initial value of 10°C and there is little east-west slope to the thermocline. Starting about year 1 after the external forcing is removed, an east-west temperature gradient develops with the formation of warm and cool patches (figure 2a). The warm patches are associated with convection in the atmosphere (wherever $T > T_C = 8.5°C$ in figure 2a) and with a deep pycnocline (figure 2b). Regions of easterly and westerly winds exist east and west of the patch, respectively (figure 2c). The zonal scale of the disturbance is mode-1 so that there is one cool region and one warm region around the equatorial belt. The disturbance propagates eastward slowly, taking 5 1/2 years to circumvent the globe. This propagation speed is much slower than any uncoupled atmospheric or oceanic wave in the model (the oceanic Kelvin wave speed is 1.7 m/s), and is good evidence of the highly coupled nature of the disturbance.

The studies of Lau (1982) and of Philander et al. (1984) provide some insight into the cause of these disturbances. The instability in their models develops in the following way. Suppose initially that $h-\bar{h}$ is weakly positive in a patch on the equator, so that a Q exists to force the

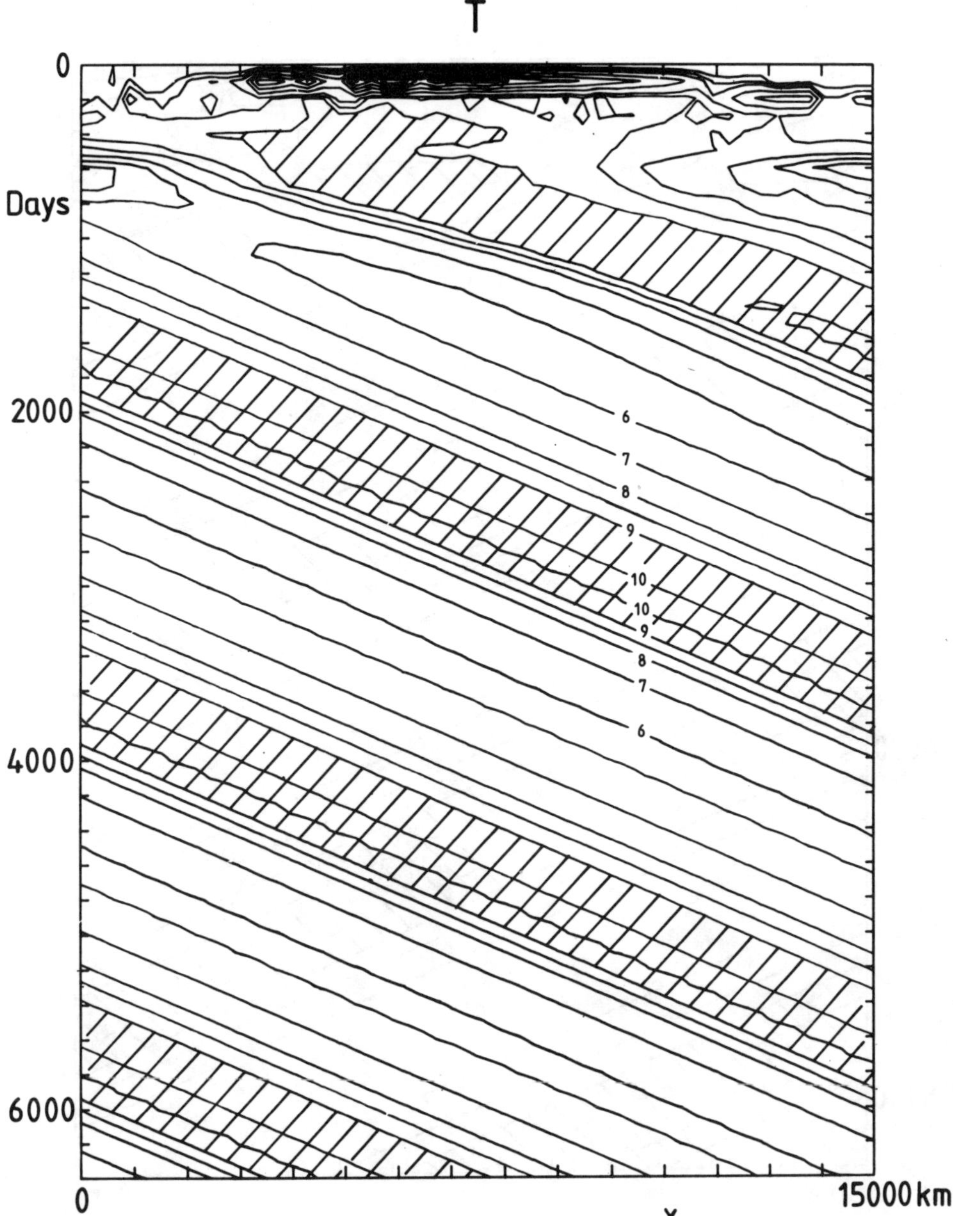

Fig. 2a: Contours of equatorial temperature vs. time for the periodic calculation illustrated schematically in configuration a of figure 1. A wind is imposed for the first 100 days. Thereafter the coupled model develops freely. The temperature is initially uniform at 10°C, but eventually an instability grows to finite amplitude and propagates eastward. The region where $T > 9°$ is crossed. Convection occurs whenever $T > T(0) - 1.5 = 8.5°C$, and the 8.5°C contour is also included.

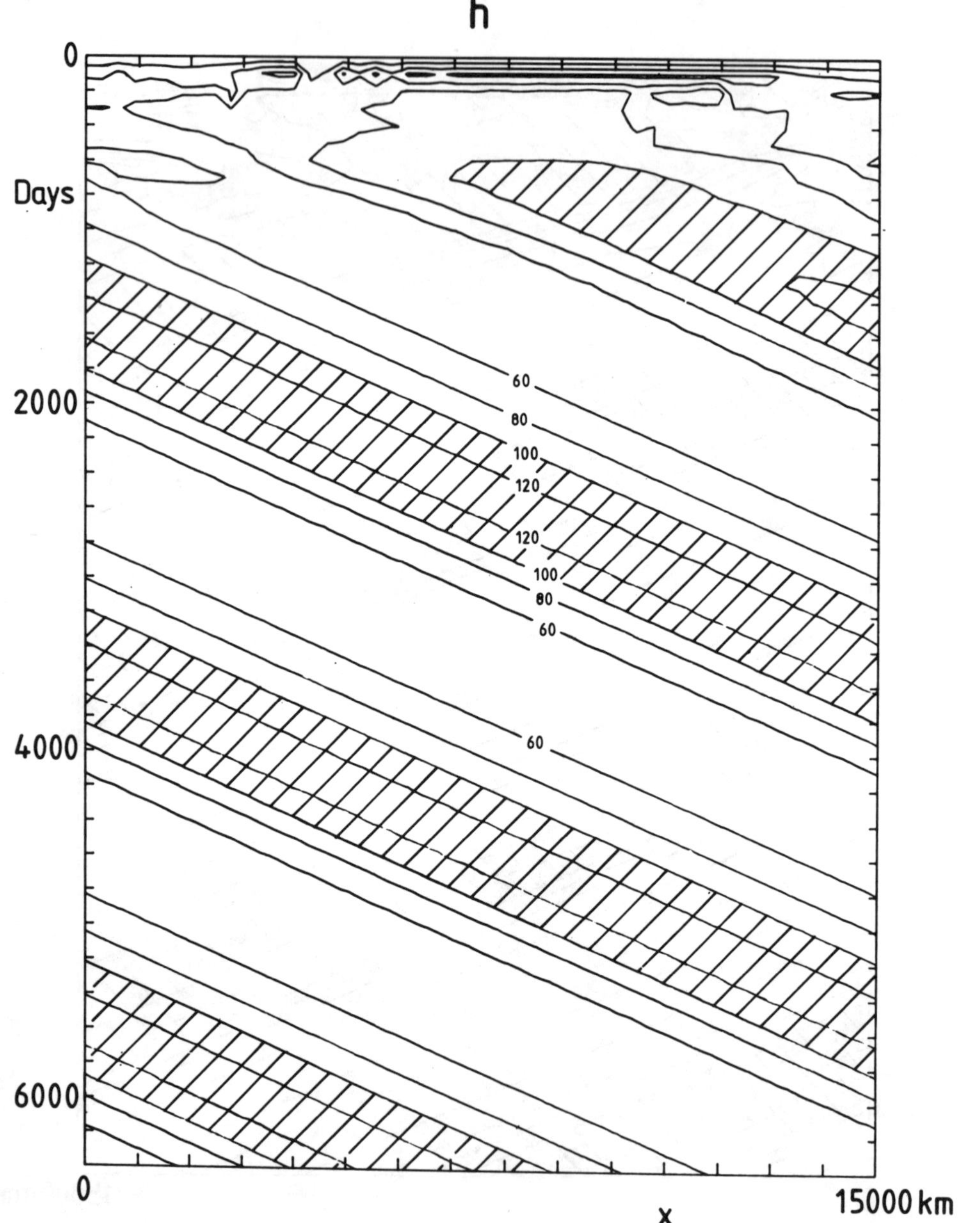

Fig. 2b: As in figure 2a, except showing contours of h. The region where h > 100 m is crossed.

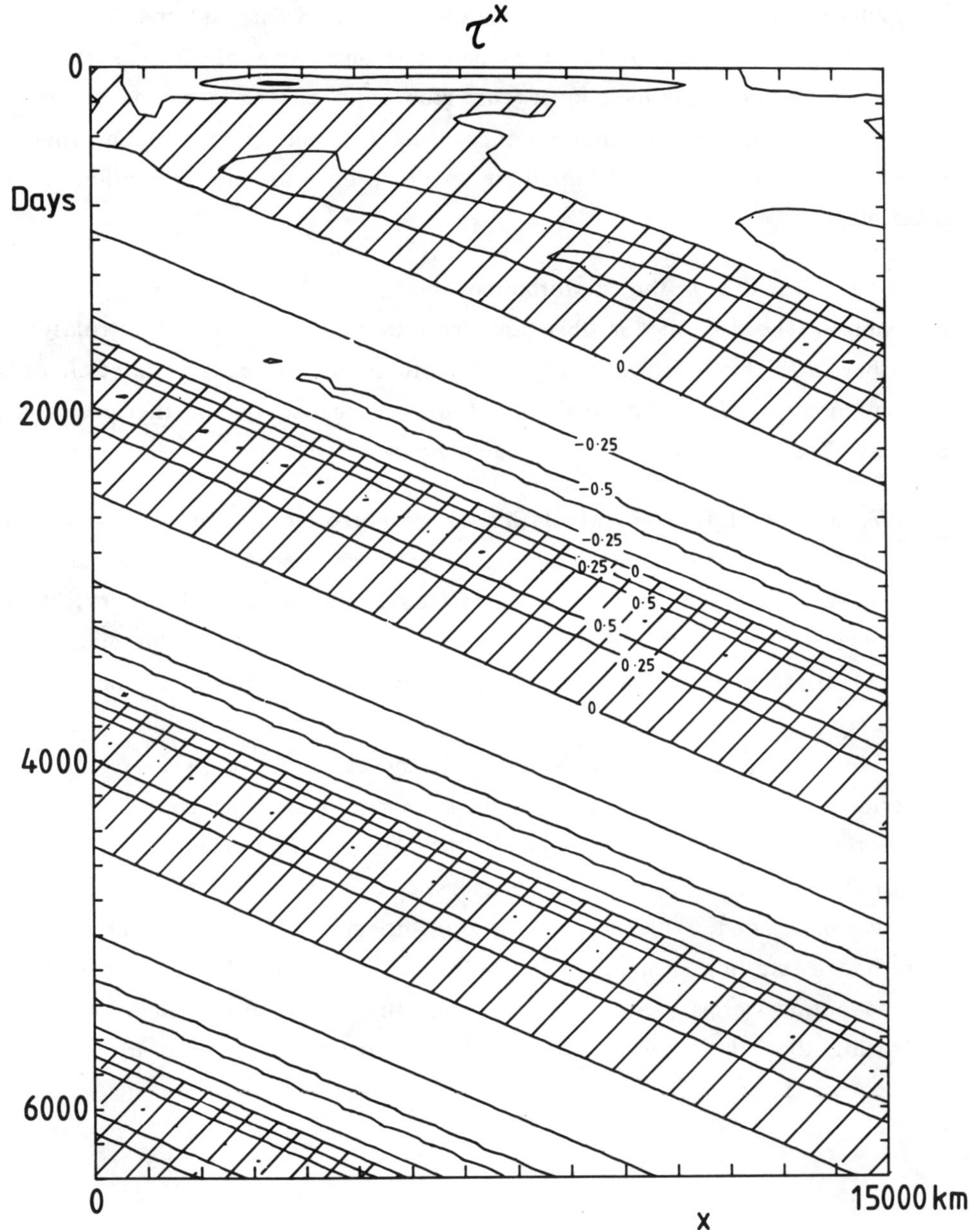

Fig. 2c: As in figure 2a, except showing contours of surface stress. Regions of positive τ^x are crossed.

atmosphere. This Q drives westerlies to the west of the patch and easterlies to the east of it. Provided the amplitude of Q is sufficiently strong, the effect of this convergent wind field on the ocean is to pile up more water in the patch, thereby increasing $h-\bar{h}$ and Q once again. There is no limit to the size of h in their models, and so the instability grows indefinitely.

The growth of the disturbance in figure 2 is caused by a similar mechanism. The reason for this similarity is that, although Q is related to T rather than h, T is generally a monotonically increasing function of h. The disturbance does not grow indefinitely because ocean thermodynamics prevents T from every increasing much beyond $\bar{T}(0)$.

A number of further calculations were performed, in order to determine the sensitivity of the coupled solutions to model parameters. The following is a brief summary of our conclusions (see Anderson and McCreary, 1985a, for a more detailed discussion). The propagation speed of the disturbance is very nearly proportional to its wavelength. There is an inverse relationship between the oscillation period and Q_o. Solutions are quite sensitive to the value of T_c. When T_c is increased to $\bar{T}(0)-1$, so that the criterion for convection is more stringent, the horizontal scale of convecting regions decreases and the period increases to 11 years. Increasing the thermodynamic parameters, δ_0, σ and w led to an increase in propagation speed, whereas decreasing them sufficiently leads to the disturbance stalling. Finally, the solution is essentially unchanged when the advection terms were dropped from the temperature equation in (1), indicating that these terms do not play an important role in the dynamics of the oscillation.

(b) Solutions with a bounded ocean

This subsection discusses when the atmosphere is cyclic with a circumference of 30,000 km, but the ocean is confined to a bounded basin with a zonal extent of 15,000 km. Land comprises the rest of the globe and is also taken to be 15,000 km wide (configurations b and c in figure 1). Over the ocean, convection is parameterized as in (7) and so Q changes as T evolves, but over land it is usually fixed in time according to (8). Thus, there is one ocean basin representing the Pacific Ocean with land extending from 'Indonesia' in the west round to 'South America' in the east. For all solutions except that in figure 4, the heating over land is modulated by the function

$$X(x) = .5\left[1 + \cos 2\pi\,(x-x_m)/x_w\right], \quad 0 < x < x_m, \tag{9}$$

where x_m = 15,000 km, x_w = 30,000 km, in order to represent the strong convection over Indonesia and the weaker convection further east (configuration b of figure 1). As for the solutions in a cyclic ocean, the ocean is initially at rest and values of h and T are given by (5). Again, for a wide range of model parameters large-scale disturbances develop and propagate slowly eastward.

Figure 3 shows time sections for 16 years of equatorial temperature, layer thickness and wind stress. After an initial period of adjustment the model settles down into an oscillatory mode with a period of about 3.5 years. Instabilities develop in the western ocean, propagate eastward, and dissipate at the eastern boundary. The speed of eastward movement of the disturbances is variable, being slowest as the warm patch approaches the eastern boundary. In mid-ocean it travels at $\sim$ 15 cm/s.

Disturbances are generated by the same mechanism as for the cyclic solutions. Essentially they can develop in any region where T remains sufficiently far above T_c. To illustrate, it is useful to follow the time development of a disturbance throughout its life cycle, from day 2800 to 4800 for example. At day 2800 the ocean begins to warm up in a region centered about 4000 km from the western boundary. This warm pool intensifies and begins to move eastward. By day 3600 it has moved to the central ocean 8000 km from the western boundary. At this time it is apparent in figures 3b and 3c that the warm pool is associated with a deepening of the pycnocline and a convergent wind field, just as for the cyclic solutions. After day 3600, the disturbance no longer increases in amplitude. Near day 4500 the anomaly has reached the eastern boundary and temperatures are a maximum there. The warm anomaly vanishes abruptly about day 4800, its dissappearance clearly associated with the growth of another warm pool in the central and western ocean.

Some features of this solution are reminiscent of those observed in the 1982/83 El Nino event. Gill and Rasmusson (1984) discuss the time development of convection, winds and SST anomalies during the event. There was an eastward movement of warm SST, convection and wind anomalies during the event. The occurrence of westerly wind anomalies west of the SST anomaly and the north-south scale of the warm pool were both very similar to those produced in the model.

Other experiments suggest the following conclusions (see Anderson and McCreary, 1985a, for a more detailed discussion). If the strength of convection over land, Q, is increased then the period lengthens; reducing Q has the opposite effect. If Q_o is increased by a factor of 1.5 then, as in the cyclic case, the period decreases; when Q_o is reduced by a factor of .5 the warm patch which develops in the west never penetrates into the eastern ocean but stalls in mid-ocean. When T_c was changed to $\overline{T}(0) - 1$ the oscillation period increases only slightly, in contrast to the cyclic case, even though the regions of convection were predictably smaller. Increasing δ_0, σ, and w increases the speed of propagation of the distrubances (and reduces their period), in agreement with the cyclic case. The atmospheric parameters, c and r, influence the ocean through their effect on the winds. If c is doubled to 120 m/s, then the model does not oscillate; the winds are weaker and the model locks into a steady state. If, however, Q_o, Q_1 and c are all doubled, then the winds are comparable with the control experiment and the behavior of the coupled system is very similar to the control run. Reducing r results in stronger winds, a very cold eastern

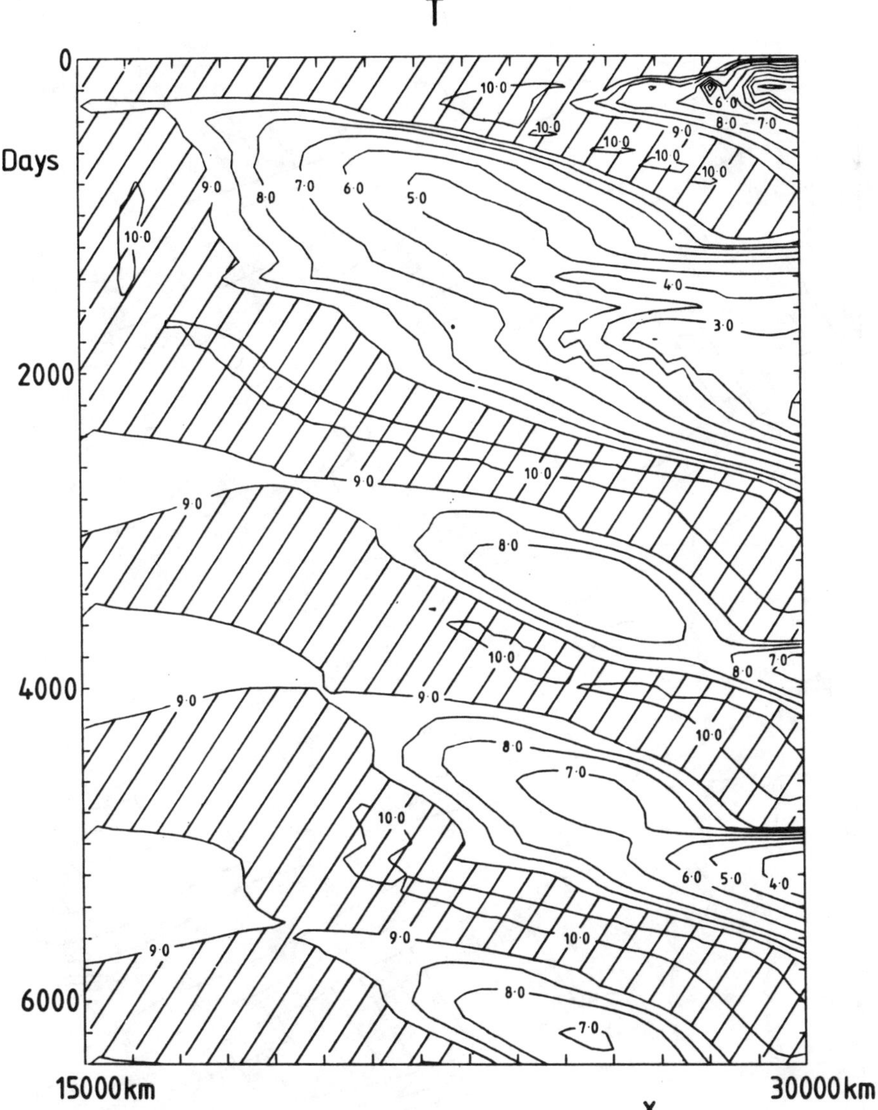

Fig. 3a: Contours of equatorial temperature vs. time for the "Pacific Ocean" configuration illustrated schematically in configuration b of figure 1. Regions of warm SST form in the west and propa-gate eastward. The ocean sometimes has a large temperature gradient between west and east, as after 2000 days. At other times the gradient is weak, resembling El Nino conditions, as after 2600 days. Regions where T is warmer than 9°C are crossed. Convection occurs wherever T > 8.5°C, and the 8.5°C contour is included.

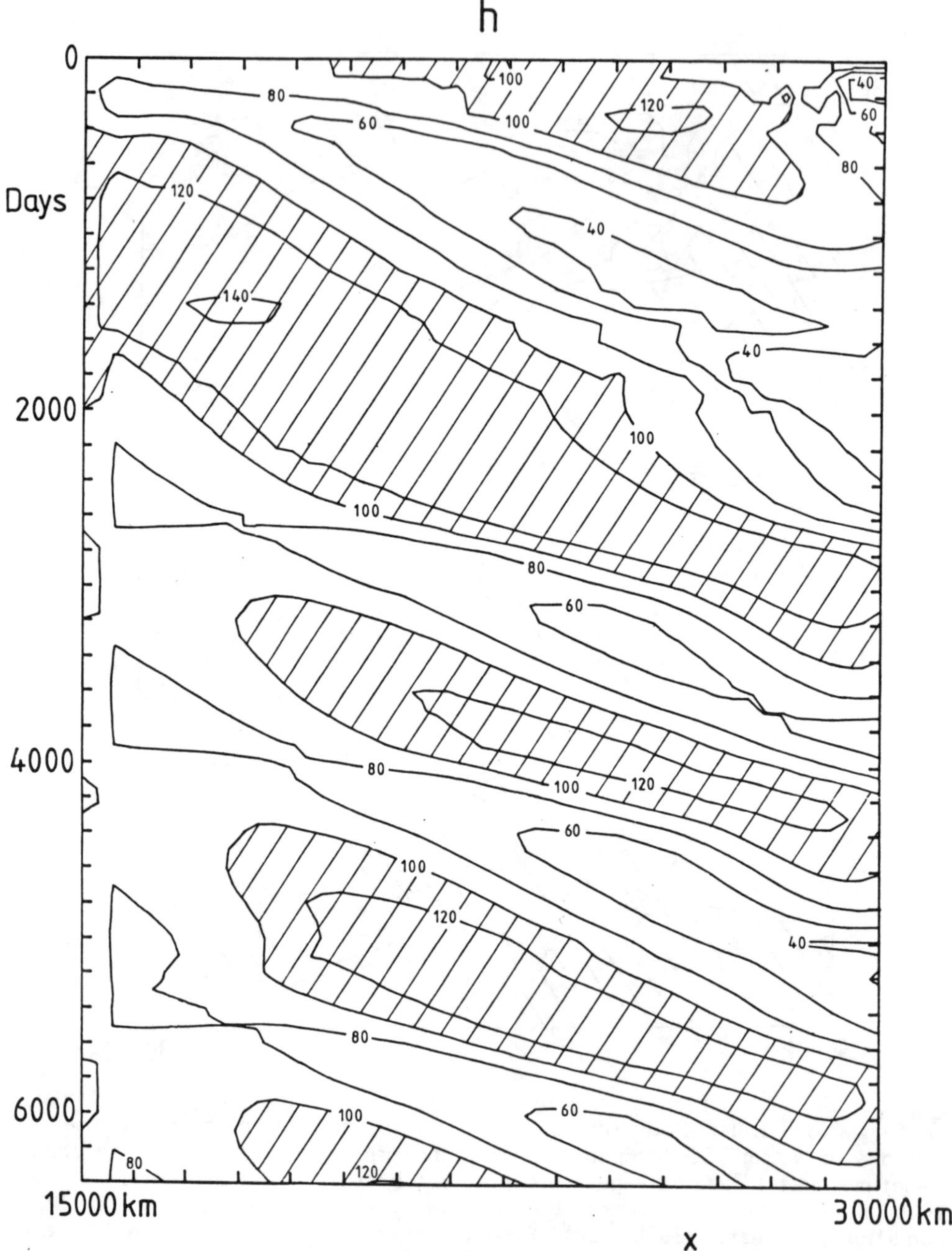

Fig. 3b: As in figure 3a, except showing contours of thermocline depth. The thermocline tilts most when the zonal temperature gradient is strong and least when the gradient is weak. Regions where h is less than 100 m are crossed.

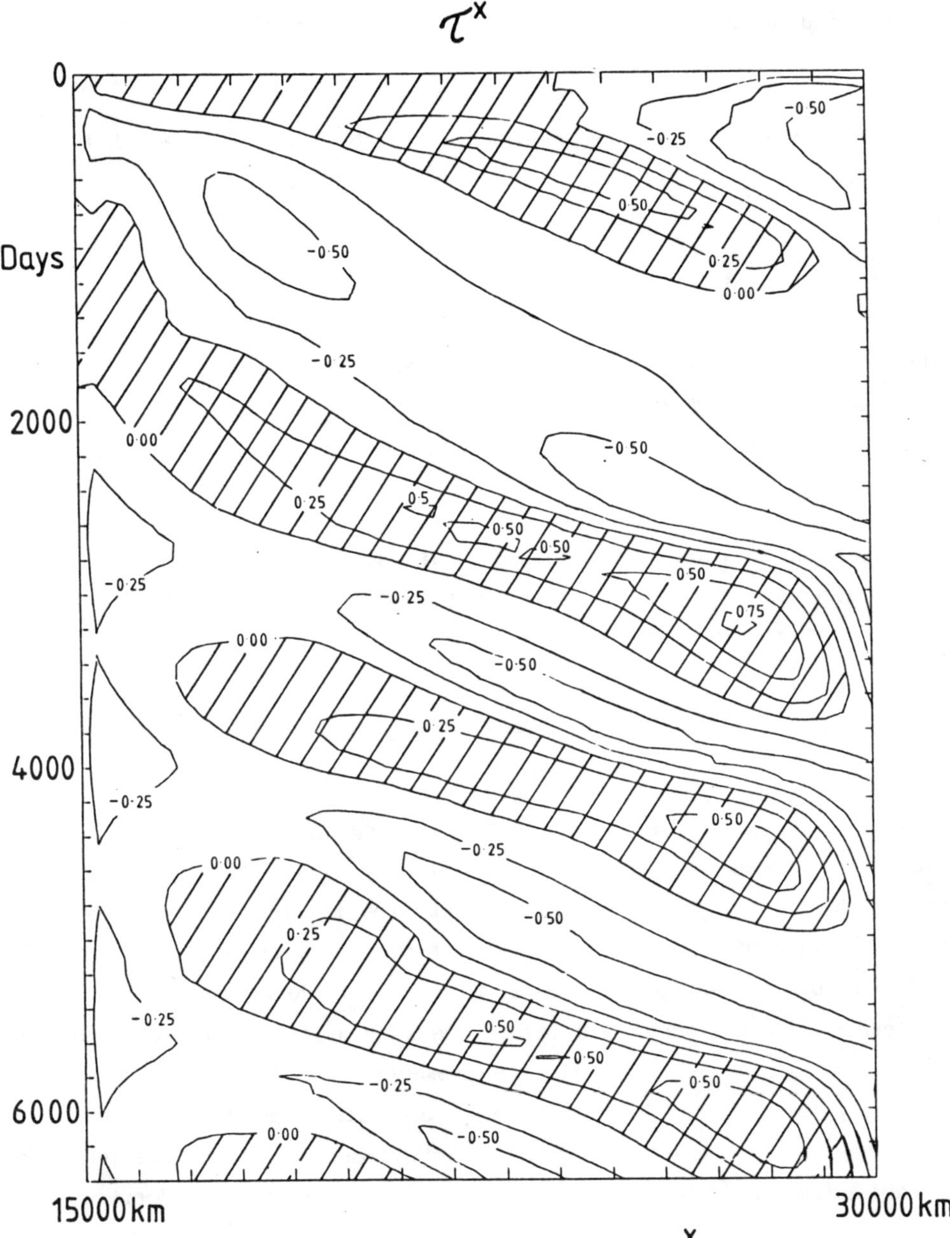

Fig. 3c: As in figure 3a, except showing contours of the zonal wind stress. The wind is eastward to the west of the pool of warm SST and westward to the east of it. In response to this wind field, the thermocline slopes up both west and east of the warm pool. Regions of positive wind stress are crossed.

Pacific and a shorter oscillation period, again a change similar to that caused by an increase in Q_o and Q_1.

A seasonal cycle can be imposed on the model by modulating Q_1 according to

$$Q_1 = Q_o \ (1 + a \sin \omega t) \ , \tag{10}$$

where $\omega = 2\pi$ years^{-1}. For values of a of order one, this change ensures that warm patches develop at the same time of year, so that the oscillation period is an integral number of years, but does not otherwise influence the instability.

Another solution investigates how the location of the convection over land influences the disturbances. The region of strong convection was shifted to the eastern end of the ocean by replacing X(x) in (8) with

$$X(x) = .5 \ (1 + \cos 2\pi x/x_w) \ , \quad 0 < x < x_m \ , \tag{11}$$

(configuraton c of figure 2). With this configuration it is useful to regard the ocean as being the Indian Ocean, where the strong convection over Indonesia occurs at the eastern boundary. Figure 4 shows the resulting time development of equatorial temperature and layer thickness. In marked contrast to the solution of figure 3, the model locks on to a state with westerlies over the ocean and with the eastern ocean warm. No disturbances develop or propagate. Evidently the presence of convection in the west is necessary for the model to be able to oscillate. The reason is likely that only when there are easterly winds over the ocean does the western ocean remain sufficiently warm for an instability to be able to develop.

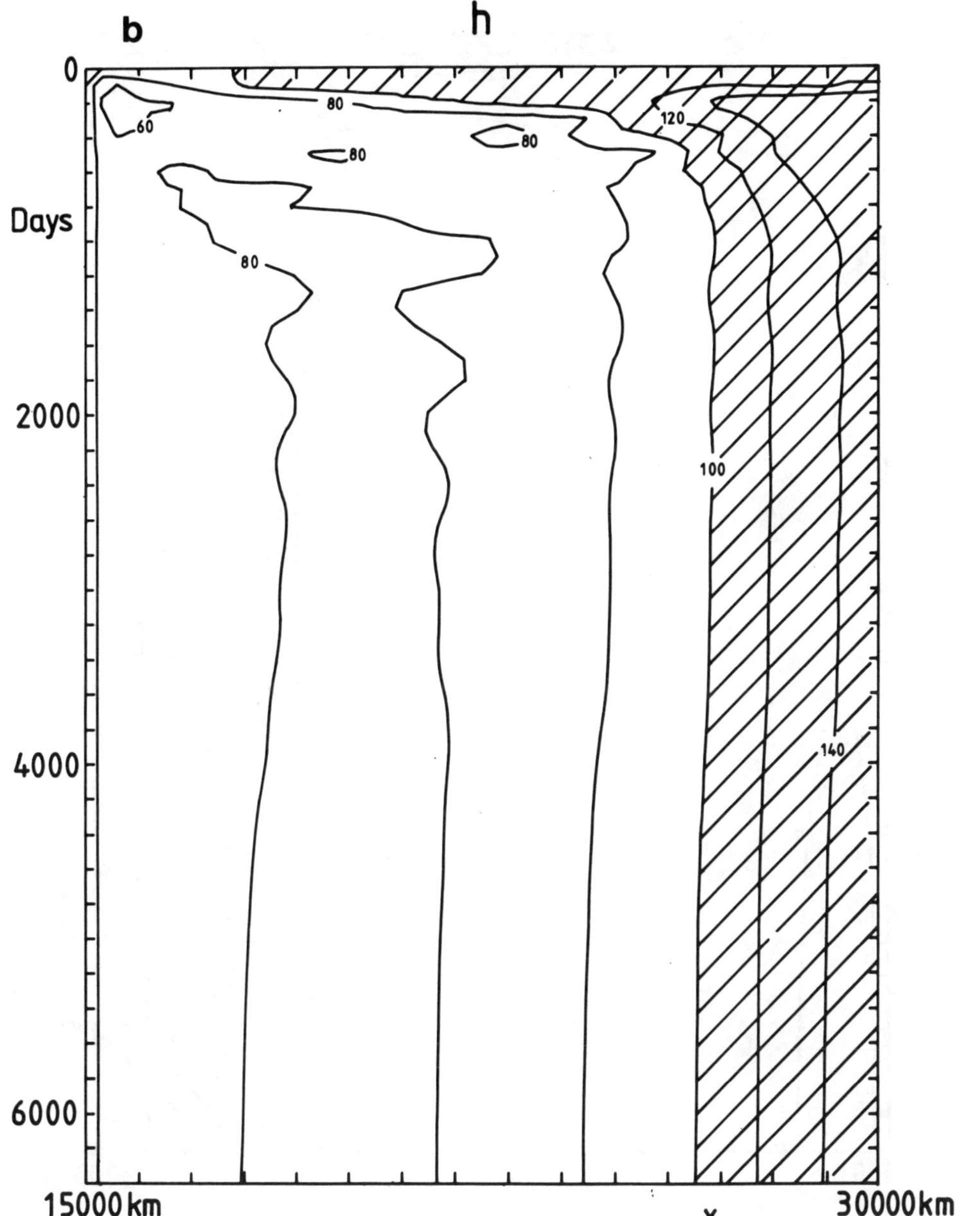

Fig. 4a: Contours of equatorial temperature and thermocline depth vs. time for the "Indian Ocean" calculation illustrated schematically in configuration c in figure 1. No instabilities develop in this case in contrast to the "Pacific Ocean" configuration of figure 3.

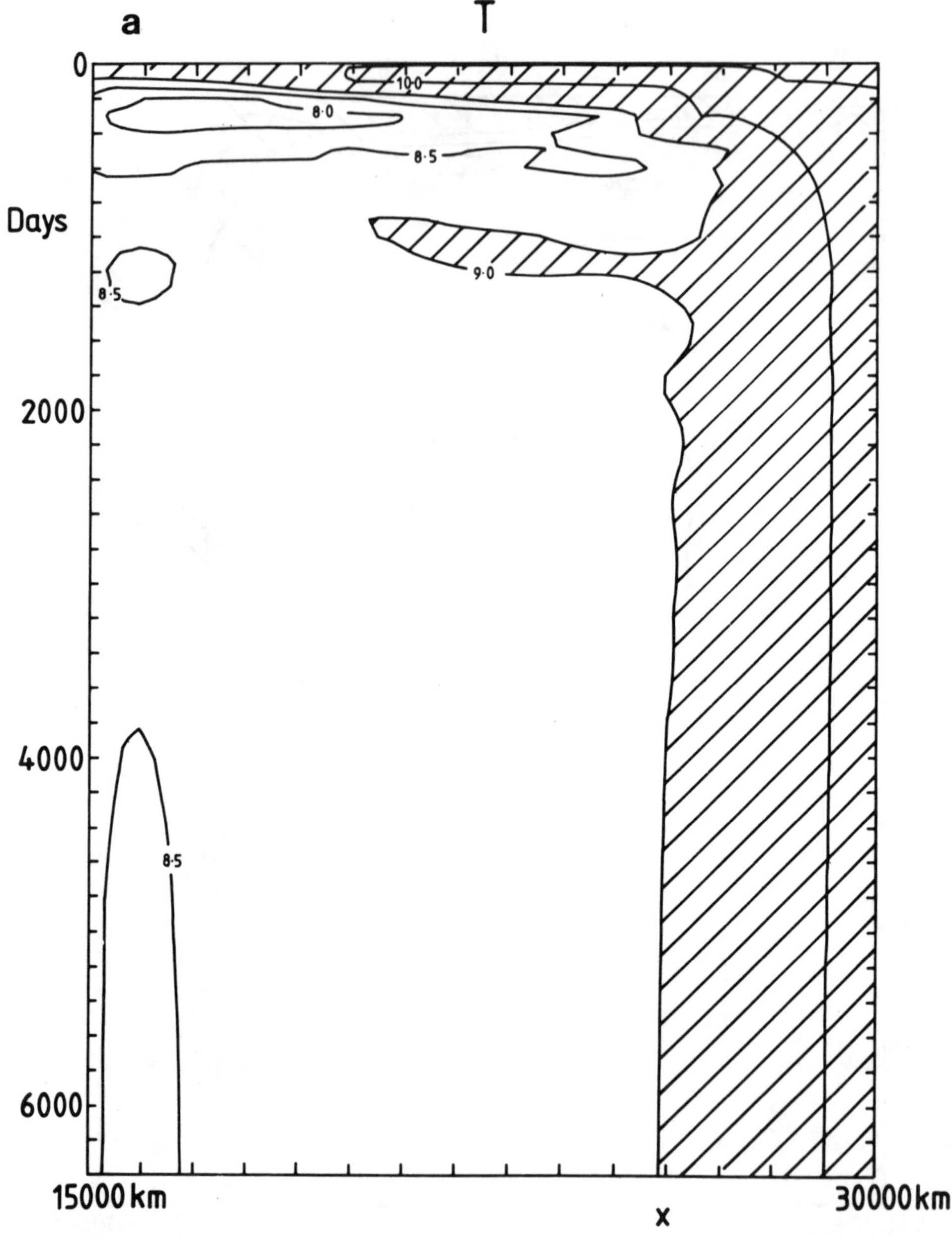

Fig. 4b:

(c) Solutions with two oceans

Two limitations of the preceding solutions are that they involve only one ocean and that convection over land is externally specified. Here both of these limitations are overcome by including a second ocean in the basin geometry. The convection that develops over the second ocean allows oscillatory solutions to exist without the necessity for any convection over land, and so land convection is set to zero throughout.

(i) Two ocean basins without land

In this subsection we consider the case of two independent oceans of width 15,000 km that are separated by land barriers of zero width (configuration d in figure 1). The atmosphere is 30,000 km wide and is cyclic. The oceans extend latitudinally to 4500 km, and solutions are symmetric about the equator. The model is forced for the first 100 days by a wind stress that converges on the barrier at x = 15,000 km, and is then allowed to run freely.

Figure 5 shows the time development of T along the equator in both oceans. Initially, a region of warm temperature (and enhanced convection) develops near x = 15,000 km. Subsequently, a slow eastward-propagating disturbance drifts eastward to the eastern edge of ocean 2, and establishes a region of warm temperature near x = 30,000 km. A similar eastward propagation then proceeds in ocean 1, and the oscillation continues in this manner.

Although the oceans are not physically connected in any way, there is a connection via the atmospheric wind field. When a warm patch approaches the eastern edge of ocean 1, say, the easterly wind field to the east of the warm patch extends over the western part of ocean 2, depressing the thermocline in the west and generating a warm patch there as well; eventually, this warm patch separates and propagates eastwards across ocean 2. Because of this strong atmospheric connection, the response in figure 5 is similar to that for the cyclic ocean case in figure 2; the propagation of the disturbance is retarded by the thin barriers, but not completely blocked.

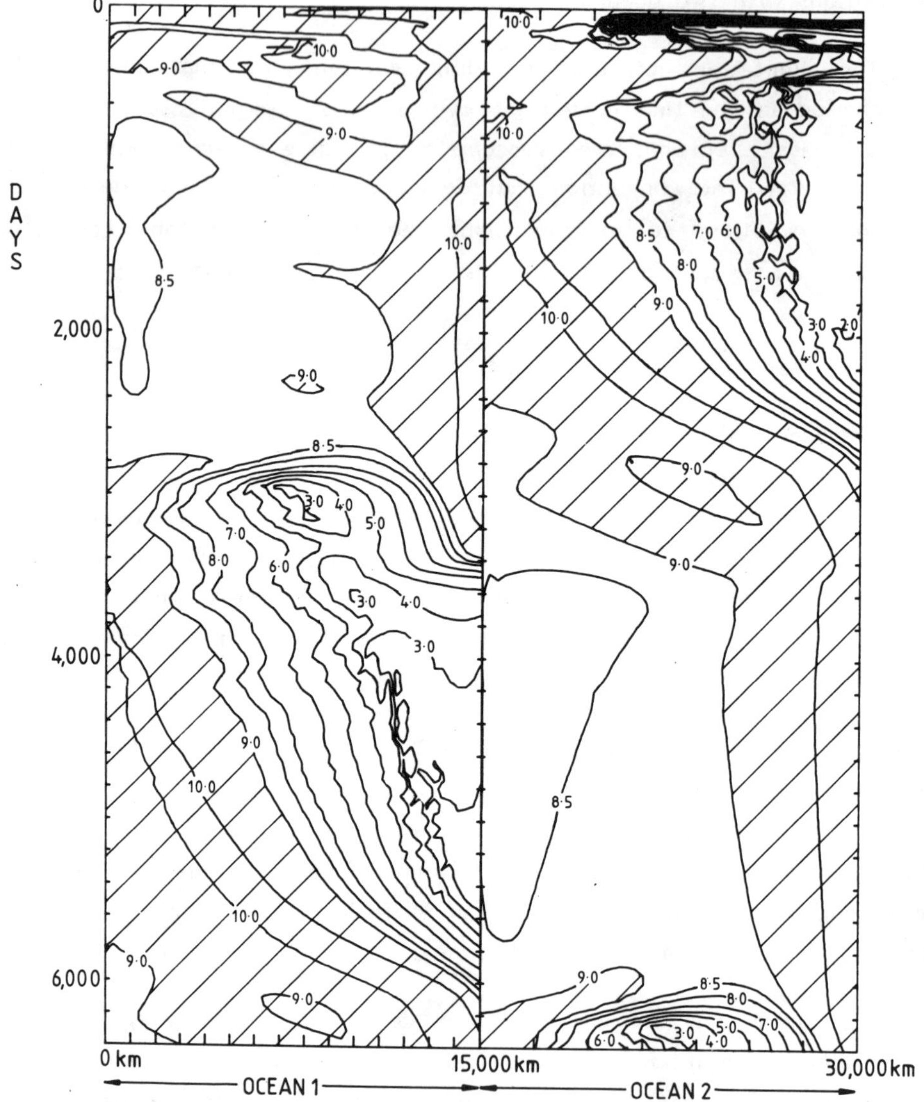

Fig. 5: Contours of T vs. time for configuration d of figure 1. There are low-frequency oscillations. A region of warm temperature develops in one ocean, and then propagates eastwards. When it approaches the eastern boundary, it begins to affect the second ocean via the wind fields. Another disturbance subsequently develops in the second ocean, then propagates eastwards, and so on. Convection occurs in the atmosphere whereever T > 8.5°C.

(ii) Two ocean basins with land separation

Here ocean 1 and ocean 2 are separated by a thin barrier on one boundary and by an extensive land mass on the other. Since the Indian Ocean is smaller than the Pacific, we consider one large ocean 15,000 km wide, a small ocean 7500 km wide, and a land separation that is also 7500 km wide. The atmosphere remains cyclic with a circumference of 30,000 km. Over land there is no convective forcing of the atmosphere. Two configurations are possible (configurations e and f in figure 1). Configuration e more closely resembles the observed Indian-Pacific Ocean configuration. The land region acts to break the atmospheric connection between the model eastern Pacific (ocean 2) and the model western Indian Ocean (ocean 1). Thus, land simulates the effects of the Andes, Atlantic Ocean and East African Highlands, which together prevent a direct atmospheric coupling between the eastern Pacific and western Indian Ocean wind fields.

Figure 6 shows the time development of equatorial SST for configuration e of figure 1. Ocean 1 (the "Indian Ocean") soon locks into a warm state with winds blowing from west to east. This state is stable, with no oscillations occurring. In contrast, ocean 2 (the "Pacific Ocean") exhibits low-frequency oscillations with a period of about 5 years, similar to the bounded-ocean solution of figure 3. This solution has several features in common with observations. For example, there is permanent convection over "Indonesia", the system oscillates at the long periods associated with ENSO, and a region of anomalous convection, SST and westerly winds propagates eastward across the "Pacific Ocean" during an El Nino event.

Figure 7 shows the time development of equatorial ocean temperature for configuration f of figure 1. In this case, a warm patch develops initially in the west of ocean 1 and then propagates to the east. Thereafter ocean 1 remains locked into a stable state with warm water in the east. Ocean 2 exhibits a weak oscillation with a very long period, a response considerably less realistic than that in figure 6.

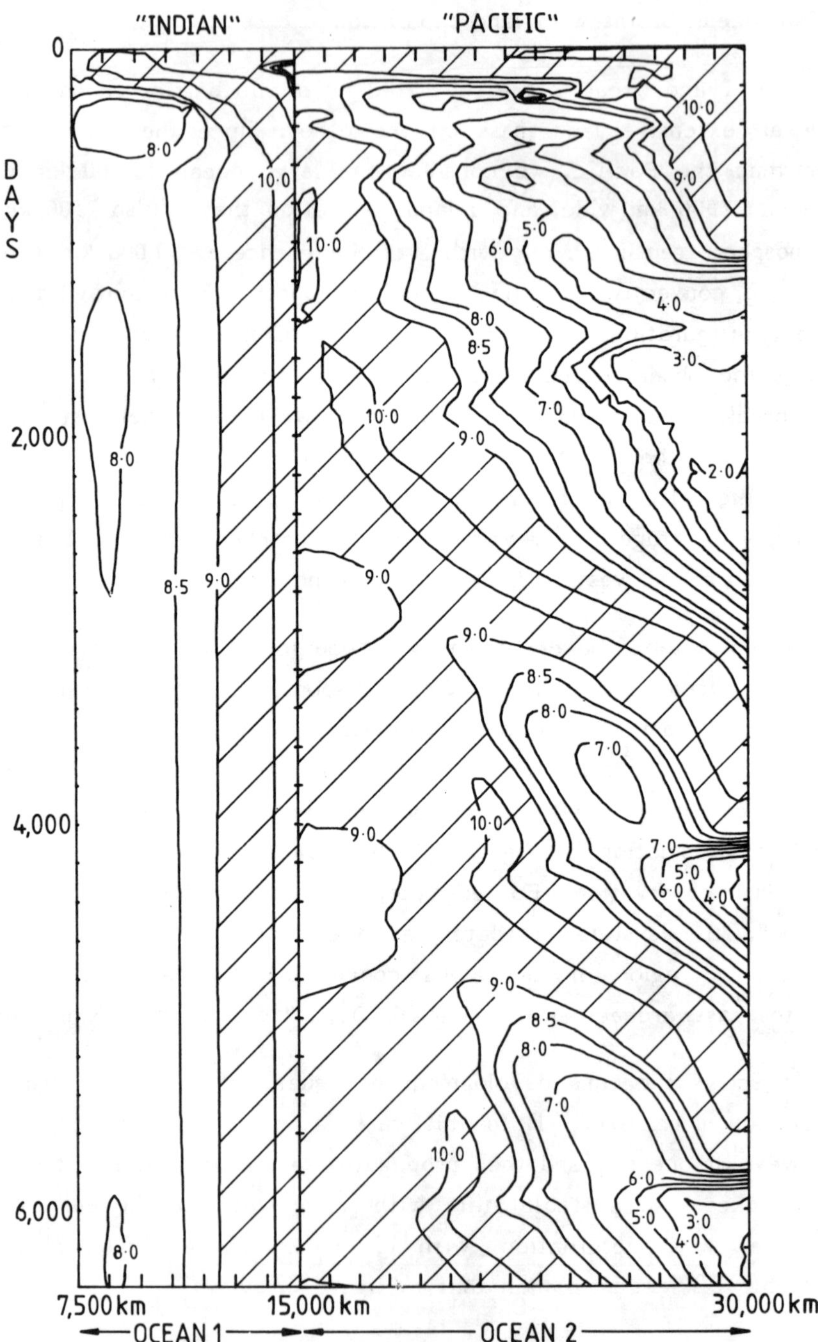

Fig. 6: Contours of T vs. time for configuration e of figure 1. The "Indian Ocean" is stable and stays in a state with warm water in the east. In contrast, the "Pacific Ocean" undergoes low-frequency oscillations.

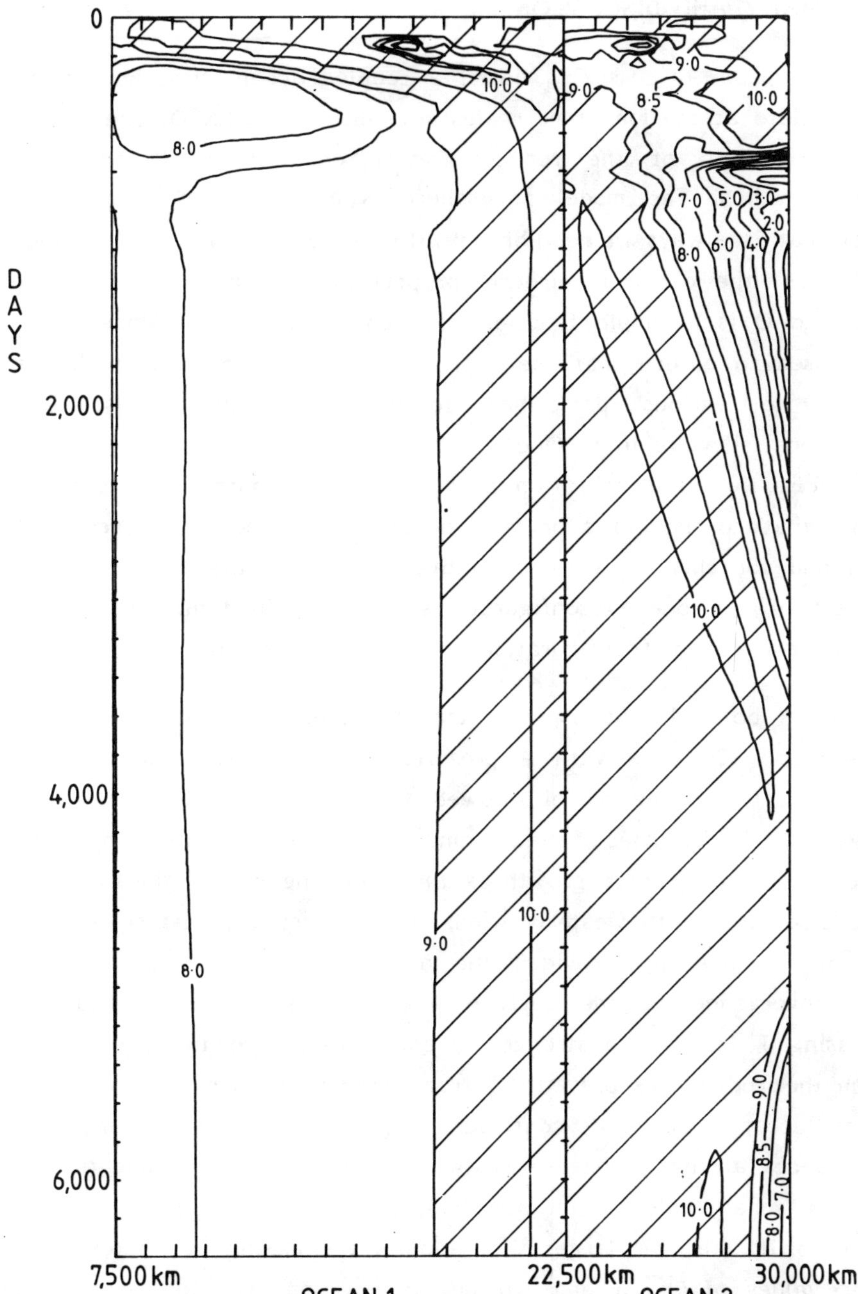

Fig. 7: Contours of T vs. time for configuraton f of figure 1. The western ocean remains stable, but the smaller eastern ocean undergoes a weak oscillation.

5. SUMMARY AND DISCUSSION

This paper discusses in detail a coupled ocean-atmosphere model that can oscillate at the long time scales associated with ENSO. The model ocean is an extension of the reduced-gravity equations that includes active thermodynamics. The model atmosphere is a linear, single-baroclinic-mode model that is always in equilibrium with a heat source, Q. The wind stress that drives the ocean is directly proportional to the atmospheric velocity field, and Q is a simple function of ocean temperature. Model parameters are chosen to ensure that the uncoupled models respond as realistically as possible to various prescribed forcings. For all the solutions the atmosphere is cyclic in x. One set of solutions is found when the ocean is also cyclic in x (configuration a in figure 1). Another set is found in a domain that consists of a bounded ocean and a land mass, each 15,000 km wide, and for this set convection over land is located either to the east or west of the ocean (configurations b and c in figure 1). A final set involves two oceans (configurations d, e and f in figure 1).

When both the atmosphere and the ocean are cyclic, the model is spun-up for 100 days with a prescribed wind stress. This wind is then switched off and the model allowed to develop without external forcing (figure 2). Instabilities develop and grow to a finite amplitude. The process underlying their growth is the following one. A localized heating, Q, is associated with westerly winds to the west and easterly winds to the east of the heating. Provided the amplitude of Q is sufficiently large, this convergence in the surface winds causes h and T to increase. Increasing T leads to a stronger Q and a rapid growth of the instability. Ocean thermodynamics prevents T from increasing much beyond $\overline{T}(O)$, and so the instability does not continue to grow indefinitely. Large-amplitude disturbances always propagate eastward. Their rate of propagation depends on their zonal scale (faster for longer disturbances), on the strength of convection (faster for larger Q_0), and on the mixing parameter (faster for larger values of δ_0, σ and w), but is not sensitive to the advection of heat.

In the bounded ocean case and with convection over land to the west (Indonesia), instabilities develop in the western or central ocean and propagate slowly eastward (figure 3). When they reach the eastern boundary they weaken and eventually vanish. Their disappearance is related to the growth of another disturbance in the west. The dependence of solutions on parameters is similar to that in the cyclic case (except for the dependence on T_C). The strength of convection over land also influences the solution. An increase in Q_1 lengthens the period between the development of warm patches, and vice versa. If convection is strongest over land to the east of the ocean (as in the Indian Ocean), no instabilities ever develop (figure 4).

Several solutions are found in a domain that contains two oceans. One advantage of having two oceans is that it is no longer necessary to specify convection over land, and so land convection is always set to zero. When there is no land present eastward-propagating disturbances easily pass from one ocean to the other, even though the oceans are separated by barriers (figure 5); they can do so because there is strong coupling between both oceans via the wind field. Configuration e of figure 1 shows the geometry most similar to that of the real Pacific and Indian Oceans. In this configuration the smaller "Indian Ocean" quickly moves to a stable state with warmest water in the east. Convection becomes strong there in association with the warm water, and remains so throughout the integration. In contrast, the larger "Pacific Ocean" exhibits low-period, eastward-propagating oscillations (figure 6). For configurations e and f of figure 1 the western ocean is stable with warm water in the east, while only the eastern ocean shows oscillations; these oscillations are stronger and faster when the larger ocean lies to the east of the smaller ocean (compare figures 6 and 7).

Solutions compare favorably with observations in several ways. The production of oscillatory solutions with long time scales is a very robust feature of the model. Disturbances always develop in the west or central ocean and propagate eastward. Finally, ENSO-type disturbances are not observed to form in the Indian Ocean, a property consistent with the solutions in figures 3, 6 and 7.

Solutions differ from observations in other ways. Observations of the wind anomalies during an El Nino event do not show easterlies to the east of the heating whereas this is a strong feature of this model atmosphere. Nobre (1983) and Gill (1983, private communication) have suggested that easterlies can occur when convection is weak but that they are suppressed when it is strong; thus, a nonlinear correction to the linear atmospheric model used here may be necessary to represent adequately the tropical atmosphere. The propagation speed of the disturbances is too slow. (Several parameters can affect this speed, and it is possible to tune the model to reproduce more faithfully the observed speed). Finally, a serious deficiency is that the model fails to simulate the rapid onset and intermittancy of El Nino events.

The model discussed in the body of the paper, as well as the others mentioned in the introduction, illustrate various possible mechanisms of tropical ocean-atmosphere interaction. Their agreement with aspects of the observations is encouraging. Their limitations are evident, but should not be regarded as discouraging; rather, they are valuable because they suggest directions for future research. All the models lack or misrepresent processes that are potentially important in the dynamics of ENSO. We are currently improving our model by developing a model atmosphere that includes a humidity equation, thereby allowing a better parameterization of convection. Our goal is to develop a coupled system that can simulate more realistically the rapid onset and intermittancy of ENSO.

ACKNOWLEDGEMENTS

This research was sponsored by the National Science Foundation under grant No. ATM79-19698 and by the National Environmental Research Council under grant No. GRS/468. The work was carried out in the summers of 1983 and 1984 while the author was visiting Oxford. The dynamical framework of the model ocean is similar to one formulated in 1978 by David Anderson and Adrian Gill, but never published. The programming assistance of Robert Wells is greatly appreciated.

REFERENCES

Anderson, D.L.T., 1984: An advective mixed layer model with applications to the diurnal cycle of the low-level East African Jet. Tellus, 36A 278-291.

Anderson, D.L.T. and J.P. McCreary, 1985a: Slowly propagating disturbances in a coupled ocean-atmosphere model, J. Atmos. Sea. (To appear.)

Anderson, D.L.T. and J.P. McCreary, 1985b: A note on the role of the Indian Ocean in a coupled ocean-atmosphere model of El Nino and the Southern Oscillation, J. Atmos. Sci. (Submitted.)

Cane, M.A. and S. Zebiak, 1985: A theory for El Nino and the Southern Oscillation. (To be submitted.)

Egger, J., 1977: On the linear theory of the atmospheric response to sea surface temperature anomalies. J. Atmos. Sci., 34, 603-614.

Gill, A.E., 1980: Some simple solutions for heat-induced tropical circulation. Q.J. Roy. Met. Soc., 106, 447-462.

Gill, A.E., 1985: Elements of coupled ocean-atmosphere models for the tropics. In Coupled Ocean-Atmosphere Models, Amsterdam: Elsevier. J.C.J. Nihoul Ed., p. 303-328.

Gill, A.E. and E.M. Rasmusson, 1984: The 1982/83 climate anomaly in the equatorial Pacific. Nature, 306, 229-232.

Haney, R.L., 1971: Surface thermal boundary conditions for ocean circulation models. J. Phys. Oceanogr., 1, 241-248.

Hirst, A.C., 1985: Free equatorial instabilities in simple coupled atmosphere-ocean models. In Coupled atmosphere-ocean models, Amsterdam: Elsevier. J.C.J. Nihoul Ed., p. 153-166

Hughes, R.L., 1979: A highly simplified El Nino model. Ocean Modelling, No. 22

Hughes, R.L., 1980: On the equatorial mixed layer. Deep Sea Res., 27A, 1067-1078.

Hughes, R.L., 1984: Developments on a highly simplified El Nino Model. Ocean Modeling, No. 54 (unpublished manuscript).

Lau, K.M., 1981: Oscillations in a simple equatorial climate system. J. Atmos. Sci, 38, 248-261.

Lau, K.-M, 1984: Subseasonal scale oscillation, biomodal climatic state and the El Nino/Southern Oscillation. In Coupled atmosphere-ocean models, Amsterdam: Elsevier. J.C.J. Nihoul Ed., p. 29-40.

Lau, N.-C., 1981: A diagnostic study of recurrent meteorological anomalies appearing in a 15-year simulation with a GFDL general circulation model. Mon. Wea. Rev., 109, 2287-2311.

Lau, N.-C., and A.H. Oort, 1985: Response of a GFDL General circulation model to SST Fluctuations observed in the tropical Pacific Ocean during the period 1962-1976. In: Coupled Ocean-Atmosphere Models, Amsterdam: Elsevier. J.C.J. Nihoul Ed., p. 289-302.

Manabe, S., and D.G. Hahn, 1981: Simulation of atmospheric variability. Mon. Wea. Rev., 109, 2260-2286.

McCreary, J.P., 1983: A model of tropical ocean-atmosphere interaction, Mon. Wea. Rev., 111 (2), 370-389.

McCreary, J.P. and D.L.T. Anderson, 1984: A simple model of El Nino and the Southern Oscillation. Mon. Wea. Rev., 112, 934-946.

Nobre, C.A., 1983: Tropical heat sources and their associated large-scale atmospheric circulation. Ph.D.thesis, M.I.T., Cambridge, Mass.

Philander, S.G.H., T.Yamagata and R.C. Pacanowski, 1984: Unstable air-sea interactions in the tropics. J. Atmos. Sci. (In press.)

Webster, P.J., 1972: Response of the tropical atmosphere to local steady forcing. Mon. Wea. Rev., 100, 518-541.

Welander, P. and B. Long, 1985: The overall mechanism of the El Nino phenomenon. (Submitted to Science.)

Yamagata, T., 1985: Stability of a simple air-sea coulped model in the tropics. In: Coupled atmosphere-ocean models, Amsterdam: Elsevier. J.C.J. Nihoul Ed., p. 637-658.

Zebiak, S.E., 1982: A simple atmospheric model of relevance to El Nino. J. Atmos. Sci., 39, 3017-2027.

Zebiak, S.E., 1984: Tropical atmosphere-ocean interaction and the El Nino/Southern Oscillation phenomenon. Ph.D. thesis. M.I.T., Cambridge, Mass.

MAXIMUM ENTROPY PRODUCTION AS A CONSTRAINT IN CLIMATE MODELS

STEPHEN D. MOBBS

Applied Mathematics Department
Leeds University
Leeds LS2 9JT

1. INTRODUCTION

The principal aims of climate modelling are to understand the physical processes governing the present day mean state of the atmosphere and oceans and also to predict the response of the atmosphere-ocean system to possible changes in imposed conditions, such as fluctuations in the solar heating or changes in atmospheric composition. There are two contrasting approaches currently being used in climate modelling research. One approach assumes that the problem will be solved if all the physical processes influencing the system are understood in sufficient detail, the only remaining difficulty (not to be overlooked!) being the inadequacy of present day computers to include all these details in numerical models. The other approach involves a search for simple "laws" which control the mean system. (Here a mean would usually be a time average over years or longer with, in addition, possible spatial averaging.) Progress towards a solution to the climate problem is likely to come about by a combination of these two methods.

The primary problem associated with the formulation of simple climate models is one of closure. (This problem also occurs in complex models but is arguably less fundamental in that context.) As explained by Paltridge and Platt (1976), models based on sound physical principles such as energy conservation usually have at least one more unknown than there are equations and so additional assumptions such as specified thermal diffusivity or specified cloud cover are needed. The present article describes an approach to the closure problem first suggested by Paltridge (1975). Paltridge hypothesized that the climate system is constrained to operate at

J. Willebrand and D. L. T. Anderson (eds.), Large-Scale Transport Processes in Oceans and Atmosphere, 281–323.

or near a local minimum of the entropy exchange rate and used this closure condition with remarkable success in an energy balance atmosphere-ocean climate model. Subsequent efforts have, however, failed to provide a complete and sound physical justification for the technique. Various applications of the minimum entropy exchange principle and related principles are described here and some possible approaches to the problem of justifying their application are described.

2. PALTRIDGE'S MINIMUM ENTROPY EXCHANGE PRINCIPLE

Paltridge (1975) developed a one dimensional, atmosphere-ocean energy balance climate model which he constrained to operate at a minimum of the entropy exchange. The meridional variation of climatic variables was represented by dividing the globe into 10 latitude zones (or "boxes") of equal area. Within each box, coupled energy balance equations for the atmosphere and ocean were constructed, along with a third relation concerning the upward flux of heat from the surface and a fourth relation giving the observed ratio of atmospheric to oceanic meridional heat fluxes. The four equations for each box were expressed in terms of five variables: the surface temperature, the fractional cloud cover, the total upward flux of heat from the surface, the meridional atmospheric heat flux and the meridional oceanic heat flux. Figure 1 shows schematically the energy transfers in a single box. Closure of the model was achieved by the following postulate:

Paltridge's Hypothesis: The climate system operates at a minimum of

$$E_1 = \sum_{\text{boxes}} F_S/F_L \, , \qquad (2.1)$$

where F_S is the absorbed solar radiation and F_L is the outgoing longwave radiation.

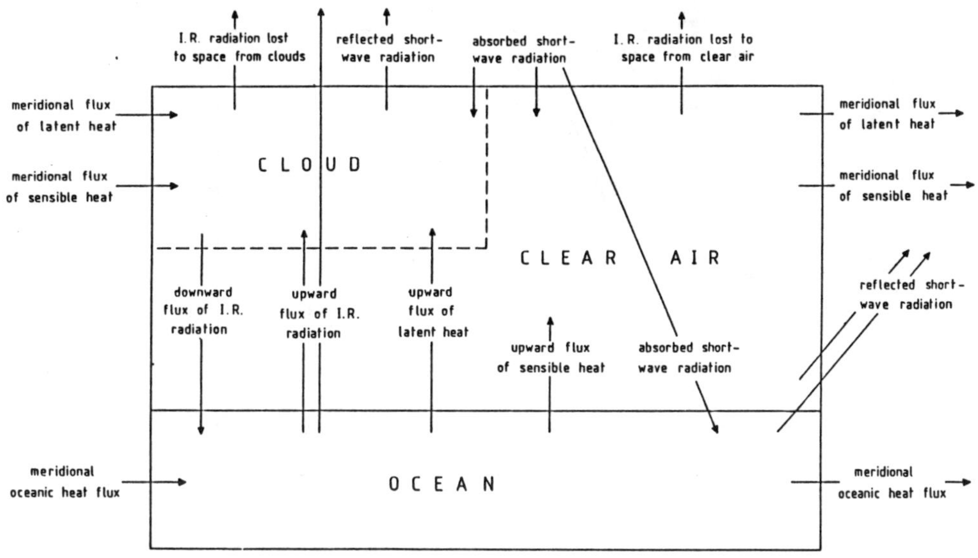

Fig. 1: Energy transfer processes included in Paltridge's (1975) energy balance climate model.

An absolute minimum of E_1 was found at values of the meridional heat fluxes close to observed values.

E_1 does not have any obvious physical interpretation so Paltridge replaced E_1 by

$$E_2 = \sum_{\text{boxes}} (F_S - F_L)/T_a \tag{2.2}$$

(where T_a is the absolute temperature) which he interpreted as the rate of entropy exchange between the climate system and its surroundings. He found an absolute minimum of E_2 at values of the heat fluxes close to observed values. Figure 2 shows the temperature, cloud cover and total meridional heat flux corresponding to minimum E_2 plotted against $x = \sin\theta$, where θ is the latitude. The graphs are drawn from results tabulated by Paltridge (1975). Also shown are observed values of these quantities taken from

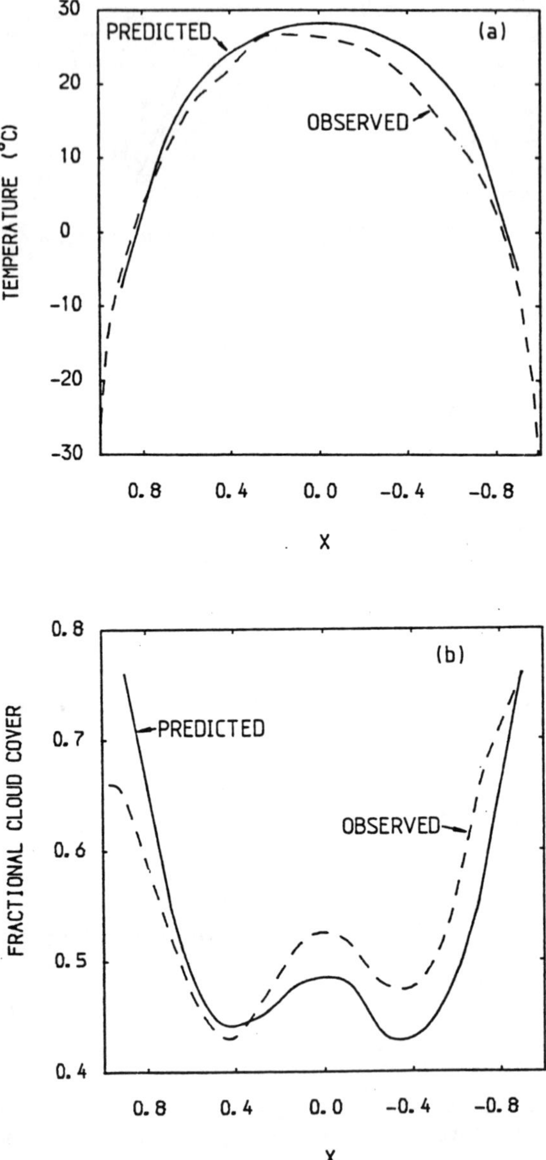

Fig. 2: Predictions of Paltridge's minimum entropy exchange model compared with observed values (from Sellers, 1965) (a) surface temperature; (b) fractional cloud cover; (c) total northward heat flux. Each is plotted against x, the sine of the latitude.

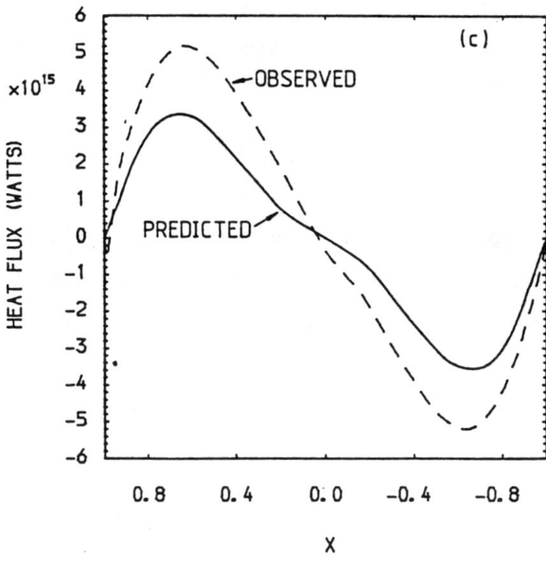

Fig. 2c

Sellers (1965). Quite clearly, the agreement between prediction and observation is remarkable, particularly for the cloud cover which is difficult to predict accurately even when the physics is more fully represented.

The results immediately raise the question as to why a minimum entropy exchange principle should be valid, if indeed the results are more than just a coincidence. Rogers (1976), whilst discussing this problem, showed that Paltridge's ideas could be expressed in a much simpler form. Writing

$$E_1 = \int_{-1}^{1} \frac{F_S(x) - F_L(x)}{F_L(x)} dx + 2 , \tag{2.3}$$

Paltridge's minimum entropy exchange principle requires E_I to be a minimum subject to the total energy of the system being conserved, i.e.

$$\int_{-1}^{1} \left[F_S(x) - F_L(x) \right] dx = 0 . \tag{2.4}$$

This minimisation problem is easily solved using the methods of the calculus of variations. Introducing a Lagrange multiplier λ , the requirement is that

$$\delta \int_{-1}^{1} \left[\frac{F_S(x) - F_L(x)}{F_L(x)} - \lambda \left(F_S(x) - F_L(x)\right) \right] dx = 0 . \tag{2.5}$$

Rogers used the simplifying assumption that the incident solar radiation F_S (which he specified from observations) is independent of the outgoing longwave radiation F_L. Then the requirement of a minimum is satisfied for

$$F_L = \left[\frac{F_S}{\lambda} \right]^{1/2} \tag{2.6}$$

and from (2.4), it follows that λ satisfies

$$\lambda^{1/2} = \frac{\int_{-1}^{1} F_S^{1/2} \, dx}{\int_{-1}^{1} F_S \, dx} . \tag{2.7}$$

Rogers demonstrated that using observed values of F_S, a good fit of the predicted meridional temperature distribution to the observed values could be obtained. Note that (2.6) implies that

$$F_L \propto F_S^{1/2} \tag{2.8}$$

and so an investigation of the validity of Paltridge's hypothesis could alternatively be expressed as an investigation of the validity of (2.8). It appears that (2.8) is a good approximation for the Earth's atmosphere-ocean system but as Rogers points out, this relationship clearly does not apply on some planets, e.g. Mercury (no atmosphere or oceans), Venus (too little temperature variation - but is F_S independent of F_L with so much cloud?).

Paltridge (1978) repeated his previous calculations without using the restrictive assumption concerning the upward heat flux from the surface and obtained equally convincing results. He also developed a two dimensional model (having zonal as well as meridional structure) with the globe divided into 400 equal area boxes. The model was constrained to be at a minimum of the entropy exchange rate E_2 and again he obtained predicted temperature, cloud cover and heat fluxes in good agreement with reality. (The meaning of the local heat flux is not entirely clear in this case, since any arbitrary non-divergent flux can be added to it without changing the other variables. I am grateful to Dr. J. Willebrand for pointing this out.) Paltridge also noted that the total rate of change of entropy of the climate system, can be expressed as

$$\frac{dS}{dt} = \frac{dS_i}{dt} + \frac{dS_e}{dt} , \tag{2.9}$$

where dS_i/dt is the internal rate of entropy production and dS_e/dt is the rate of exchange of entropy with the surroundings. Thus, for a steady state, i.e. $dS/dt = 0$, a minimum of the entropy exchange rate corresponds to a maximum entropy production rate.

Paltridge's (1975) one dimensional results were essentially reproduced by Grassl (1981). Grassl also noted that the full entropy balance of the climate system should include the flux of entropy carried by the incident and emitted radiation but he did not exploit this fact. A discussion of the full entropy balance equation is given in section 6.

3. ATTEMPTS TO JUSTIFY PALTRIDGE'S HYPOTHESIS

An initial justification of Paltridge's minimum entropy exchange principle could be made on the grounds that it works; it gives answers in good agreement with reality. However, this is not sufficient, because any useful climate model must be able to predict changes in the climate state which may be induced by changes in the external or internal parameters of the system. The success of the method for present day parameter values does not mean that equally good results would be obtained for other values. What is needed, therefore, is some a priori justification of the method based on sound physical principles. In this section, some attempts to find such a verification are described. Unfortunately, a totally convincing explanation is still awaited.

Paltridge (1979 and 1981) developed a qualitative argument involving the main energy conversions and feedback mechanisms of a zero dimensional climate model and arrived at the following conclusions:

If (1) the rate of conversion of available potential energy P to kinetic energy K is a super-linear function of P, (i.e. the gradient of the graph of conversion rate against P increases with P),

(2) the rate of dissipation is an increasing function of K,

(3) the system has several possible steady states and

(4) there is random variability in the rate of energy input which is sufficient to cause the climate system to move from one steady state to another,

then, it can be expected that the system will eventually occupy the state with the maximum dissipation (i.e. the maximum internal rate of entropy production).

The idea is that the climate system jumps from one steady state to another due to random fluctuations in the forcing, but that jumps to higher dissipation are more likely than jumps to lower dissipation. Assumption (2) is almost certainly true and there is good theoretical evidence for (3) from simple thermodynamic and dynamical climate models (e.g. North et al. (1981), Charney and DeVore (1979)). Assumptions (1) and (4) would be more difficult to prove in practice. There is, however, another difficulty with this argument. The minimum entropy exchange (or maximum entropy production) principle, as applied by Paltridge and Grassl, states that:

> of all the energetically possible steady states in the neighbourhood of the realised one, the realised one has the maximum (minimum) entropy production (exchange).

Paltridge's (1979, 1981) later argument asserts that the entropy producton is likely to increase with time, which is clearly different. Nicolis and Nicolis (1980) noted this difference. Paltridge (1981) also discussed the possible inclusion of vertical temperature structure in his (1978) zonally averaged, one dimensional model. He concluded that the minimum entropy exchange concept did not work in that case.

Mobbs (1982) approached the problem from the point of view of the consistency of the maximum entropy production idea with the observational data. He derived a variational principle for the zonally averaged atmospheric temperature from the thermodynamic energy equation. This required the use of restricted variations; in other words, certain terms in the functional could not be varied arbitrarily. By assuming parameterizations of the meridional and vertical heat fluxes which are known to give a reasonable fit to observations in mid-latitudes, results from the calculus of variations were used to obtain a sufficient condition for the existence of a maximum of the entropy production. Observational data for the meridional and vertical temperature structure then allowed regions where the entropy production was a maximum to be identified. Figures 3 and 4 show the regions where the condition for maximum entropy production is satisfied. These are precisely the regions where the parameterizations of the eddy heat fluxes hold. Outside these regions, the test for a maximum was inconclusive. Thus there is strong evidence that the mean atmospheric state is consistent with a maximum in the entropy production. This does not prove that the system is in a state of maximum entropy

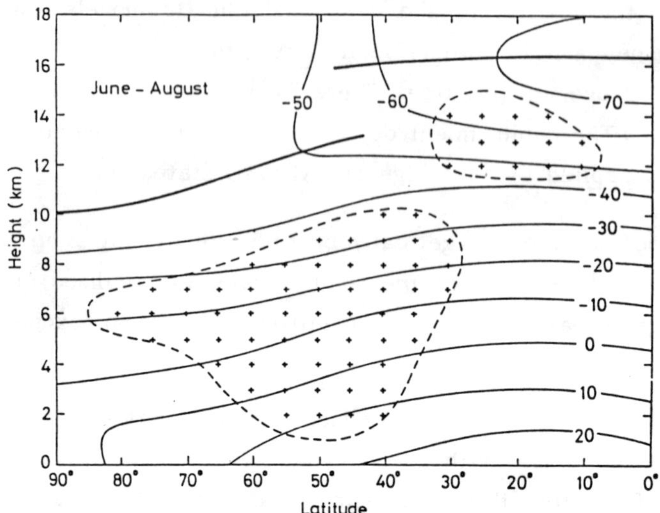

Fig. 3: Cross section of the northern hemisphere troposphere averaged for the period June - August, showing the region (indicated by crosses) where the mean conditions are consistent with the entropy production being at a maximum. The thin solid lines are isotherms (in °C) and the heavy lines are the tropopause. (Reproduced form Mobbs (1982)).

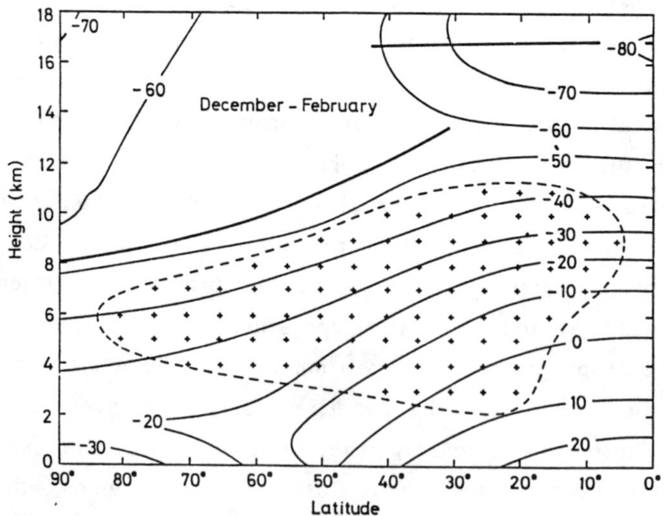

Fig. 4: As figure 3, but averaged for period December-February.

production. However, the method could, but does not, disprove the hypothesis. A fundamental assumption of this study is that the observational data and the parameterized eddy heat fluxes are consistent with each other. The effects of the inevitable departures from this situation are unclear.

Noda and Tokioka (1983) applied the principle of minimum entropy exchange to a zonally averaged, 10 latitude, 10 level energy balance climate model. In contrast to Paltridge's simple model, Noda and Tokioka found multiple minima, with some climate states corresponding closely to present day conditions but others differing considerably. A disturbing conclusion of the study (from the point of view of the success of the minimum entropy exchange principle) was that the details were very sensitive to the particular parameterizations of the cloud physics, humidity and radiation processes. In fact, when the absorption of long-wave radiation by water vapour was explicitly included, no minimum could be found. A significant point to note here is that Noda and Tokioka (and Paltridge in the case of his simple model) looked for absolute minima with respect to all the free variables in the model (subject to the energy balance constraint). It may in fact be that the climate system is at a minimum of the entropy exchange rate only with respect to a restricted number of variables (e.g. the temperature). This could explain why some parameterizations appear to allow minima whilst others do not. Further discussion of the point is given in section 4.

4. APPLICATION OF THE MINIMUM ENTROPY EXCHANGE PRINCIPLE IN ONE DIMENSIONAL ENERGY BALANCE MODELS

(a) Formulation of a simple model

Some qualitative understanding of the behaviour and status of the minimum entropy exchange principle can be gained by applying it to simple, one dimensional energy balance climate models. The model used here is essentially the Budyko-Sellers-North model (Budyko 1969, Sellers 1969, North 1975). Figure 5 shows the energy fluxes considered in the model. Incident solar radiation Q Wm^{-2} is partially reflected (αQ Wm^{-2}, where α is the albedo) and the remainder absorbed. Long-wave radiation I Wm^{-2} is emitted and the energy balance is achieved by a meridional flux R. The energy balance of an element of the climate system between latitudes θ and $\theta + \delta\theta$ requires that

$$(Q - \alpha Q - I)2\pi r_o \cos\theta \; r_o \; \delta\theta = -R + (R + \delta R)$$

where r_o is the radius of the Earth. Hence

$$Q(1 - \alpha) - I = \frac{1}{2\pi r_o^2} \frac{dR}{dx} \tag{4.1}$$

where $x = \sin\theta$.

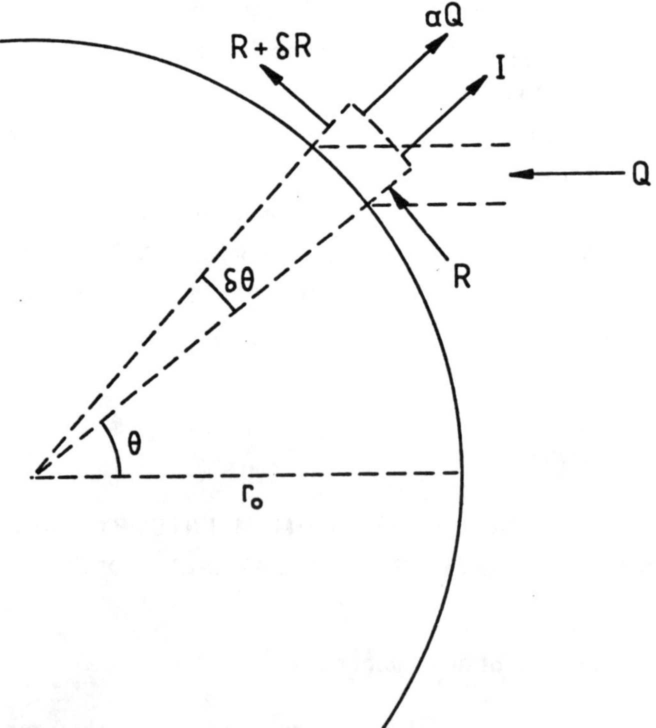

Fig. 5: The energy transfer processes incorporated in the simple one dimensional energy balance climate model.

The terms on the left of equation (4.1) are parameterized as follows:

(i) The incident solar radiation Q contains a geometrical factor due to the sphericity of the Earth and can be written as

$$Q = Q_o S(x) = Q_o(1 + S_2 P_2(x)) \qquad (4.2)$$

where Q_o is $^1/_4$ of the solar constant, $P_2(x)$ is the second order Legendre Polynomial $((3x^2-1)/2)$ and S_2 is a constant. Values of the constants are taken from North (1975) to be $Q_o = 334.4 \text{ Wm}^{-2}$ and $S_2 = -0.482$. According to North, the geometrical factor in equation (4.2) is accurate to better than 2%.

(ii) It is convenient to consider the co-albedo $a(x) = 1-\alpha(x)$. The simple model used in this section takes

$$a(x,x_s) = \begin{cases} a_o & \text{for ice-free areas } (x<x_s) \\ b_o & \text{for ice-covered areas } (x>x_s) \end{cases}$$

and following Sellers (1969), we take $a_o \approx 0.70$ (value actually used was 0.697 from North (1975) and from Budyko (1969), $b_o \approx 0.38$. These co-albedo values take into account (insomuch as they agree with observations) the mean cloud cover. The condition for ice to form is again taken from Budyko (1969) as

$$\begin{array}{ll} T > -10° \text{ C} & \text{ice forms} \\ T < -10° \text{ C} & \text{no ice} \end{array} \qquad (4.3)$$

where T is the surface temperature in degrees Celcius.

(iii) Following Budyko (1969), the long-wave radiation is parameterized by

$$I = A + BT \tag{4.4}$$

where A and B are constants. This could be regarded as a linearization of the Stefan-Boltzmann T^4 law but in fact observations show that the effective emissivity is strongly temperature dependent (due to different cloud heights, for example) so that formula (4.4) is better regarded as being fitted directly to observational data. Recent values for A and B which are used here are given by North et al. (1981) as

$$A = 203.3 \ \mathrm{Wm}^{-2},$$
$$B = 2.09 \ \mathrm{Wm}^{-2}\ {}^\circ\mathrm{C}^{-1}$$

combining the parameterizations gives

$$Q_o(1 + S_2P_2)a(x,x_s) - (A + BT) = \frac{1}{2\pi r_o^2}\frac{dR}{dx} \, . \tag{4.5}$$

The model equation (4.5) should be regarded as an equation for the surface temperature T. Clearly, however, it cannot be solved as it stands, since the meridional heat flux R is undetermined. This illustrates the fundamental problem of climate modelling; namely one of <u>closure</u>. It is instructive to consider the most extreme solutions of (4.5) before considering how realistic closure may be achieved.

(b) Closure Methods

(i) First extreme: $R = 0$. This corresponds to a climate system which is unable to transport heat horizontally and hence from (4.5)

$$A + BT = Q_o S(x) a(x).$$

Representing the co-albedo by

$$a(x) = a_o + a_2 P_2(x)$$

with $a_o = 0.681$ and $a_2 = -0.202$, which is a reasonable fit to observed values (North et al. 1981), the temperature is easily obtained. Figure 6 shows the temperature plotted against x for this no transport extreme. As expected, the equator is much too hot and the poles too cold.

(ii) Second extreme: infinite diffusion. In this extreme example it is assumed that heat can be transported infinitely quickly around the globe, so that an isothermal state results. The temperature is then given by:

$$A + BT = Q_o \int_0^1 S(x) a(x)\, dx$$

and has the value 14.97°C for the above parameter values (North et al. 1981). (Note that here and throughout this section, symmetry between the hemispheres has been assumed.)

Figure 6 also shows the observed mean surface temperature (from Sellers 1965). As expected, the real climate system lies between the two extremes. The above exercise is instructive because it indicates the limits between which the temperature may lie and gives a way of assessing the accuracy of a particular closure scheme. For instance, at the equator, temperatures between 15°C and 60°C seem to be possible. If the deviation of a predicted temperature from the observed value is small compared to this 45°C range, then in some sense the closure scheme is a good one.

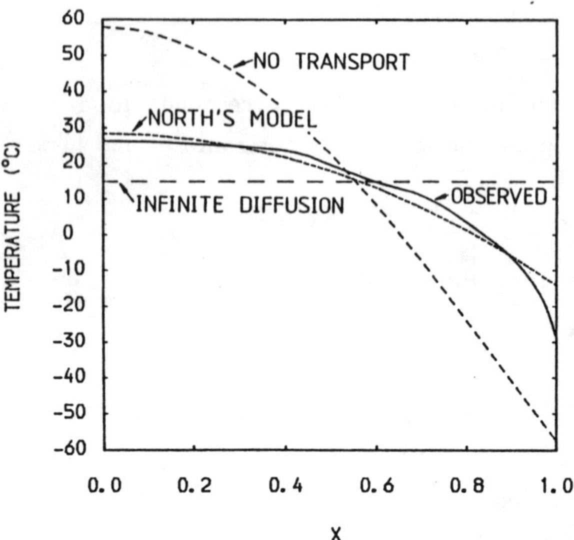

Fig. 6: The variation of surface temperature with $x = \sin\theta$ corresponding to the extreme solutions of the one dimensional energy balance model. Also shown are the observed temperature (from Sellers, 1965) and the results of North's (1975) diffusive heat flux model.

Figure 7 shows the poleward heat flux, R, plotted against x for the two extreme cases and for the observed present day conditions (taken from Sellers, 1965). The observed case is for the northern hemisphere.

There are essentially two types of closure scheme for this model:

(i) The conventional method is to assume a linear diffusion law for R so that

$$R \propto \nabla T$$

or

$$R = -D(1 - x^2)\frac{dT}{dx} ,$$

where D is the diffusivity (e.g. North 1975). North calculates D by constraining the model to give the observed value for the ice-line

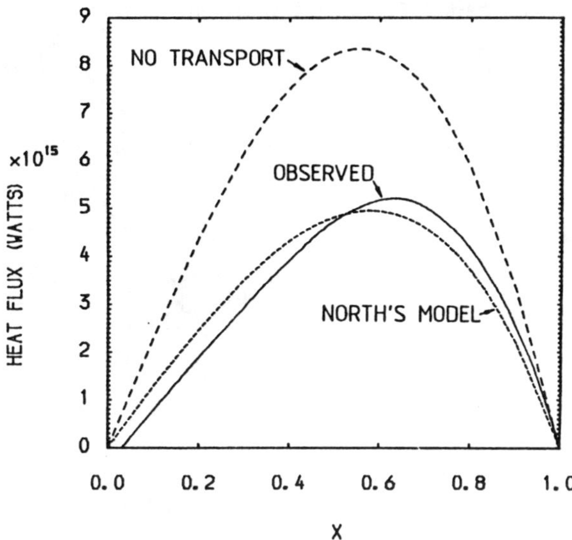

Fig. 7: As figure 6, but showing the poleward heat flux. The observed case (from Sellers, 1965) is for the northern hemisphere.

latitude x_S, which is about 0.95. Results from North (1975) are shown in Figure 6. By the above criterion for assessing a closure scheme, the fit is clearly good. The limitation of the method, however, is that it does not allow x_S to be predicted for other values of the external forcing, unless another way of estimating D (based, perhaps, on baroclinic instability theory) is found.

(ii) The model may be closed by using a minimum entropy exchange constraint or some related extremal condition. Five examples of this technique are now presented.

(c) Examples of Extremal Closure Conditions

Example 1

Following Paltridge (1975), suppose that the climate is at a minimum of

$$F = \int_0^1 \frac{F_S - F_L}{F_L}\,dx$$

or in the present model

$$F = \int_0^1 \frac{Q_o S(x) a(x,x_s) - I(x)}{I(x)}\,dx\ . \tag{4.6}$$

The requirement of global energy balance is, from equation (4.5)

$$\int_0^1 \left[Q_o S(x) a(x,x_s) - I(x) \right] dx = 0\ . \tag{4.7}$$

Suppose that the solution is approximated by $I = I_o + I_2P_2(x)$, or equivalently, by $I = I_o + I_2x^2$. (This is the simplest polynomial allowing variations of I which are symmetrical about the equator.) Then the Budyko ice-line condition (4.3) becomes, after using (4.4),

$$I_0 + I_2x^2 = 182.4 \ \mathrm{Wm}^{-2} \ . \tag{4.8}$$

Equations (4.7) and (4.8) can be regarded as having 3 variables I_0, I_2 and x_s and can therefore be solved for two variables I_2 and x_s^2 for example, in terms of I_0. Substituting in equation (4.6) then allows F to be expressed as a function of I_0. Figure 8 shows a graph of F plotted against I_0. The graph has 2 branches with 4 local minima. The discontinuities in gradient occur when x_s decreases to zero (i.e. the temperature becomes less than -10°C everywhere) or increases to one (i.e. the temperature becomes greater than -10°C everywhere). One of the minima (marked 4) is so close to a

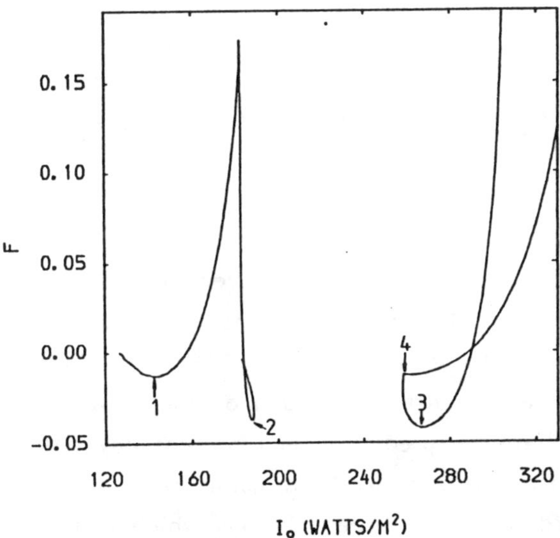

Fig. 8: The functional F plotted against I_0, subject to the energy balance condition (4.7) and the Budyko ice-line condition (4.8). The four minima of F are indicated.

discontinuity that it is ignored in the following (since a small change in parameter values could result in its disappearance). Figure 9 shows x_s plotted against I_0 for the same model. Note the presence of multiple solutions for given I_0.

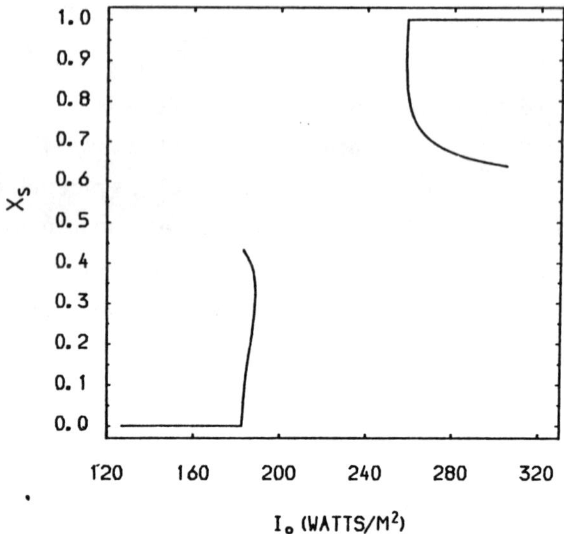

Fig. 9: The sine of the latitude of the edge of the polar ice cap plotted against I_0, corresponding to F shown in figure 8.

Closure of the model is now achieved by choosing I_0 to correspond to one of the minima of F. These occur at values of x_s of 0.712, 0.276 and zero (corresponding to latitudes of 45°, 16° and 0°). Figure 10 shows the temperature plotted against x for each of these cases. For $x_s = 0.712$, the correspondence between prediction and present day observations is fair but in the other cases it is poor. In a qualitative sense, the three solutions can be identified with the three solutions which can be obtained using the diffusion law closure scheme (see North et al. 1981). The solution with the largest value of x_s is identified as corresponding to present day conditions. The solution with $x_s = 0$ corresponds to a completely ice-covered state. At least for the diffusive case, North et al. (1979) have shown that these two solutions are stable, whilst the third, intermediate one is unstable and therefore has no physical significance. Figure 11 shows the meridional heat flux for the three solutions, plotted against x. Agreement with the observations is not good in any of the cases but is best for $x_s = 0.712$. Note that the discontinuity in the gradient of R

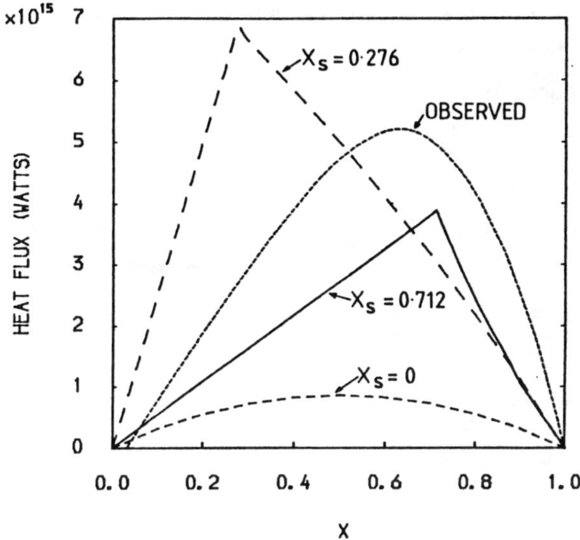

Fig. 10: The surface temperature as a function of the sine of the latitude, x, corresponding to the minima 1 (x_S = 0), 2 (x_S = 0.276) and 3 (x_S = 0.712) of F shown in figure 8. The observed surface temperature (from Sellers, 1965) is also shown.

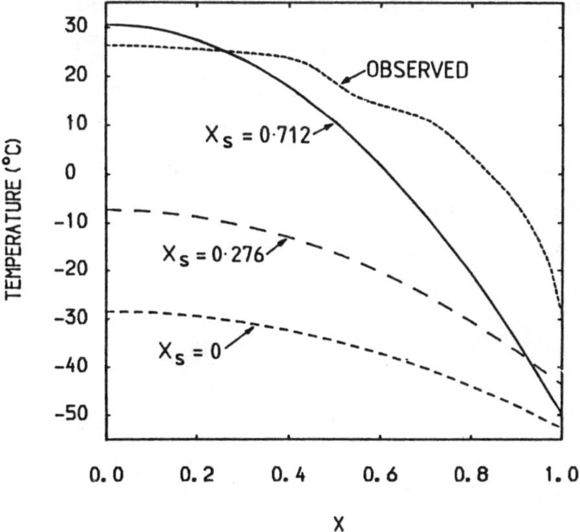

Fig. 11: As figure 10, but showing the variation of poleward heat flux with x. The observed values (from Sellers, 1965) are for the northern hemisphere.

is due to the discontinuity in the albedo at the edge of the ice caps. It is quite evident from Figures 10 and 11 that this method of closure gives poor results compared with the diffusion law method.

Example 2

This example differs from the previous one only in the way in which the minimisation of F is performed. The energy balance constraint (4.7) can be applied using the method of Lagrange multipliers by minimising

$$F = \int_0^1 \left[\frac{Q_0 S(x) a(x,x_s) - I(x)}{I(x)} - \mu^2 \left(Q_0 S(x) a(x,x_s) - I(x)\right) \right] dx , \qquad (4.9)$$

where μ^2 is a Lagrange multiplier. Suppose now that F is minimised with respect to I. This means that if I is perturbed about the solution to the problem (subject to the energy balance condition being satisfied), F will increase. However, this perturbation is carried out whilst holding x_s constant and this is the essential difference between Example 1 and Example 2. Note that x_s does not have to be specified; the requirement is merely that F is to be a minimum with respect to I only and, in particular, not with respect to x_s. This type of variational principle is often referred to as a restricted variational principle. With these conditions, the Euler-Lagrange equation for an extremal of F is

$$- \frac{Q_0 S(x) a(x,x_s)}{I^2} + \mu^2 = 0$$

and hence

$$I = \frac{[Q_0 S(x) a(x,x_s)]^{1/2}}{\mu} . \qquad (4.10)$$

Note the similarity between (4.10) and Rogers' (1976) result (equation (2.6)). The Lagrange multiplier μ^2 and x_s can be found by substituting expression (4.10) into the energy balance equation (4.7) and by using the Budyko ice-line condition. The model is found to have only one solution with $x_s = 0.935$ (latitude 69°). Figure 12 shows the temperature plotted against x for this solution. Agreement with observations is clearly good, as is the predicted value of x_s (present day observed value is about 0.95).

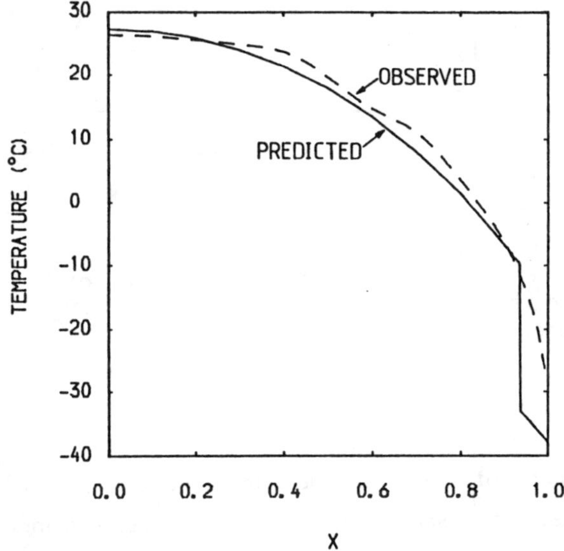

Fig. 12: Predicted surface temperature flux as a function of the sine of the latitude, x_s, corresponding to an extremum of F (equation (4.9)). The observed values are from Sellers (1965).

Figure 13 shows the corresponding meridional heat flux. Here the agreement is poor.

The reason for the apparently good results for the temperature but poor heat flux results lies in their relative sensitivities to the parameterizations and other model details. Examination of observational data shows that the maximum value of R (at about 40° latitude) amounts to less than 10% of the total radiation received by either hemisphere (Sellers

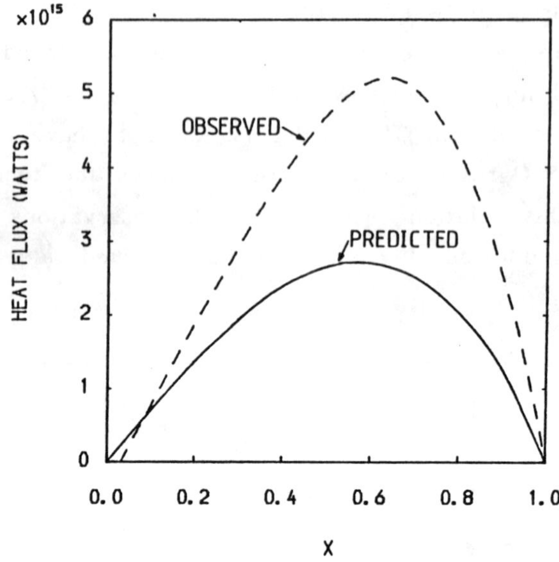

Fig. 13: Predicted poleward heat flux as a function of the sine of the latitude, x_S, corresponding to an extremum of F (equation (4.9)). The observed values are from Sellers (1965) for the northern hemisphere.

1965), so that small errors in T (and hence in F_L) can lead to large errors in R. From this point of view, the meridional heat flux can be regarded as a much better (i.e. more sensitive) test of the performance of a climate model than the temperature.

There is no doubt that the results from Example 2 are better than those from Example 1, particularly for the temperature. This suggests that in looking for a justification for Paltridge's minimum entropy exchange hypothesis, it may be necessary to search for a restricted variational principle.

Example 3

Following Paltridge (1978), suppose that the climate model is at an extremum of the entropy exchange rate, subject to global energy balance, or in other words

$$F = \int_0^1 \left[\frac{Q_0 S(x) a(x,x_s) - I(x)}{T_a(x)} - \mu^2 (Q_0 S(x) a(x,x_s) - I(x)) \right] dx \qquad (4.11)$$

is at an extremum. T_a is the absolute temperature and the long-wave parameterization (4.4) can be written as

$$I = A' + BT_a ,$$

where $A' = -367\ \mathrm{Wm}^{-2}$. By varying T_a but holding x_s fixed (as in Example 2), the Euler-Lagrange equation

$$-\frac{Q_0 S(x) a(x,x_s)}{T_a^2} + \mu^2 = 0$$

is obtained and hence

$$T = \frac{1}{\mu} \left[\frac{Q_0 S(x) a(x,x_x) - A'}{B} \right]^{1/2} . \qquad (4.12)$$

Again x_s and μ can be found using the global energy balance and Budyko ice-line conditions. There is a single solution, this time with $x_s = 0.948$ (latitude 71°). Figures 14 and 15 show the temperature and heat flux for this case. The most striking feature is the similarity between this case and Example 2. As noted by Paltridge (1975), the two extremal conditions give very similar results. Presumably this is because if the expressions for F in equations (4.9) and (4.11) were linearized about the mean absolute

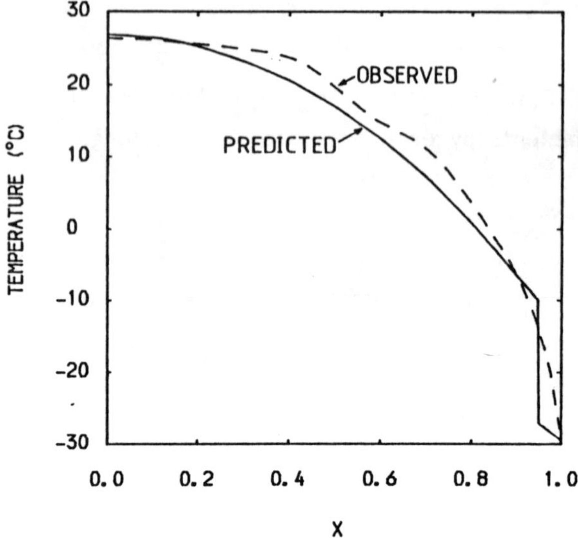

Fig. 14: As figure 12, but with F given by equation (4.11).

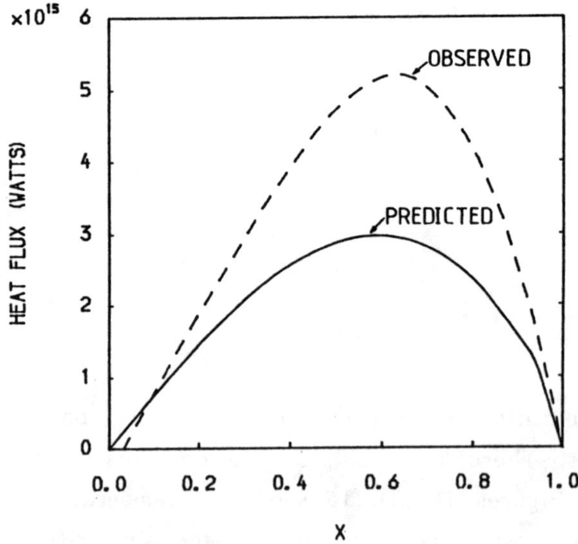

Fig. 15: As figure 13, but with F given by equation (4.11).

temperature, they would reduce to almost the same quantity (apart from a constant factor), since the global variations in absolute temperature (≈ 60 K) are small compared with the mean absolute temperature (≈ 287 K).

Example 4

It has been conjectured by Lorenz (1955, 1960) that the atmosphere may be generating available potential energy at the maximum possible rate. (The available potential energy is defined in this context as the maximum potential energy which could be released by a hypothetical adiabatic rearrangement of the fluid.) The rate of generation of available potential energy is given to good accuracy (Lorenz 1955) by

$$G = \frac{1}{g} \int_0^{\bar{p}_0} \frac{\Gamma_d}{\Gamma_d - \bar{\Gamma}} \frac{\overline{q' T'_a}}{\bar{T}_a} \, dp \, ,$$

where q is the net diabatic heating, p the pressure, g the acceleration due to gravity, Γ the lapse rate and Γ_d the dry adiabatic lapse rate. An overbar represents an average over an isentropic surface and a prime the deviation from such an average. Clearly G depends on the vertical structure of the atmosphere and has no direct meaning in the context of the one dimensional energy balance model used here. An analogous quantity, however, is

$$\int_0^1 \left[Q_0 S(X) a(x, x_s) - I(x) \right] T(x) \, dx$$

and so, with the global energy balance constraint applied, the relevant functional is

$$F = \int_0^1 \left[(Q_0 S(x) a(x,x_s) - I(x)) T(x) - \mu^2 (Q_0 S(x) a(x,x_s) - I(x)) \right] dx \,. \quad (4.13)$$

Lin (1982) considered a similar quantity but solved the problem by expanding I(x) as $I_0 + I_2 P_2(x)$. Lin found that present day conditions were close to a maximum of his functional. The condition for F to be a maximum is (from 4.13)

$$T = \frac{Q_0 S(x) a(x,x_s) - A - \mu^2 B}{2B} \quad (4.14)$$

From the global energy balance constraint and the Budyko ice-line condition, it follows that there is one solution with $x_s = 0.960$ (latitude 74°). Figures 16 and 17 show the temperature and heat flux for this example. Again agreement with observations is good for the temperature but

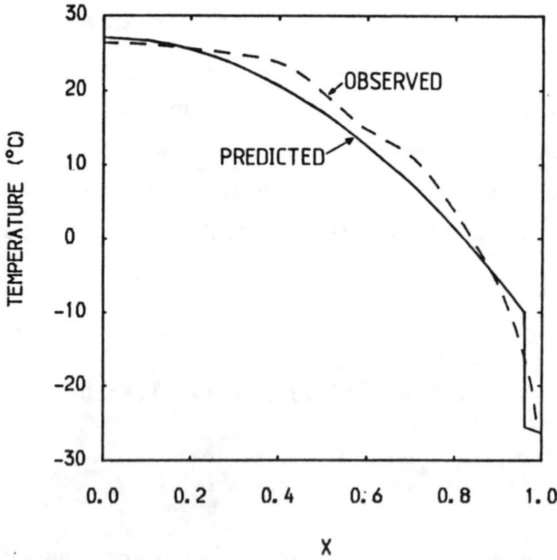

Fig. 16: As figure 12, but with F given by equation (4.13).

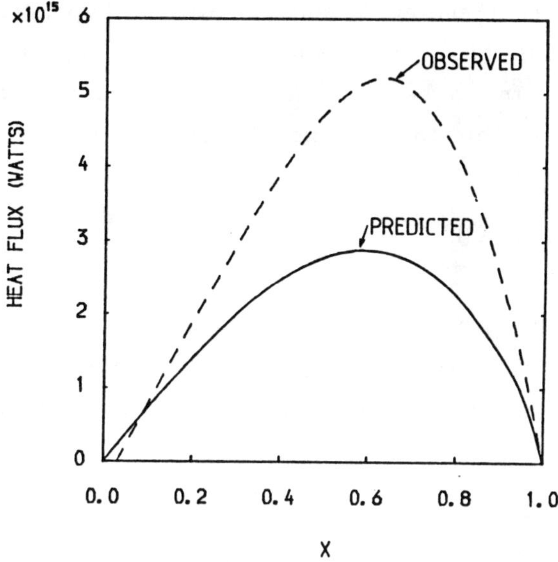

Fig. 17: As figure 13, but with F given by eqaution (4.13).

more significantly, the results are almost identical to those from Examples 2 and 3. The main conclusion to be drawn from these results, therefore, is that any search for an explanation of Paltridge's results should not be confined to a study of the entropy balance; the controlling parameter may well be something which, superficially at least, has a quite different physical interpretation.

Example 5

As a final example, consider minimising

$$F = \int_0^1 \left[\frac{Q_0 S(x) - I(x)}{I(x)} - \mu^2 \left(Q_0 S(x) a(x, x_s) - I(x)\right) \right] dx \qquad (4.15)$$

subject to the global energy balance constraint. Expression (4.15) is just (4.9) with the absorbed solar radiation replaced by the total incident radiation in the term to be minimised. No justification for this procedure will be given at present. The condition for an extremum is

$$I = \frac{[Q_0 S(x)]^{1/2}}{\mu} \tag{4.16}$$

and by applying the global energy balance and Budyko ice-line conditions, it is found that there are (for $x_s \neq 0$ or 1) two solutions with $x_s = 0.936$ (latitude 69°) and $x_s = 0.272$ (latitude 16°). Figures 18 and 19 show the temperature and meridional heat flux corresponding to these two solutions.

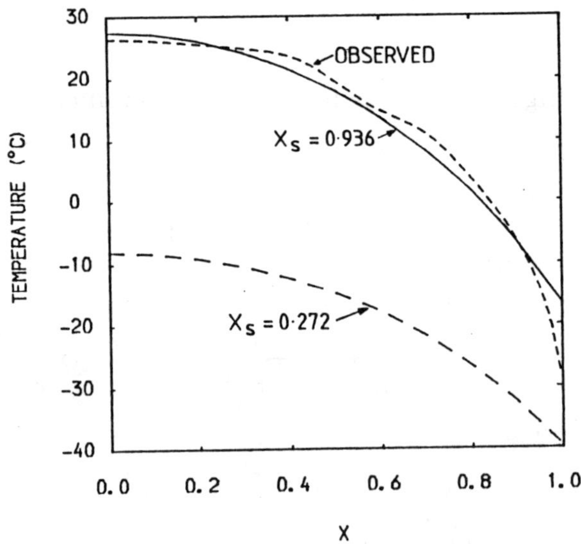

Fig. 18: As figure 12, but with F given by equation (4.15).

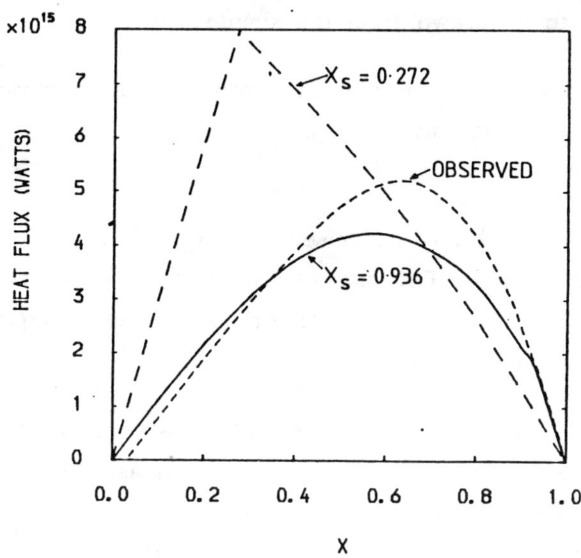

Fig. 19: As figure 13, but with F given by equation (4.15).

For the first solution (x_S = 0.936), the agreement between the predicted and observed temperature is very good and the results for the heat flux are considerably better than those for the previous examples. It seems unlikely that the improvement occurs because the incident (rather than absorbed) radiation is a controlling factor; indeed, the climate system never detects the reflected radiation. A possible explanation is that the discontinuous albedo function has been removed from the minimised quantity. In the real climate system, although the albedo varies very abruptly between the ice-free and ice-covered regions, the effect of this "step" in the albedo on, for example, the temperature, will be smeared over a range of latitudes. Removal of the "step" from the minimised quantity in some sense approximates this smoothing.

(d) Conclusions to be drawn from the simple model

(i) The extremal condition most appropriate as a closure condition for climate models may be expressable only as a restricted variational principle.

(ii) The minimised (or maximised) quantity may not be related to the entropy balance. For example, a quantity analogous to the rate of generation of available potential energy gives equally good results in the simple model.

(iii) Better results are obtained when allowance is made for the non-local effect of the albedo contrast between ice-free and ice-covered regions.

5. RELATED STUDIES

A detailed discussion of the concept of available potential energy was given by Dutton and Johnson (1967). They also considered various extreme states of the atmosphere such as maximum or minimum potential energy and discussed the possible application of Hamilton's principle to atmospheric dynamics. Dutton (1973) considered the total entropy of the atmosphere (rather than its rate of production or exchange). He showed that, associated with the actual atmospheric state with total entropy S, there is always a unique equilibrium state with the same mass and energy but which is at rest and has entropy S_0. Dutton suggested that $S - S_0$ may be minimised. From observations, $(S - S_0)/S_0$ is about 0.015, i.e. $S - S_0$ is small compared to S. However, this does not imply that $S - S_0$ must be close to a minimum. There does not appear to be any obvious connection between Dutton's hypothesis and that due to Paltridge.

Schulman (1977) applied Lorenz's maximum generation of available potential energy concept to a multi-level energy balance model of the atmosphere. He presented results for 5-level, 5-latitude and 5-level,

9-latitude models, using a number of different parameterization schemes for the radiative and sensible heat processes. Predicted temperature distributions resembling those observed were obtained, although the corresponding rates of generation of available potential energy were at least twice the observed values. The results, however, suggest that the atmosphere may indeed be near its state of maximum generation of available potential energy.

Golitsyn and Mokhov (1978) applied the constraint of minimum entropy exchange to various simple Budyko-Sellers type climate models, rather as in section 4. However, as an alternative to varying the temperature as in section 4, they varied the meridional heat flux or the diffusivity coefficient for the meridional heat flux. Generally, the predicted temperatures were in poor agreement with observation, particularly for North's (1975) model which closely resembles the one used in the previous section. In the case of a five latitude model due to Budyko, however, reasonable agreement was obtained.

Mokhov and Golitsyn (1978) reformulated several Budyko-Sellers models as variational principles involving the entropy production by meridional heat fluxes. These were used to discuss the stability properties of the solutions. While such an approach gives a neat and useful mathematical way of analysing the stability of models, it does not address the quite different problem of closure. It is still necessary to use another condition (such as the diffusive heat flux assumption) to complete the solution. A similar approach was used by North et al. (1979) who reformulated North's (1975) model as a variational principle which did not involve the entropy production. Their stationary functional had two minima corresponding to the two stable states and one saddle point corresponding to the unstable solution.

Nicolis and Nicolis (1980) used thermodynamic arguments to conclude that the entropy production could increase or decrease with time for different climate regimes. They also noted that the time dependence of the entropy production (as considered by Paltridge 1979 and 1981) is not directly relevant to the maximum entropy production constraint used by

Paltridge (1975 and 1978). This point was discussed in section 3. Nicolis and Nicolis reproduced Golitsyn and Mokhov's minimum entropy exchange results for North's climate model. Like Golitsyn and Mokhov, they found the predicted temperature to be too high at the equator and too low at the poles, in contrast to the results presented here. However, the predicted meridional heat flux was very similar to that found in Examples 2, 3 and 4 of section 4 (i.e. too low).

Shutts (1981) applied a principle of maximum entropy production to a quasi-geostrophic, two level model of the atmosphere. The main difference between this study and the others described here is that it involves a dynamical model rather than a thermodynamic one. Shutts defined entropy in a dynamical context using methods from statistical mechanics and in common with the thermodynamic approaches, it was necessary to apply a constraint of energy conservation. A constraint of no angular momentum transfer between the atmosphere and the Earth was also found to be needed. Expressions for the zonal-mean surface wind in reasonable agreement with those derived using more conventional parameterizations were obtained. Shutts also applied the concept to a model of the wind-driven ocean circulation, obtaining results which had some realistic features.

6. THE ENTROPY BALANCE EQUATION

The entropy balance equation for the climate system is a reformulation of the first law of thermodynamics:

$$TdS = de + pd\alpha - \sum_k \mu_k dc_k \quad , \tag{6.1}$$

where s is the specific entropy, θ is the specific internal energy, p is the pressure and α is the inverse of the density ρ (i.e. $\alpha = 1/\rho$). μ_k is the chemical potential of the kth constituent of the system and c_k its mass fraction (i.e. ρ_k/ρ, where ρ_k is the density of the kth constituent). Restricting attention to only those changes in composition due to the phase changes of water, k then represents water vapour (V), liquid water (W) or ice (I).

The thermodynamic energy equation for the system takes the form (Gill 1982)

$$\frac{\partial}{\partial t}(\rho e) + \nabla\cdot(\rho e\underline{u}) + \nabla\cdot(\underline{F}_S + \underline{F}_L + \underline{F}_H) = Q_H - \rho\nabla\cdot\underline{u}\ , \tag{6.2}$$

where $\underline{u}$ is the velocity of the centre of mass of a fluid element of the system and Q_H denotes latent heating due to the phase changes of water. $\underline{F}_S$, $\underline{F}_L$ and $\underline{F}_H$ are the fluxes of solar radiation, long-wave radiation and sensible heat respectively. Equation (6.1) can be applied to an arbitrary element of the system giving

$$T\frac{Ds}{Dt} = \frac{De}{Dt} + p\frac{D\alpha}{Dt} - \sum_k \mu_k \frac{Dc_k}{Dt}$$

(Where $D/Dt \equiv \partial/\partial t + \underline{u}\cdot\nabla$ is the material derivative). With the use of the continuity equation

$$\frac{\partial\rho}{\partial t} + \nabla\cdot(\rho\underline{u}) = 0$$

and equation (6.2), this can be written as

$$T\frac{\partial}{\partial t}(\rho s) = -T\nabla\cdot(\rho s\underline{u}) - \Sigma\cdot(\underline{F}_S + \underline{F}_L + \underline{F}_H) + Q_H - \sum_k \mu_k\left[\frac{\partial\rho_k}{\partial t} + \nabla\cdot(\rho_k\underline{u})\right] = 0. \tag{6.3}$$

Now the balance equation for k takes the form

$$\frac{\partial \rho_k}{\partial t} + \nabla\cdot(\rho_k \underline{u}_k) = \sigma_k , \tag{6.4}$$

where σ_K is the rate of formation of k per unit volume. In particular, denoting the rate of melting of ice by σ_{IW}, the rate of evaporation by σ_{WV} and the rate of direct conversion of water vapour to ice by σ_{VI}, then $\sigma_V = \sigma_{WV}-\sigma_{VI}$, $\sigma_W = \sigma_{IW} - \sigma_{WV}$ and $\sigma_I = \sigma_{VI}-\sigma_{IW}$. Noting also that the total mass of water of all phases within an element of the system must be conserved, it follows that

$$\sum_k \frac{Dc_k}{Dt} = 0 . \tag{6.5}$$

It is easily shown using equations (6.4) and (6.5) that

$$\sum_k \mu_k \left[\frac{\partial \rho_k}{\partial t} + \nabla\cdot(\rho_k \underline{u})\right]$$

$$= (\mu_V - \mu_W)\sigma_{WV} + (\mu_I - \mu_V)\sigma_{VI} + (\mu_W - \mu_I)\sigma_{IW}$$

$$+ (\mu_I - \mu_V)\nabla\cdot\underline{F}_V + (\mu_I - \mu_W)\nabla\cdot\underline{F}_W , \tag{6.6}$$

where $\underline{F}_V$ and $\underline{F}_W$ are the diffusive fluxes of water vapour and liquid water given by

$$\underline{F}_V = \rho_V(\underline{u}_V - \underline{u})$$

and $$\underline{F}_W = \rho_W(\underline{u}_W - \underline{u}) .$$

Note that the expression in (6.6) would be zero if all the phase changes were to take place at equilibrium, since then all the chemical potentials would be equal. However, this is not the case. For example, evaporation from the sea surface usually occurs when the vapour pressure is considerably less than its saturation value. Equation (6.6) can be combined with (6.3) to give the required expression for $\partial(\rho s)/\partial t$. The rate of change of the total entropy of the system is then given by

$$\begin{aligned}\frac{dS}{dt} &= \int \frac{\partial}{\partial t}(\rho s)\, dV \\ &= -\int \frac{\nabla\cdot(\underline{F}_S + \underline{F}_L)}{T}\, dV \\ &\quad + \int \frac{Q_H - (\mu_V - \mu_W)\sigma_{WV} - (\mu_I - \mu_V)\sigma_{VI} - (\mu_W - \mu_I)\sigma_{IW}}{T}\, dV \\ &\quad + \int \underline{F}_H\cdot\nabla\left(\frac{1}{T}\right) dV + \int \underline{F}_V\cdot\nabla\left(\frac{\mu_I - \mu_V}{T}\right) dV + \int \underline{F}_W\cdot\nabla\left(\frac{\mu_I - \mu_W}{T}\right) dV\end{aligned} \tag{6.7}$$

after using Gauss' theorem and noting that

$$\underline{F}_H\cdot d\underline{a} = \underline{F}_V\cdot d\underline{a} = \underline{F}_W\cdot d\underline{a} = \underline{u}\cdot d\underline{a} = 0$$

on the boundary of the system ($d\underline{a}$ is the outward vector element of area). If attention were restricted to the atmosphere alone, for example, it would then be necessary to take into account the non-zero values of $\underline{F}_V \cdot d\underline{a}$ and $\underline{F}_W \cdot d\underline{a}$ at the sea surface. Noda and Tokioka (1983) showed that the contribution of the flux of water vapour into the atmosphere due to evaporation at the sea surface is a significant term in the entropy balance equation for the atmosphere.

The first term on the right hand side of equation (6.7) can be identified as the entropy exchange rate. More accurately, it is the entropy production due to the exchange of radiation. The remaining terms are the rate of internal production of entropy. If phase changes are assumed to take place at equilibrium, the entropy production rate dS_i/dt reduces to

$$\frac{dS_I}{dt} = \int \frac{Q_H}{T}\, dV + \int \underline{F}_H \cdot \nabla\left(\frac{1}{T}\right) dV \ .$$

An estimate of the order of magnitude of these two terms, using globally averaged values for the variables and length scales, suggests that the first term is of the order of 10 times larger than the second term.

Essex (1984a) has noted that equation (6.1) neglects the entropy of the radiation interacting with the climate system. The flux of entropy associated with an energy flux $\underline{F}$ of blackbody radiation is

$$\underline{H} = \frac{4}{3}\left(\frac{\underline{F}}{T}\right) ,$$

where T is the temperature of the matter in equilibrium with the radiation. Thus, for the Earth,

$$\underline{H} = \frac{4}{3}\left(\frac{F_S}{T_s} + \frac{F_L}{T}\right) \quad ,$$

where T_S is the effective temperature of the sun and T the effective temperature of the Earth. The correct entropy balance of the climate system is obtained by adding the term

$$- \int \nabla \cdot \underline{H} \, dV$$

to the right hand side of equation (6.7). The extra term can be regarded as an extra contribution to the entropy exchange rate. For the simple climate model used in section 4, the modified minimum entropy exchange condition would require

$$\int_0^1 \frac{Q_0 S(x) a(x,x_s) - I(x)}{T(x)} dx - \frac{4}{3} \int_0^1 \left[\frac{Q_0 S(x) a(x,x_s)}{T_s} - \frac{I(x)}{T(x)} \right] dx$$

to be a minimum. An attempt to use this condition to close the model failed to give any solutions at all! Essex (1984a) has been critical of climate modellers for neglecting the contribution of the radiation to the entropy balance. However, even if there exists an extremal condition for the climate system which involves the entropy, there appears to be no fundamental reason why this <u>must</u> involve the radiation contribution. Indeed, the failure of the modified minimum entropy exchange principle to close the simple model of section 4 suggests that the radiation contribution may be irrelevant.

Essex (1984b) had proved that for an atmosphere in <u>radiative equilibrium</u>, the entropy production should be a minimum. Unfortunately, the

Earth's atmosphere is very far from radiative equilibrium and so it seems unlikely that this result will be applicable to the Earth's climate system. It may, however, have applications in the study of some other planetary atmospheres.

7. CONCLUSIONS

(i) There is evidence from energy balance climate models of varying complexity that the Earth's climate system obeys the following law:

Subject to the constraint of global energy balance, the entropy exchange is at a local minimum with respect to the temperature.

This result is based only on the success of its applications and as yet does not have any sound, *a priori*, physical basis.

(ii) Study of simple climate models suggests that other extremal conditions may work equally well. It may be, therefore, that the underlying, fundamental constraint does not involve the entropy explicitly.

(iii) In applications, the minimum entropy exchange principle should be regarded as a closure condition. It is an alternative to closure by parameterization of, for example, the eddy meridional heat flux.

(iv) *How could the result be proved*? Suggested approaches are:

(a) The law is probably some kind of stability condition. Investigations of the global stability characteristics of climate models may help to solve the problem.

(b) Another possibility is an extension of Essex's (1984b) minimum entropy production theorem for radiative equilibrium, although this would appear to be less likely than (a).

REFERENCES

Budyko, M.I., 1969: The effect of solar radiation variations on the climate of the Earth, Tellus 21, 611-619.

Charney, J.G. and J.G. DeVore, 1979: Multiple flow equilibria in the atmosphere and blocking, J. Atmos. Sci. 36, 7, 1205-1216.

Dutton, J.A., 1973: The global thermodynamics of atmospheric motion, Tellus 25, 2, 89-110.

Dutton, J.A. and D.R. Johnson, 1967: The theory of available potential energy and a variational approach to atmospheric energetics, Advances in Geophysics, Vol. 12, ed. H.E. Landsberg and J. van Mieghem, Academic Press, New York, pp. 334-436.

Essex, C., 1984a: Radiation and the irreversible thermodynamics of climate, J. Atmos. Sci. 41, 12, 1985-1991.

Essex, C., 1984b: Maximum entropy production in the steady state and radiative transfer, Astrophys. J. 285, 279-293.

Gill, A.E., 1982: Atmosphere-Ocean Dynamics, Academic Press, New York.

Golitsyn, G.S. and I.I. Mokhov, 1978: Stability and extremal properties of climate models, Izvestiya Atm. Oc. Physics 14, 4, 271-277.

Golitsyn, G.S., 1983: Almost empirical approaches to the problem of climate, its variations and fluctuations, Advances in Geophysics, Vol. 25, ed. B. Saltzman, Academic Press, New York , pp. 85-115.

Grassl, H., 1981: The climate at maximum entropy production by meridional atmospheric and oceanic heat fluxes, Quart. J. R. Met. Soc. 107, 153-166.

Lin, C.A., 1982: An extremal principle for a one-dimensional climate model, Geophys. Res. Lett. 9, 6, 716-718.

Lorenz, E.N., 1955: Available potential energy and the maintenance of the general circulation, Tellus 7, 2, 157-167.

Lorenz, E.N., 1960: Generation of available potential energy and the intensity of the general circulation, Dynamics of Climate, ed. R.L. Pfeffer, Pergamon, Oxford, pp. 86-92.

Mobbs, S.D., 1982: Extremal principles for global climate models, Quart. J. R. Met. Soc. 108, 535-550.

Mokhov, I.I. and G.S. Golitsyn, 1978: Variational estimate of climate system stability in elementary models, Izvestiya Atm. Oc. Physics 14, 6, 426-431.

Nicolis, G. and C. Nicolis, 1980: On the entropy balance of the earth-atmosphere system, Quart. J. R. Met. Soc. 106, 691-706.

Noda, A. and T. Tokioka, 1983: Climates at minima of the entropy exchange rate, J. Met. Soc. Japan 61, 6, 894-908.

North, G.R., 1975: The theory of energy balance climate models, J. Atmos. Sci. 32, 11, 2033-2043.

North, G.R., L. Howard, D. Pollard and B. Wielicki, 1979: Variational formulation of Budyko-Sellers climate models, J. Atmos. Sci. 36, 2, 255-259.

North, G.R., R.F. Cahalan and J.A. Coakley, Energy balance climate models, Rev. Geophys. Space Phys. 19, 1, 91-121.

Paltridge, G.W., 1975: Global dynamics and climate - a system of minimum entropy exchange, Quart. J. R. Met. Soc. 101, 475-484.

Paltridge, G.W., 1978: The steady-state format of global climate, Quart. J. R. Met. Soc. 104, 927-945.

Paltridge, G.W., 1979: Climate and thermodynamic systems of maximum dissipation, Nature 279, 630-631.

Paltridge, G.W., 1981: Thermodynamic dissipation and the global climate system, Quart. J. R. Met. Soc. 107, 531-547.

Paltridge, G.W. and C.M.R. Platt, 1976: Radiative Processes in Meteorology and Climatology, Elsevier, Amsterdam.

Rogers, C.D., 1976: Comments on Paltridge's 'minimum entropy exchange' principle, Quart. J. R. Met. Soc. 102, 455-457.

Schulman, L.L., 1977: A theoretical study of the efficiency of the general circulation, J. Atmos. Sci. 34, 4, 559-580.

Sellers, W.D., 1965: Physical Climatology, Chicago University Press, Chicago.

Sellers, W.D., 1969, A global climate model based on the energy balance of the earth-atmosphere system, J. Appl. Met. 8, 392-400.

Shutts, G.J., 1981: Maximum entropy production states in quasi-geostrophic dynamical models, Quart. J. R. Met. Soc. 107, 503-520.

HEAT TRANSFER BY THERMAL CONVECTION IN A ROTATING FLUID SUBJECT TO A HORIZONTAL TEMPERATURE GRADIENT

RAYMOND HIDE

Geophysical Fluid Dynamics Laboratory
Meteorological Office (Met O 21)
Bracknell, Berkshire RG12 2SZ
England, UK

1. INTRODUCTION

Laboratory studies and associated analytical and numerical investigations of thermally-induced motions in a rapidly-rotating fluid of low viscosity are of intrinsic interest in fluid dynamics, and they also bear on a wide variety of natural flow phenomena in atmospheres and oceans. Advective heat flow perpendicular to the rotation axis $r = 0$ (where (r, Φ, z) are the cylindrical polar coordinates of a general point in a frame of reference that rotates with angular speed Ω about $r = 0$) requires inter alia that u_r, the r-component of the Eulerian relative flow velocity $\underline{u}$, shall be non-zero. Since $u_r = (2\Omega \bar{\rho} r)^{-1} \partial p/\partial\Phi$ (where p denotes pressure and $\bar{\rho}$ mean density) nearly everywhere when Ω is so large that the geostrophic relationship (see equation (2) below) holds throughout most parts of the fluid, in rapidly rotating systems advective heat transfer must be associated with departures from axial symmetry in the pattern of flow. Theory indicates and experiments amply confirm that strong departures from axial symmetry are "generic" features of the patterns of thermally-induced flow in rapidly rotating fluids, even when the boundary conditions are axisymmetric.

There have been many experiments on thermal convection in a fluid annulus that rotates about a vertical axis and is subject to axisymmetric impressed differential heating and cooling with strong horizontal components and, in many cases, no impressed vertical component. These include careful determinations of the dependence of heat transfer across the annulus on Ω

J. Willebrand and D. L. T. Anderson (eds.), Large-Scale Transport Processes in Oceans and Atmosphere, 325–336.

and other factors, including sloping endwalls which produce effects that are analogous in some respect to the latitudinal variation of the vertical component of Coriolis parameter that features prominently in theories of large-scale flow phenomena in atmospheres and oceans. We note here in passing that systematic studies of effects due to departure from axial symmetry in the boundary conditions have also been instructive. So far as heat transfer is concerned, the most dramatic enhancement is produced simply by introducing a rigid radial barrier connecting the inner wall of the annulus to the outer wall at all levels. Any tendency for rotation to reduce heat transfer is largely annulled by the presence of the barrier, across which at any station (r,z) can be supported the non-zero geostrophic pressure drop

$$\int_{\Phi_2}^{2\pi-(\Phi_2-\Phi_1)} 2\Omega\,\bar{\rho}\,r\,u_r\,d\Phi \qquad (1)$$

(where $\Phi_1 < \Phi < \Phi_2$, is the range of azimuth occupied by the barrier) and concomitant temperature drop, associated with non-zero geostrophic values of u_r that do not depend strongly on Φ.

It was impossible in a single lecture to do more than indicate in general terms how rotation affects thermal convection in a rotating fluid subject to a horizontal temperature gradient. Details can be found by consulting the literature cited at the end of this summary.

2. GEOSTROPHY

It is useful to consider certain general properties of the motion of a fluid of low viscosity that departs but little from solid body rotation with steady angular velocity Ω when typical timescales of the relative motion greatly exceed $2\,\Omega^{-1}$. Such motion is "geostrophic" nearly everywhere, satisfying

$$2\,\rho\,\underline{\Omega} \times \underline{u} + \nabla p - \rho\nabla V = 0. \tag{2}$$

Here $\underline{u}$ is the Eulerian relative flow velocity, ρ denotes density, p pressure, and ∇V is the acceleration due to gravity and centripetal effects. Equation (2) is the leading approximation to the full equation of motion

$$2\rho\,\underline{\Omega} \times \underline{u} + \nabla p - \rho\,\nabla\,V = \underline{A} \tag{3}$$

where

$$\underline{A} \equiv -\rho\,D\underline{u}/Dt - \rho\,\underline{r} \times d\underline{\Omega}/dt + \underline{F} \tag{4}$$

the "ageostrophic" contribution. It is valid in regions where the Coriolis term $2\rho\,\underline{\Omega} \times \underline{u}$ greatly exceeds the relative acceleration term $\rho\,D\underline{u}/Dt \equiv \rho\,[\partial\underline{u}/\partial t + (\underline{u}\cdot\nabla)\,\underline{u}]$ (where t denotes time), the "precessional" term $\rho\,\underline{r} \times d\underline{\Omega}/dt$ (where $\underline{r}$ is the position vector of a general point in a frame which rotates with angular velocity $\underline{\Omega}(t)$ relative to an inertial frame), and the term $\underline{F}$ which represents the viscous and other forces (e.g. Lorentz forces in magnetohydrodynamic systems) per unit volume.

Now equation (2) is mathematically degenerate, for its order is lower than that of the full equation of motion (3), and it cannot therefore be solved under the complete set of boundary conditions. For this to be possible it is necessary to include ageostrophic terms in the analysis, so that the flow cannot be geostrophic everywhere. Hence:

> regions of highly ageostrophic flow occuring not only on the boundaries of the system but also in localised regions (detached shear layers, jet streams, etc.) of the main body of the fluid are necessary concomitants of geostrophic motion. (5)

Within these highly ageostrophic regions, $\underline{A}$ is comparable in magnitude with $2\rho\underline{\Omega} \times \underline{u}$ and the corresponding relative vorticity $\underline{\xi} \equiv \nabla \times \underline{u}$ can be comparable with or even exceed 2Ω in magnitude. Many examples of such vorticity concentrations are found in nature and in laboratory systems, such as those considered below. They are often associated with steep gradients of temperature ("thermal fronts"), as in the case of jet-streams and western boundary currents found in atmospheres and oceans.

Equations (2) and (3) lead directly to another important finding (see above), namely that:

> the motion of fluid of low viscosity that departs only slightly from steady rapid rigid-body rotation will not in general be symmetric about the rotation axis, even when the boundary conditions are axisymmetric. (6)

This simple result elucidates of the occurence of large-scale non-axisymmetric disturbances in the Earth's atmosphere and other natural systems, and it receives direct verification from the experiments outlined below. It can be deduced as follows. In cylindrical coordinates (r,Φ,z) where $\underline{\Omega} = (0,0,\Omega)$, the Φ-component of equation (3) gives

$$\rho\, u_r = (2\Omega)^{-1}\,\{-r^{-1}\,\partial p/\partial\Phi + A_\Phi\} \tag{7}$$

if $u = (u_r, u_\Phi, u_z)$ and (A_r, A_Φ, A_z), since $\partial V/\partial\Phi = 0$ by the assumption of axial symmetry in the boundary conditions. Now over any cylindrical surface of radius r, the rate of advective transport M(r,t;Q) or any quantity Q (per unit mass), such as heat, angular momentum,etc. is given by

$$M(r,t;Q) \equiv \int_{z_1}^{z_2} \int_{0}^{2\pi} \rho\, r\, u_r\, Q\, d\Phi\, dz$$
$$= \frac{1}{2\Omega} \int_{z_1}^{z_2} \int_{0}^{2\pi} \{- \frac{\partial p}{\partial \Phi} + r\, A_\Phi \}\, Q d\Phi dz \,, \qquad (8)$$

where $z_1 < z < z_2$ is the axial extent of the surface. Since the ageostrophic contribution A_ϕ to equation (7) decreases rapidly with increasing Ω, advective transport perpendicular to the axis of rotation, as measured by M(r,t;Q) will be negligible unless the flow pattern departs significantly from axial symmetry. In the axisymmetric case we have $\partial p / \partial \Phi = 0$, and M(r,t;Q) is consequently of the order of the small ageostrophic contribution.

This argument is the basis of (6). There may be singular cases when the flow remains axisymmetric and in consequence any advective transport perpendicular to the rotation axis is negligible. Indeed, such cases can be realised in the laboratory by taking certain special precautions, but the general conclusion to be drawn from the laboratory studies outlined below is that (6) is a correct inference from the geostrophic equation.

Order of magnitude estimates of $\underline{A}$ given by equation (4) in cases when $d\Omega/dt = 0$ and $\underline{F} = \nu\rho \nabla^2 \underline{u}$, where ν denotes kinematic viscosity, indicate that geostrophic balance can be expected in thermally-driven systems when Ω is sufficiently large to satisfy both

$$\Omega >> \nu/l^2 \text{ and } \Omega >> \left(\frac{g\, \alpha\, \Delta T}{l} \right)^{1/2} \left(\frac{\nu}{\kappa} \right)^{1/2} . \qquad (9a,b)$$

Here l is a typical length, ΔT a typical impressed temperature contrast, α the thermal coefficient of cubical expansion and κ the coefficient of thermometric conductivity or it's radiative equivalent.

3. REGIMES OF THERMAL CONVECTION IN A ROTATING FLUID ANNULUS

The earliest labóratory experiments on thermal convection in a rotating liquid cylindrical annulus which rotates with angular speed Ω about its vertical axis of symmetry and is subject to axisymmetric applied differential heating showed that the general character of the flow is largely determined by two dimensionless parameters, namely

$$\Theta \equiv g\, d\, \Delta\rho / \bar{\rho}\, \Omega^2 (b-a)^2 \tag{10}$$

and

$$I \equiv 4\, \Omega^2\, L^4 / \nu^2 \tag{11}$$

Here g denotes the acceleration of gravity, d the fluid depth, b and a are the radii of the outer and inner side-walls; L depends on b, a and d and has the dimensions of length (and is equal to $(b-a)^{5/4}/d^{1/4}$ over a wide range of conditions). $\Delta\rho$ is a measure of the impressed horizontal temperature contrast associated with the applied differential heating and cooling, which can be taken as the difference between the maximum and the minimum density associated with the impressed temperature field T_0.

The side-walls are held at temperatures T_b and T_a respectively and the fluid is subject to diabatic heating per unit volume at a rate $q\bar{\rho}$ times the specific heat c. The radial variation of the impressed temperature $T_0(r)$ takes the simple form

$$T_0(r) = (b-a)^{-1} \left[(r-a)\, T_b + (b-r)\, T_a\right] - q\, \frac{(r-a)\,(r-b)}{2\kappa} \tag{12}$$

(where κ is the thermometric conductivity of the fluid) when $(b-a) \ll \frac{1}{2}(b+a)$ and q is constant. When there are no heat sources or sinks within the fluid, we have the most extensively studied case of all, often referred to as the "wall heated" case. Then $q = 0$, equation (12) becomes

$$T_0(r) = (b-a)^{-1} \left[(r-a)T_b + (b-r)T_0 \right] \quad \left\{ \begin{array}{l} \text{wall heated} \\ \text{and} \\ \text{wall cooled} \end{array} \right\} ; \tag{13}$$

and $\Delta\rho$ is $\left[\rho(T_b) - \rho(T_a)\right]$.

When $q \neq 0$ we have "internally heated" systems (which when $q < 0$ correspond to "internally cooled" systems). There are three particularly interesting limiting cases of internally heated systems, namely the "inner wall cooled" case when the outer wall is a thermal insulator, so that $dT_0/dr = 0$ at $r = b$ and equation (12) becomes

$$T_0(r) = T_a + \frac{q(r-a)(2b-a-r)}{2\kappa} \quad \left\{ \begin{array}{l} \text{internal heating;} \\ \text{inner wall cooled} \end{array} \right\} ; \tag{14}$$

the "outer wall cooled" case, when the inner wall is a thermal insulator so that $dT_0/dr = 0$ at $r = a$ and equation (12) becomes

$$T_0 = T_a - \frac{q(r-a)^2}{2\kappa} \quad \left\{ \begin{array}{l} \text{internal heating;} \\ \text{outer wall cooled} \end{array} \right\} ; \tag{15}$$

and the "both walls cooled" case, when heat is removed at the same rate <u>via</u> both outer and inner side-walls simultaneously, so that dT_0/dr at $r = a$ is equal to $-dT_0/dr$ at $r = b$ and equation (12) becomes

$$T_0 = T_a - \frac{q(r-a)(r-b)}{2\kappa} \quad \left\{ \begin{array}{l} \text{internal heating;} \\ \text{both walls cooled} \end{array} \right\} . \tag{16}$$

In the first two of these internally heated cases, $\Delta\rho$ is the same as in the wall-heated case, $\left[\rho(T_b) - \rho(T_a)\right]$, but in the "both walls cooled" case we have $\Delta\rho = \left[\rho(T_0(r = \tfrac{1}{2}(a+b)) - \rho(T_a)\right]$.

Careful studies of the principal spatial and temporal characteristics of flows over a wide range of precisely specified and carefully controlled experimental conditions led to the discovery of several fundamentally different free types of flow, only one of which is symmetrical about the axis of rotation. When J is less than a certain critical value of about 2×10^5, viscosity ensures that the motion is essentially axisymmetric for all values of Θ. However, when J exceeds this critical value there exists a range of Θ, namely $\Theta_R > \Theta > \Theta_L$ (where Θ_R and Θ_L depend on J) within which highly non-axisymmetric flow occurs. These non-axisymmetric motions are either "regular" or "irregular" depending on the values of Θ and J. The regular flows, which occur when $\Theta_R > \Theta > \Theta_I$ are spatially periodic in ϕ and often exhibit periodic temporal "vacillation" in amplitude, shape or even wavenumber, but under certain conditions these periodic variations are so slight that, apart from a steady azimuthal drift of the flow pattern relative to the walls of the apparatus, the flow is virtually steady. In sharp contrast to this behaviour, irregular flows (which occur within the range $\Theta_I > \Theta > \Theta_L$) exhibit complicated aperiodic fluctuations in both space and time. Axisymmetric flow occurs when $\Theta > \Theta_R$ or $\Theta < \Theta_L$.

The non-axisymmetric flow regimes are manifestations of "sloping" or "slantwise" convection and the observed dependence of Θ_R and Θ_L on J and other parameters has been successfully interpreted on the basis of baroclinic instability theory suitably modified to take viscous boundary layers into account. Barotropic instability is the likely cause of the transition from regular to irregular non-axisymmetric flow that occurs at $\Theta = \Theta_I$.

Heat flow determinations show that the tendency for advective heat transfer to decrease with increasing Ω is pronounced within the axisymmetric and irregular non-axisymmetric flow regimes found, respectively, where $\Theta > \Theta_R$ and where $\Theta_I > \Theta > \Theta_L$. However, within the regular non-axisymmetric regime (where $\Theta_R > \Theta > \Theta_I$) advective heat transfer is virtually independent of Ω and typically about 0.9 times the "non-rotating" value.

4. PATTERNS OF REGULAR NON-AXISYMMETRIC FLOW

The mathematical equations governing the flows we are considering are the equation of motion, equation (3), together with the equations of continuity and state for a liquid, respectively

$$\nabla \cdot \underline{u} = 0 \tag{17}$$

and

$$\rho = \hat{\rho}\ (1-\alpha\ (T-\hat{T})) \tag{18}$$

(where T denotes temperature, α thermal coefficient of cubical expansion (for convenience here taken as constant), and is the density at the reference temperature T), and the equation of heat transfer

$$\frac{\partial T}{\partial t} + (\underline{u}\cdot\nabla)\ T = \kappa\nabla^2\ T + q\ . \tag{19}$$

Equations (12) to (16) are, of course, solutions of equation (19) when $\underline{u} = 0$. Across any cylindrical vertical surface constant, the rate of heat transfer (made up of a conductive component and a convective (or "advective") component when radiative effects are negligible) is given by

$$H\ (r,t) = \int_{-1/2d}^{1/2d} \int_{0}^{2\pi} \rho c \left[\kappa \frac{\partial T}{\partial r} + u_r\ T \right] r\ d\phi\ dz \tag{20}$$

if the fluid extends in the axial direction from $z = -\frac{1}{2}d$ to $z = \frac{1}{2}d$. As anticipated above (see equation (8)), the geostrophic contribution to the advective heat flow term on the right-hand side of equation (20) would vanish if the flow were axisymmetric, since the geostrophic part u_r is proportional to $\partial p / \partial \Phi$. As we have already indicated, this result points to the <u>raison d'être</u> of the non-axisymmetric regimes of flow found when is sufficiently large; geostrophic flow cannot convey heat perpendicularly to the axis of rotation unless the flow is non-axisymmetric!

The boundary conditions on $\underline{u}$ under which the governing equations must be satisfied are that $\underline{u} = 0$ at a rigid bounding surface, and that the stress should vanish at a free surface. The thermal boundary conditions require continuity of heat flow which, at a bounding surface, is purely conductive and proportional to $\kappa \nabla T$. When the boundary conditions on the side-walls at $r = r^*$ where $r = a$ or b are combined with the geostrophic relationship given by equation (2) and used in conjunction with the standard relationship for the radial flow in the Ekman boundary layers on $z = -\frac{1}{2} d$ and $z = \frac{1}{2} d_X$ to evaluate the radial heat flow at $r = r^*$ (see equation (20)), expressions for $H(r^*t)$ can be obtained. These show that

$$H(r^*,t) = (\text{negative definite quantity}) \cdot \Gamma(r^*, \tfrac{1}{2}d, t) \tag{21}$$

if

$$\Gamma(r^*,z,t) \equiv \int_0^{2\pi} U_\Phi (r^*,\Phi,z,t)\, r\, d\phi , \tag{22}$$

where U_Φ (r^*, Φ,z,t) is the value of u_Φ evaluated just outside the viscous boundary layer on $r = a$ or $r = b$ as the case may be.

This relationship between the heat flow at a side-wall and the line integral of the tangential velocity near the side-wall embodies the arguments that have been used to provide a general interpretation of the upper-level flow pattern in the case when $q = 0$ everywhere (see equation (13)), heat being introduced into the system <u>via</u> one of the side-walls and removed <u>via</u> the other side-wall. The corresponding impressed radial temperature gradient has the same sign at all values of r and the upper level pattern of motion in the regular flow regime consists of a single jet-stream meandering in a wavy pattern between the bounding cylinders, with a positive (i.e. "westerly") azimuthal component when heat enters <u>via</u> the outer side-wall and leaves <u>via</u> the inner side-wall (so that the impressed radial temperature gradient is positive), and negative ("easterly") when the overall radial heat transfer is in the opposite direction, from the inner to the outer cylinder.

As a further test of (21), experiments have been carried out using internal heating, so that the term q in equation (19) is not equal to zero. This was done by passing an alternating electric current through the fluid. Heat could be removed via the inner side-wall, the outer side-wall, or both side-walls. The observed upper surface flow patterns were found to be in good agreement with predictions for these three cases made on the basis of equation (22). In the cases where heat is removed via one side-wall only, $H(r^*,t)$ vanishes at the other side-wall and, by equation (22), the quantity $\Gamma(r^*,\frac{1}{2}d,t)$ vanish. For this to happen, $U_\Phi(r^*, \Phi,\frac{1}{2}d,t)$ will be positive at some values of Φ and negative at others (or zero at all values of Φ, as in the axisymmetric regime that occurs when $\Theta > \Theta_R$). This requirement can be satisfied by adding closed eddies to the wavy pattern found in the cases when $q = 0$.

The most striking case of all studied in the experiments is that when heat is removed via both side-walls, implying that the impressed radial temperature gradient changes sign near mid-radius. The corresponding upper level flow consists of several separate closed eddies, each circulating anticyclonically (when $q > 0$), in accordance with equation (22), with the horizontal flow confined to a narrow jet-stream at the periphery of each eddy.

REFERENCES

The results summarized in this lecture are but a few of the extensive findings of many investigations of thermal convection in rotating fluids carried out during the past thirty-five years and described in numerous reports in the literature. References to these reports can be found in the bibliographies of the following recent papers:

Hide, R., 1985: Thermal convection in a rotating fluid subject to a horizontal temperature gradient, pp. 159-171 in: Turbulence and predictability in geophysical fluid dynamics and climate dynamics. - LXXXVIII. Corso, Soc. Italiana di Fisica - Bologna - Italy.

Hide, R,. 1984: Presidential address: the giant planets; Galileo Galilei to Project Galileo. - Quart. J. R. Astron. Soc. 25, 232-247.

Hide, R., 1984: Rotating fluids in geophysics and planetary physics. - Geo Journal 8.4 , 365-377.

Hide, R., 1984: On the dynamics of rotating fluids and planetary atmospheres: a summary of some recent work, pp. 79-85 in: Predictability of fluid motions, eds. G. Holloway and B.J. West, New York, American Institute of Physics (xvi + 612 pages).

Hignett, P., 1985: Characteristics of amplitude vacillation in a differentially heated rotating fluid annulus. - Geophys. Astrophys. Fluid Dyn. 31, 247-281.

Hignett, P., A.A. White, R.D. Carter, W.D.N. Jackson and R.M. Small, 1985: A comparison of laboratory measurements and numerical simulations of baroclinic wave flows in a rotating cylindrical annulus. - Quart. J. R. Met. Soc. 111, 131-154.

Read, P.L. and R. Hide, 1983: Long-lived eddies in the laboratory and in the atmospheres of Jupiter and Saturn. - Nature 302, 126-129.

Read, P.L. and R. Hide, 1984: An isolated baroclinic eddy as a laboratory analogue of the Great Red Spot on Jupiter. - Nature 308, 45-48.

SOME ASPECTS OF TURBULENT DIFFUSION

MARCEL LESIEUR

Institut de Mécanique de Grenoble
Domaine Universitaire
38402 Saint-Martin d'Héres Cedex, France

1. INTRODUCTION

One of the most striking manifestations of turbulence in a flow is the considerable enhancement of its diffusion properties, compared with molecular diffusion. The quantities transported by the flow can be passive or active. If one is interested in ocean or atmosphere dynamics, these quantities are for instance the temperature, momentum, potential vorticity, Lagrangian tracers, error (this entity will be defined later),and others.

Turbulent diffusion in the atmosphere or the oceans plays a role at all scales, from the smallest three-dimensional dissipative scales to the largest quasi-two-dimensional, meso- or synoptic scales. The diffusion laws differ of course according to scale and to the diffusing quantity considered.

The scale of the motion can generally be associated with characteristic values of nondimensional parameters such as the Rossby number $U/2 \cdot \Omega \cdot L$ and the Froude number $U/N \cdot L$. We choose to classify the different turbulent diffusion problems one will envisage here, from large to small Rossby and (or) Froude numbers. This lecture will then focus on the four following mathematical models, corresponding to approximations of the Navier-Stokes equations when these parameters decrease from high to low values, and the "integral scale" L goes from small three-dimensional scales to large planetary scales.

J. Willebrand and D. L. T. Anderson (eds.), Large-Scale Transport Processes in Oceans and Atmosphere, 337–357.

(a) Three-dimensional Navier-Stokes equations (constant density)

$$\frac{\partial \vec{u}}{\partial t} + \vec{u} \cdot \vec{\nabla} \vec{u} = - \frac{1}{\rho_o} \vec{\nabla} p + \nu \nabla^2 \vec{u} \; ; \; \vec{\nabla} \cdot \vec{u} = 0$$

$$\frac{D \rho'}{Dt} = \kappa \nabla^2 \rho'$$

Notice that here the density fluctuation ρ' behaves like a passive scalar since it does not enter the dynamical equation. These equations apply to small scale three-dimensional turbulence in both atmosphere and ocean.

(b) Boussinesq equations

$$\frac{\partial \vec{u}}{\partial t} + \vec{u} \cdot \vec{\nabla} \vec{u} = - \frac{1}{\rho_o} \vec{\nabla} p' + \frac{\rho'}{\rho_o} \vec{g} + \nu \nabla^2 \vec{u} ; \; \vec{\nabla} \cdot \vec{u} = 0$$

$$\frac{D \rho'}{Dt} + w \frac{d \bar{\rho}}{dz} = \kappa \nabla^2 \rho'$$

Now the density fluctuation is <u>not</u> a passive scalar. These equations apply in particular to stably stratified turbulence which plays a fundamental role in the vertical mixing in the ocean and in mesoscale atmospheric processes.

(c) Quasi-geostrophic equation

$$\frac{D_H}{D_t}\left(-\nabla_H^2 \psi - \mu^2 \frac{\partial^2\psi}{\partial z^2} + \beta y\right) = \nu \nabla_H^2 (-\nabla_H^2 \psi)$$

The quantity conserved[1] following the fluid motion is the potential vorticity (active). These equations apply to large scale dynamics in the atmosphere and the ocean.

(d) Two-dimensional Navier-Stokes equations

$$\frac{D_H}{Dt}(-\nabla_H^2 \psi) = \nu \nabla_H^2 (-\nabla_H^2 \psi)$$

$$\frac{D_H}{Dt}\theta = \kappa \nabla_H^2 \theta$$

Two quantities are conserved following the fluid motion: the vorticity (active) and a passive one θ, which could be the temperature, or the density of some tracer. These equations are a simplification of the quasi-geostrophic equation if the horizontal kinetic energy supply due to baroclinic instability is modelled by some external forcing at the internal radius of deformation.

We will consider successively the kinematics, phenomenology, and analytical theories of isotropic turbulence, and see how it applies to the problems a), b), c) and d) listed above.

---

1) The word "conserved" applies only to nonlinear terms: indeed there is a potential vorticity dissipation due to molecular dissipation.

2. KINEMATICS OF ISOTROPIC TURBULENCE

Let us take the velocity field $\vec{u}(\vec{x},t)$ to be a random function statistically homogeneous and isotropic, and let <...> be the ensemble average operator. Defining

$$U_{ij}(\vec{r},t) = \langle u_i(\vec{x},t)\, u_j(\vec{x}+\vec{r},t) \rangle$$

as the velocity correlation tensor and

$$\hat{U}_{ij}(\vec{k},t) = (1/2\pi)^d \int e^{i\vec{K}\cdot\vec{r}}\, U_{ij}(\vec{r},t)\, d\vec{r}$$

as the spectral tensor ("d" denotes the dimension of space) the following properties can be derived

$$\langle \hat{u}_i(\vec{k}',t)\, \hat{u}_j(\vec{k},t) \rangle = \hat{U}_{ij}(\vec{k},t)\, \delta(\vec{k}+\vec{k}')$$

$$\hat{U}_{ij}(\vec{k},t) = (1/2)\left[(\delta_{ij} - k_i k_j/k^2)\, E(k,t)/2\pi k^2 + i\varepsilon_{ij\lambda} k_\lambda\, H(k,t)/2\pi k^4\right]$$

where E(k,t) is the kinetic energy spectum

$$(1/2) \langle \vec{u}^2(\vec{x},t) \rangle = \int_0^\infty E(k,t)\, dk ,$$

$k^2 E(k,t)$ is the enstrophy spectrum

$$(1/2) \langle (\vec{\nabla}\times\vec{u})^2 \rangle = \int_0^\infty k^2 E(k,t)\, dk$$

and H(k,t) is the helicity spectrum

$$(1/2) \langle \vec{u}\cdot(\vec{\nabla}\times\vec{u}) \rangle = \int_0^\infty H(k,t)\, dk$$

Obviously helicity is zero in 2 dimensions, or if we assume invariance with respect to mirror reflection. Helicity is known to play an important role in MHD turbulence for the generation of magnetic fields (dynamo effect, Moffatt, 1978), and could also be of some importance in tornado dynamics (Lilly, 1983).

The evolution rates of energy and enstrophy are given by

$$(d/dt) \int_0^\infty E(k,t)\, dk = -2\nu \int_0^\infty k^2 E(k,t)\, dk$$

$$(d/dt) \int_0^\infty k^2 E(k,t)\, dk = -2\nu \int_0^\infty k^4 E(k,t)\, dk + \int_0^\infty k^2 T(k,t)\, dk$$

where T(k,t) is the energy transfer function. The last term on the r.h.s. of the enstrophy balance represents the production of enstrophy due to the mean stretching of vortex filaments by the turbulence. This term is zero in two dimensions.

3. PHENOMENOLOGY OF THREE-DIMENSIONAL TURBULENCE

This analysis applies to case a) of section 1. As we saw before the kinetic energy dissipation rate ε is given by

$$\varepsilon = 2\nu \int_0^\infty k^2 E(k)\, dk$$

It can be shown by dimensional arguments that in the kinetic energy cascade, the energy spectrum is

$$E(k) = f(k, \varepsilon) = \text{const.}\ \varepsilon^{2/3} k^{-5/3}$$

which is the Kolmogorov law (Kolmogorov, 1941). This can also be derived, using the so-called Obukhov theory, by introducing a constant flux of kinetic energy equal to $\varepsilon = k\, E(k)/\tau(k)$, where $\tau(k)$ is a nonlinear time characteristic of the relaxation of triple velocity correlations towards a quasi-equilibrium state. The simplest expression for $\tau(k)$, also obtained phenomenologically, is

$$\tau(k) = [k^3 E(k)]^{-1/2}$$

This theory applies also to the helicity and to a passive scalar. If η is the passive scalar flux and $E_\Theta(k)$ the scalar variance spectrum, we have

$$E_\theta(k) \approx \eta \varepsilon^{-1/3} k^{-5/3}$$

This law is valid in the so-called "inertial convective range", when molecular viscous and conductive effects can be neglected. The same kind of analysis can be performed in the physical space as well. Let us consider an "eddy" with typical rotational velocity v_r and radius r, then write that the energy dissipation rate ε is proportional to $v_r^2/(r/v_r)$. We obtain

$$v_r \approx (\varepsilon r)^{1/3}$$

which is exactly the Kolmogorov law if we replace r by k^{-1} and r v_r^2 by E(k). If a pair of Lagrangian tracers were separated by r, they would disperse with a dispersion coefficient σ given by

$$\sigma \approx (1/2)\,(d/dt)\, r^2 \approx r v_r \approx \varepsilon^{1/3} r^{4/3}$$

The latter law is known as Richardson's law (Richardson, 1926). We should mention that Kolmogorov's law is extremely well verified experimentally (Grant et al., 1962; Gargett et al., 1984).

4. PHENOMENOLOGY OF TWO-DIMENSIONAL TURBULENCE

This section applies to case d) of section 1. Since there is no enstrophy production (in the absence of external forces), enstrophy is bounded by its initial value, and the energy dissipation rate goes to zero with ν. A transfer of energy towards the small scales must then be accompanied by a much higher transfer of the quantity towards the large scales, in order to preserve the "conservation" of enstrophy. This has been shown in particular by Fjortoft (1953). The energy cannot cascade to the small scales and is trapped in the large scale. Therefore an ultraviolet cascade (towards large k) can only be envisaged for the enstrophy: indeed the vorticity of the "meso-scale" eddies, which is conserved following their motion, will be transferred to the small scales because of the shearing action of the large eddies. Indicating with ß the enstrophy dissipation rate, the same phenomenological arguments used in the three-dimensional case for the energy can now be used for the enstrophy cascade in two dimensions. One obtains:

$$E(k) = f(k,\beta) \approx \text{const } \beta^{2/3} k^{-3}$$

Actually, two types of homogeneous two-dimensional turbulences can be considered, depending on the existence of a forcing:

i) If there is a stationary injection of energy and enstrophy at a given fixed wavenumber, the energy injected at the forcing with the rate ε will not be able to dissipate by viscosity appreciably in the small scales (no direct cascade), and will then be transferred to larger and larger scales. Assuming then a stationary quasi-equilibrium energy spectrum with a constant flux of energy ε, one obtains again by dimensional arguments the so-called inverse energy cascade: the spectrum is still $\approx \varepsilon^{2/3} k^{-5/3}$, but the energy flux is now negative (Kraichnan, 1967; Leith, 1968; Pouquet et al., 1975).

ii) The freely evolving case has been studied by Batchelor (1969) and Rhines (1975). A recent EDQNM analysis (Lesieur et al.,1985, see section 5) using the fact that $E(k) \approx k^3$ as $k \to o$, and

$E(k) \approx \beta^{2/3}\ k^{-3}\ [\mathrm{const} + \ln(k/k_I)]^{-1/3}$ (where k_I is the wavenumber characteristic of the peak of the energy spectrum), justifies Batchelor's choice (1969) $k_I \approx (\nu t)^{-1}$ (see Lesieur et al., 1985).

Let us now consider the inertial convective ranges of the passive scalar. In the enstrophy cascade an à-la-Obukhov analysis shows that $E_\theta\ (k) \approx \eta\beta^{-1/3}\ k^{-1}$ or equivalently $E_\theta\ (k) \approx (\eta/\beta)\ k^2\ E(k)$. This shows that the scalar spectrum is proportional to the spectrum of the cascading quantity, i.e. the enstrophy. This is not surprising since the enstrophy is the spectrum of the vorticity, which is conserved following the motion as is the passive scalar.

In the inverse energy cascade one can analogously write

$$E_\theta\ (k) \approx (\eta/\varepsilon)\ E(k) \approx \eta\varepsilon^{-1/3}\ k^{-5/3}$$

It is then worth while asking the question of whether the passive scalar follows the inverse energy cascade towards larger scales or on the contrary cascades to small scales. Indeed, unlike the velocity field, the passive scalar is not constrained by enstrophy conservation. EDQNM calculations of this problem have shown that the actual direction of the scalar cascade is directly towards the small scales (Lesieur et al., 1985). This behaviour presents analogies with the interaction between baroclinic and barotropic energies in the quasi-geostrophic system, as applied to the dynamics of planetary scale motions in the earth atmosphere: in that case the temperature is injected at low wave-numbers due to differential heating, cascades along a direct $k^{-5/3}$ inertial range up to the internal radius of deformation, where it is partially converted into barotropic energy; the latter flows back through an inverse $k^{-5/3}$ cascade, and the final state consists of two cascades of opposite direction for the energy and the temperature. Such dynamics has been clearly displayed with the aid of the EDQNM theory applied to a two-layer quasi-geostrophic model (Hoyer et al., 1982).

Finally, and in analogy with the three-dimensional case, we will look at the pair dispersion coefficient, still defined as $\sigma \approx r\ v_r$. In the inverse energy cascade, one still obtains the 3-D Richardson law which yields $r \approx t^{3/2}$ when ε is constant. In the enstrophy cascade, $v_r \approx r/\tau$, with $\tau \approx \beta^{-1/3}$ (independent of k), and the dispersion coefficient is given by $\sigma \approx \beta^{1/3} r^2$ as shown by Lin (1972). This yields an exponential dispersion $r^2 \approx \exp(\beta^{1/3} t)$ provided β is constant (i.e. with a stationary injection). The r^2-law for the dispersion coefficient can be shown to be valid for the freely-decaying case also (Larchevêque and Lesieur, 1981).

Actually, the preceeding derivation can be generalized to obtain a relationship between the slope of the energy spectrum and the dispersion coefficient: let us assume a $k^{-\alpha}$ energy inertial range, to which one can associate a local velocity at $r = k^{-1}$ equal to $v_r = r^{(\alpha-1)/2}$ and a dispersion coefficient $r\, v_r = r^{(\alpha+1)/2}$. This allows to calculate, in two and three dimensions, the exponent of the inertial range from the dispersion data: for instance, a $r^{4/3}$ dispersion coefficient will correspond to a $k^{-5/3}$ energy spectrum, while a r^2-law would imply a k^{-3}-spectrum. Such a correspondence has nevertheless to be taken carefully, since it has been shown in two dimensions (Babiano et al., 1985) that any energy spectrum decreasing faster than k^{-3} would imply the r^2-dispersion law. This latter exponent should then appear to be, at least in two dimensions, the upper bound for the dispersion coefficient power law with respect to the separation r.

5. ANALYTICAL THEORIES OF ISOTROPIC TURBULENCE

Now we are going to build up analytical statistical tools which permit to verify these phenomenological predictions. These theories will be complemented by direct numerical simulations.

Our objective in developing such theories is mainly to obtain evolution equations for the various spectra we have introduced before. These theories can be introduced either from the closure or the stochastic model standpoints. They correspond more or less to the same physics, that is, that the fourth order cumulants linearly damp the third order moments. It seems that they are good tools to study turbulence as long as the departure from Gaussian behaviour is not too high. The closure point of view, leading in particular to the Eddy-Damped Quasi-Normal Markovian approximation (EDQNM) can be found in Orszag (1970) and André & Lesieur (1977). The stochastic models point of view is presented in Frisch et al. (1974). For a more detailed description of these theories, the reader is referred to Lesieur (1984). In 3-D isotropic turbulence, the EDQNM theory leads to the following evolution equation for the energy spectrum

$$\frac{\partial}{\partial t} E(k,t) = T(k,t) - 2\nu k^2 E(k,t) + f(k,t)$$

where

$$T(k,t) = \int_{\Delta} \theta_{kpq} \frac{xy+z^3}{q} \cdot \left[k^2 E(p)E(q) - p^2 E(q)E(k)\right] dp\, dq$$

is the transfer function (in three dimensions) involving the integration over a domain Δ such that the wave vectors $\vec{k}$, $\vec{p}$, $\vec{q}$, should form a triangle ($\vec{k} + \vec{p} + \vec{q} = \vec{0}$). θ_{kpq} is the relaxation time of triple correlations, approximately equal to the time $\tau(k)$ introduced in section 3, and x, y, and z are the cosines of the triangle (k,p,q) interior angles. The spectrum of possible external forces is f(k,t) (which is zero in freely decaying turbulence). An equivalent equation can be obtained for the passive scalar spectrum

$$\frac{\partial}{\partial t} E_\theta(k,t) = T_\theta(k,t) - 2\kappa k^2 E_\theta(k,t) + f_\theta(k,t)$$

where $T_\theta(k,t)$ is the scalar transfer.

These two evolution equations allow us (in three dimensions) to calculate "eddy" viscosities and diffusivities in spectral space, which can be used for subgridscale modelling purposes (in three dimensions). Let us first consider a spectrum with a "gap" (i.e. a separation of scales) at some wavenumber k_C. The energy transfer through k_C is due to triads such that $k \ll k_C \ll p \approx q$

This yields eddy-viscous and conductive transfers:

$$T^{EV}(k) \approx -2\,\nu_e k^2 E(k)$$

$$T^{EC}_{\theta}(k) \approx -2\,D_e k^2 E_{\theta}(k)$$

where the eddy viscosity and diffusivity are given by

$$\nu_e = (1/15) \int_{K_c}^{\infty} \theta_{oqq} \left[5\,E(q) + q \frac{\partial E}{\partial p} \right] dq$$

$$D_e(k) = (2/3) \int_{K_c}^{\infty} \theta_{oqq} E(q)\, dq$$

(see Kraichnan, 1976; Chollet and Lesieur, 1981, and Herring et. al, 1982). These eddy-quantities can be generalized to the case of a continuous spectrum (no gap), and used as subgridscale diffusion operators for Large-Eddy-Simulation purposes (LES). It yields eddy viscosities and diffusivities proportional to $(E(k_c)/k_c)^{1/2}$.

These subgridscale operators allows us to obtain self-similar free evolutions of the large scales (explicitly simulated) in very good agreement with the predictions of the phenomenological theories (Chollet, 1984, Comte-Bellot and Corrsin, 1966). A promising (but more costly) approach has also been developed by coupling a direct numerical simulation of the large eddies with a complete EDQNM statistical modelling of the small scales (Chollet, 1984): the resulting energy spectrum is shown in Fig. 1. Such subgridscale modelling can be shown (Chollet, 1984), in the three dimensional isotropic case, to be equivalent to an "averaged" eddy viscosity in the real space of the form given by Smagorinsky (1963). This is no longer true for anisotropic and inhomogeneous turbulence where Smagorinsky's eddy viscosity represents the mean shearing action of the large eddies, while eddy viscosities and diffusivities proportional to $(E(k_c)/k_c)^{1/2}$ are characteristic of the local shear in the neighbourhood of the numerical grid mesh. It is then possible that these latter subgridscale procedures could prove to be more useful than Smagorinsky's procedure in inhomogeneous three-dimensional simulations where the grid mesh corresponds to quasi-isotropic eddies cascading along a Kolmogorov inertial range.

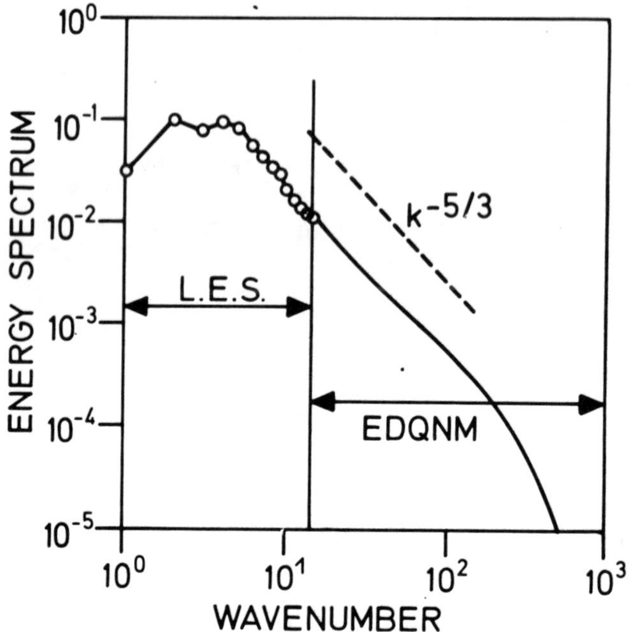

Fig. 1: Energy spectrum of a large-eddy-simulation (o) coupled with an EDQNM modelling of the subgrid scales Lesieur, 1984.

The subgridscale modelling problem arises in fairly different terms for two-dimensional or quasi-two-dimensional turbulence. Nonlocal fluxes of energy across the cutoff wave number can also be evaluated in the enstrophy cascade and they correspond to two terms. One of them represents the shearing action of the large scales (wave number q) on wave number k through interactions ($q \ll k < k_C < q$), and the second is a "negative eddy-viscosity" term which sends the energy transferred in the subgridscales (due to the large scale shearing) back into these large scales. The total balance between the two terms can be shown to be exactly zero in the enstrophy cascade so that there is no energy flux through this cascade. The same analysis applied to the enstrophy flux shows that only the first mechanism is important. A subgridscale modelling based on this concept and transferring enstrophy at the right rate ß through the cutoff, while no energy is being transferred, has been developed in Basdevant et al. (1978): it leads to a k^{-3}-spectrum extending up to the cutoff, but seems to affect the spatial coherence of the large scales, as was shown by Basdevant and Sadourny (1983). Other promising approaches could be the use of high order Laplacian dissipative operators, or the "anticipated vortex" method proposed by Basdevant and Sadourny (1983).

To conclude this section, let us mention the results on the diffusion of "error" obtained with the EDQNM theory. Here the problem considered is the so-called "statistical predictability" problem where one is interested in the increase (due to nonlinear interactions in the Navier-Stokes equations) of decorrelation between two velocity fields initially decorrelated only in the small scales. One can introduce the error spectrum E_Δ, density per wavenumber of the energy of the difference between the two fields. Fig. 2 shows the evolution of the energy and error spectra for two-dimensional unforced turbulence (Métais and Lesieur, 1985). The energy displays a k^{-3} enstrophy cascade. Both spectra behave like k^3 when $k \to 0$ due to nonlocal interactions in Fourier space. In the enstrophy cascade, the error spectrum seems to be proportional to the enstrophy spectrum, an analogy with the inertial convective range of the 2-D passive scalar studied in the second section (see also Lesieur and Herring, (1985).

These unforced calculations show an increase of predictability of about 30 to 40 %, compared to the case of a stationary forced two-dimensional turbulence with the same initial conditions. The latter problem had already been studied with the aid of the statistical theories of turbulence (Quasi-Normal and EDQNM) (Lorenz, 1969; Leith and Kraichnan, 1972). Applied to atmospheric predictability spectra, it yields predictability times of about ten days. These times are the theoretical limit for a deterministic prediction of the state of the atmosphere, when it is approximated by statistically stationary two-dimensional Navier-Stokes equations on a plane. The increased predictability due to freely-evolving situations could have implications in dynamical meteorology or oceanography, whenever such systems are encountered.

6. DIFFUSION OF TEMPERATURE IN STRATIFIED TURBULENCE

This section will be devoted to the study of the decay of a three-dimensional initially isotropic turbulence under the effect of a stable stratification. Some results taken from a large eddy simulation of the homogeneous problem (Métais, 1985) will be presented.

The phenomenology of the initial stage has been given by Riley et al. (1981). Initially, the Froude number U/NL characterizing the relative importance of inertial forces and buoyant forces is high and the turbulence decays as in the unstratified case; the ratio U/L decreases like t^{-1}, and the Froude number decreases to values of order 1 in a few times N^{-1}. At that time, corresponding to the appearance of internal gravity waves, the increasing integral scale of turbulence L has collapsed on the decreasing buoyant scale $L_B = (\varepsilon/N^3)^{1/2}$. The calculation by Métais (1985) is a homogeneous calculation on the basis of the Boussinesq equations presented in section 1. The code is pseudo-spectral with a 32^3-resolution. The initial state is an evolved isotropic field of isotropic turbulence to which gravity is suddenly applied, and the initial Froude number is of order 3. The subgridscale modelling used is the isotropic one presented in section 5. Indeed it can be shown that such a subgridscale parametrization is valid as long as the buoyant scale remains inferior to the spatial grid mesh. Fig. 3 shows the time evolution of the following normalized quantities: horizontal kinetic energy in the unstratified (a) and stratified (b) cases; vertical kinetic (c) and potential energy (d) in the stratified case. It follows that the horizontal kinetic energy is nearly unaffected by stratification. Due to internal waves, there is a periodic exchange between the vertical kinetic and potential energies, but their sum (the total vertical energy) decays like the horizontal kinetic energy, proportionally to $t^{-1.18}$.

The possible appearance of two-dimensional turbulence after the collapse of three-dimensional turbulence under stratification was first

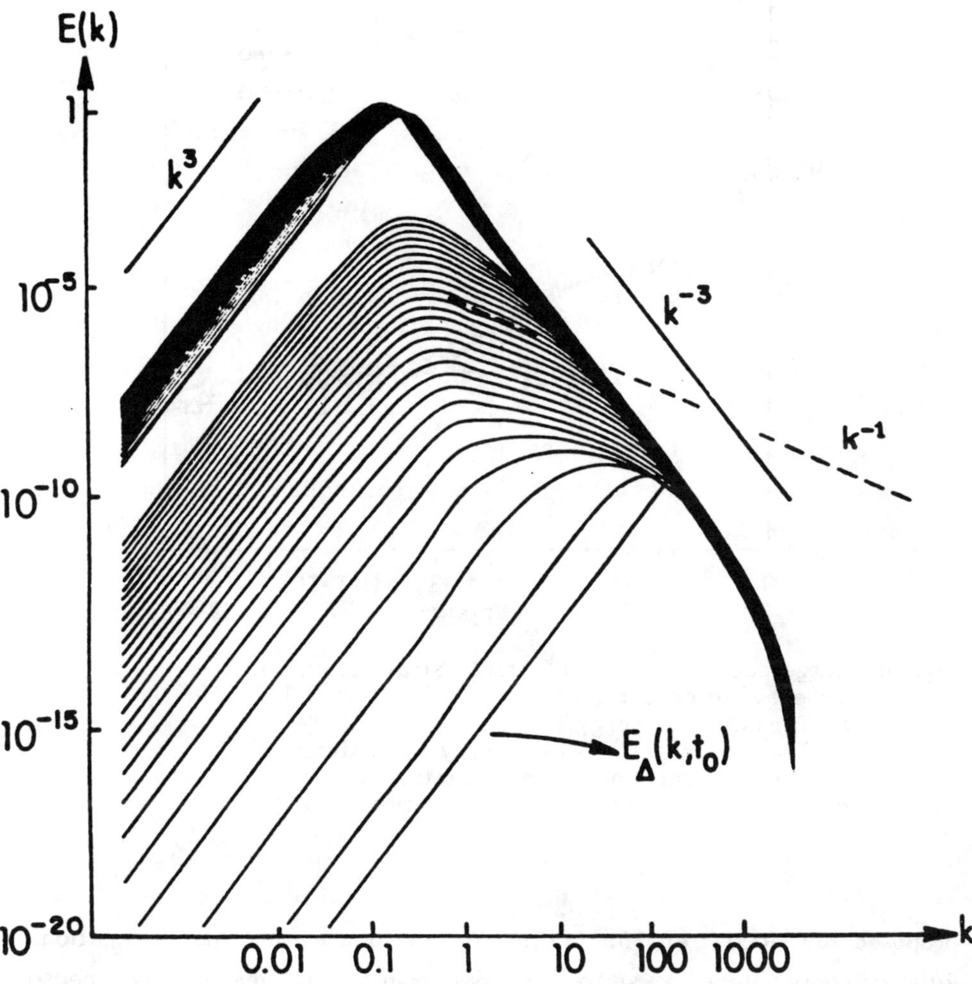

Fig. 2: EDQNM calculation showing the evolution with time of the energy and error spectra in freely-evolving 2-D turbulence Kraichnan, 1976.

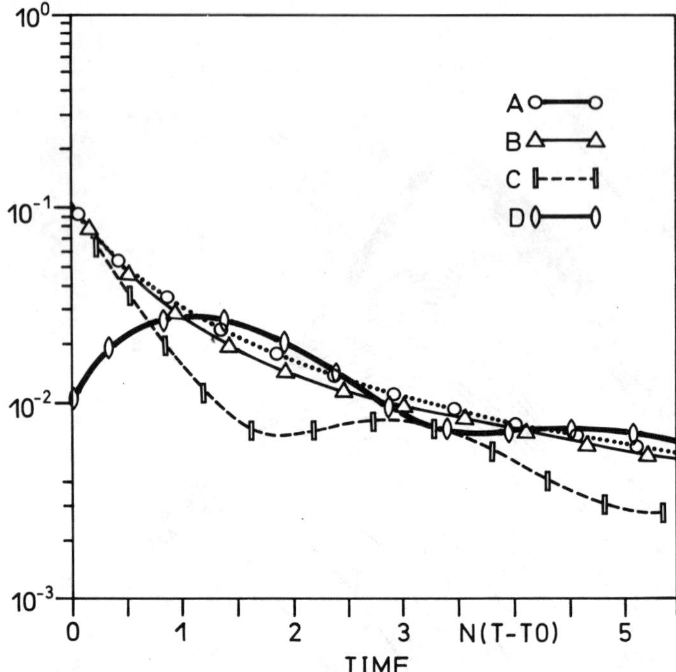

Fig. 3: Large-eddy-simulation of stably stratfied turbulence.
Time evolution of:
a- The kinetic energy (no stratification)
b- The horizontal kinetic energy (stratification)
c- The vertical kinetic energy (stratification)
d- The potential energy (stratification)

proposed by Riley et al. (1981). This would imply no dissipation of kinetic energy, and a strong inverse transfer in the energy spectrum. Experimentally one actually observes a two-dimensional collapse of the wake of an obstacle pulled through a stably stratified fluid. In the calculation of Métais (1985), there cannot be such an effect since the horizontal kinetic energy decays as in three dimensions. This fact is confirmed by Fig. 4 which shows the horizontal kinetic energy spectra in the unstratified and stratified cases; there is no appreciable inverse transfer of energy due to stratification, and therefore no tendency towards two-dimensional turbulence.

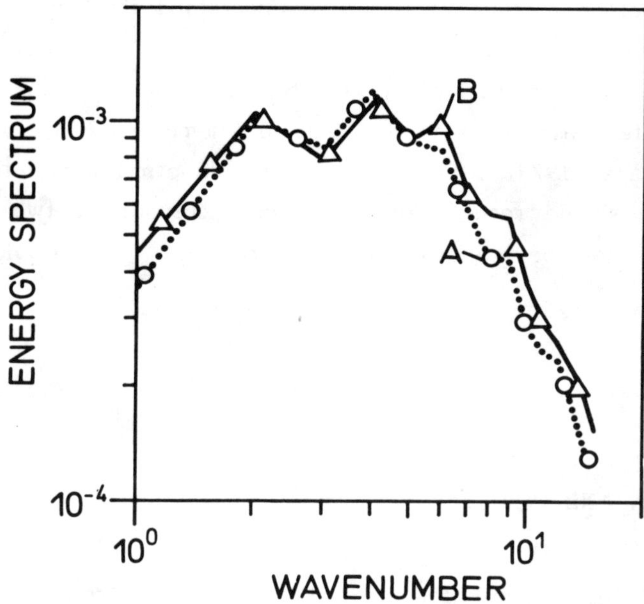

Fig. 4: Same calculation as in Fig. 3; the horizontal energy spectra at time $6.N^{-1}$ are shown:
a- in the stratified case
b- in the unstratified case

A possible way to reconcile the apparent discrepancy between the numerical calculation on one hand, and the experimental evidence on the other, could lie in the necessary time for the two-dimensional turbulence to arise. It seems that in the experiments, bidimensionalization occurs at about 10 Brunt-Väisälä periods, that is 60 N^{-1}. This is 10 times longer than the calculation performed by Métais (1985). Notice however that the latter reaches a final Froude number of about .25 which is already low, and the corresponding velocity field ought to display already some two-dimensional characteristics, which is actually not the case.

This discussion about the relevance of the concept of two-dimensional turbulence as the eventual state of a turbulence in a stably stratified fluid is extremely important. In order to understand the vertical fluxes in the small scales of the ocean (two-dimensional turbulence would imply no vertical flux), and the variability of the atmosphere in the meso-scale range (from a few kilometers to hundreds of kilometers) it has indeed been conjectured (Gage, 1979) that the mesoscale kinetic energy spectrum in the atmosphere should result from an inverse cascade of two-dimensional turbulence due to the collapse of smaller scale 3-D turbulence under stratification.

7. CONCLUSION AND DISCUSSION

We have tried to present a wide class of problems related to turbulent diffusion and having possible oceanic and atmospheric applications in the small, meso or synoptic scales. We have also presented the statistical analytical tools (closures) allowing us to predict the time evolution of the various spectra involved in these problems.

It seems that closures are a good tool to study the strong nonlinearities of turbulence, at least for what concerns the energy transfers between various scales. They also allow us to calculate eddy-exchange coefficients which can be used in 3-D (unstratified and stratified) turbulence for large eddy simulation purposes.

8. ACKNOWLEDGEMENTS

The author is indebted to J.P. Chollet and O. Metais for letting him include their numerical results in this paper, and to M. Alves, P. Cessi and R. Vautard whose lecture notes constituted the first version of this paper. Useful discussions with J. Riley on the "collapse" problem are also acknowledged. This work was sponsored by the CNRS (ATP "Recherches Atmosphériques" et "Atmosphère Moyenne").

REFERENCES

André, J.C. and Lesieur, M., 1977: Influence of helicity on the evolution of isotropic turbulence at high Reynolds number. J. Fluid Mech., 81, 187.

Babiano, A., C. Basdevant and R. Sadourny, 1985: Structure functions and dispersion laws in two-dimensional turbulence. J. Atmos. Sci., 42, 941-949.

Basdevant, C., M. Lesieur and R. Sadourny, 1978: Subgrid-scale modeling of enstrophy transfer in two-dimensional turbulence. J. Atmos. Sci., 35, 1028-1042.

Basdevant, C. and R. Sadourny, 1983: Two-dimensional turbulence. J. Méc. Théor. Appl. Numéro spécial, 243-249.

Batchelor, G.K., 1969: High speed computing in fluid dynamics. Phys. Fluids, 12 (suppl.), 233-239.

Chollet, J.P. and Lesieur, M., 1981: Parameterization of small scales of three-dimensional isotropic turbulence utilizing spectral closures. J. Atmos. Sci., 38, 2747-2757.

Chollet, J.P., 1984: Thèse de Doctorat d'Etat, Grenoble University.

Comte-Bellot,G. and S. Corrsin, 1966: The use of a contraction to improve the isotropy of grid-generated turbulence. J. Fluid Mech., 24, 657-682.

Fjortoft, R., 1953: On the changes in the spectral distribution of kinetic energy of two-dimensional non-divergent flow. Tellus, 5, 225-230.

Frisch, U., Lesieur, M. and Brissaud, A., 1974: A Markovian random coupling model for turbulence. J. Fluid Mech., 65, 145-152.

Gage, K.S., 1979: Evidence for a $k^{-5/3}$ law inertial range in mesoscale two-dimensional turbulence. J. Atmos. Sci., 36, 1950-1954.

Gargett, A.E., Osborn, T.R. and Nasmyth, P.W., 1984: Local isotropy and the decay of turbulence in a stratified fluid. J. Fluid Mech., 144, 231-280.

Grant, H.L., Stewart, R.W. and Moilliet, A. 1962: Turbulence spectra from a tidal channel. J. Fluid Mech., 12, 241-263.

Herring, J.R., Schertzer, D., Lesieur, M., Newman, G.R., Chollet, J.P., and Larchevêque, M., 1982: A comparative assessment of spectral closures as applied to passive scalar diffusion. J. Fluid Mech., 124, 411-437.

Hoyer, J.M. and R. Sadourny, 1982: Closure modeling of fully developed baroclinic instability. J. Atmos. Sci., 39, 707-721.

Kolmogorov, A.N. 1941: The local structure of turbulence in incompressible viscous fluids for very large Reynolds numbers. Dokl. Akad. Nauk SSSR, 30, 301-305.

Kraichnan, R.H., 1967: Inertial ranges in two-dimensional turbulence. Phys. Fluids., 10, 1417-1423.

Kraichnan, R.H., 1976: Eddy viscosity in two and three dimensions. J. Atmos. Sci., 33, 1521-1536.

Larchevêque, M. and M. Lesieur, 1981: The application of eddy-damped Markovian closures to the problem of dispersion of particle pairs. J. Mécanique, 20, 113-134.

Leith, C.E., 1968: Diffusion approximation for two-dimensional turbulence. Phys. Fluids, 11, 671-673.

Leith, C.E. and R.H. Kraichnan, 1972: Predictability of turbulent flows. J. Atmos. Sci., 29, 1041-1058.

Lesieur, M. and Herring, J.R. 1985: Diffusion of a passive scalar in two-dimensional turbulence, J. Fluid Mech., 161, 77-95.

Lesieur, M., 1984. Fully developed turbulence and statistical theories. Application to the coherent structures". In "Combustion and nonlinear phenomena", P. Clavin, B. Larroutuou and P. Pelcé, Eds. Les éditions de Physiques, Les Houches.

Lilly, D.K., 1983: Stratified turbulence and the mesoscale variabilty of the atmosphere. J. Atm. Sci., 40, 749-761.

Lin, J.T., 1972: Relative dispersion in the enstrophy cascading inertial range of homogeneous two-dimensional turbulence. J. Atmos. Sci., 29, 394-396.

Lorenz, E., 1969: The predictability of a flow which possesses many scales of motion. Tellus, 21, 289-307.

Mètais, O., and Lesieur, M., 1985. Statistical predictability of decaying turbulence". Submitted to J. Atm. Sci.

Mètais, O. Evolution of three-dimensional turbulence under stratification. Turbulent Shear Flows V, Cornell University, 1985.

Moffatt, H.K., 1978: Magnetic field generation in electrically conducting fluids. Cambridge University Press.

Orszag, S.A., 1970: Analytical theories of turbulence. J. Fluid Mech., 41, 363-386.

Pouquet, A., Lesieur, M., André, J.C. and Basdevant, C., 1975: Evolution of high Reynolds number two-dimensional turbulence. J. Fluid Mech., 72, 305-319.

Rhines, P., 1975: Waves and turbulence on a beta-plane. J. Fluid Mech., 69, 417-443.

Richardson, L.F. 1926: Atmospheric diffusion shown on a distance-neighbour graph. Proc. Roy. Soc. London, A 110, 709-737.

Riley, J.J., Metcalfe, R.W. et Weissman, M.A., 1981. In "Nonlinear properties of internal waves", B.J. West, ed., La Jolla Institute.

Smagorinsky, J.S., 1963: General circulation experiments with the primitive equations. I. The basic experiment. Mon. Wea. Rev., 91, 99-164.

TURBULENT DIFFUSION IN LARGE-SCALE FLOWS

ROBERT SADOURNY

Laboratoire de Météorologie Dynamique
École Normal Supérieure
24 Rue Lhomond, 75231 Paris, France

1. INTRODUCTION

We shall be concerned here with the turbulent diffusion processes associated with the large-scale quasi-two-dimensional motion in the atmosphere and in the ocean, and the way to parameterize them in numerical models. By turbulent diffusion, we mean the mean effect of small scales on the larger scales: these "small" scales are barotropic and baroclinic transient eddies which we may not necessarily want to resolve in our numerical simulations - especially in the framework of long-term climate modelling. They are to some extent unpredictable, but we may consider that their average effect on the larger scales is well defined by the structure of the larger scales themselves. This assumption is physically reasonable because the small scale instabilities and the resulting eddy fluxes are processes by which the large scale flow tends to relax its own stresses or strains. Naturally, it is also a general prerequisite for the feasibility of large eddy simulations to assume that the statistical effect of the sub-grid scale eddies is indeed parameterizable.

We shall begin our approach to turbulent diffusion modelling by the simplest case of a passive scalar, for which the derivations and interpretations are most straightforward. Then we shall show that the methods developed for the passive scalar diffusion apply equally well to the more involved cases of barotropic vorticity dynamics, and baroclinic dynamics in general, including frontogenesis - in spite of the large differences of flow behaviour encountered in these various cases.

J. Willebrand and D. L. T. Anderson (eds.), Large-Scale Transport Processes in Oceans and Atmosphere, 359–373.

2. DIFFUSION OF A PASSIVE SCALAR IN TWO-DIMENSIONAL FLOW

We shall admit in this section that two-dimensional turbulence is "nonlocal" in the sense that the strain exerted on a structure of scale l is mainly due to velocity shears whose scale l' is much larger than l. This dominance of large-scale shear occurs when the kinetic energy spectrum is steep enough. More precisely, let $K(k)$ be the one-dimensional spectrum of kinetic energy, so that

$$K = \int_0^\infty K(k)\, dk$$

is the mean kinetic energy per unit mass, and

$$Z = \int_0^\infty k^2\, K(k)\, dk$$

is its mean enstrophy per unit mass; $Z^{\frac{1}{2}}$ is a gross measure of the shear in root-mean-square sense. The strain exerted on a structure of scale $l = k^{-1}$ can be grossly measured as a frequency $\tau^{-1}(k)$, with

$$\tau^{-2}(k) = \int_0^k k'^3\, K(k')\, d\mathrm{Ln}k' \; . \tag{1}$$

Using this measure, we see that nonlocality - or dominance of large-scale shear - holds when the integrand in (1) is a decreasing function of wavenumber, or in other words, when the kinetic energy spectrum is steeper than k^{-3}. This is actually the case for two-dimensional, or quasi two-dimensional, flows, as we shall see later.

The dynamics of a passive scalar can be studied using the same phenomenological methods as used in classical turbulence theory - see for instance Lesieur, Sommeria and Holloway (1981). To fix ideas, let us consider the problem of a pollutant ξ, whose variance per unit mass is noted X:

$$X = \int_{a}^{\infty} X(k)\, dk \ ,$$

and is injected at a constant rate α at a given scale k_I. The governing equation for ξ is the transport equation

$$\frac{\partial \xi}{\partial t} + \frac{\partial(\psi,\xi)}{\partial(x,y)} = \nu\ \Delta\xi + \text{injection} \ , \qquad (2)$$

where Ψ is the streamfunction of the flow, and ν is molecular viscosity, which plays an active role only at very small scales: $k > k_D >> k_I$. In the stationary regime, we expect X to cascade from the injection range to the dissipation range at the constant rate α. Then simple phenomenological arguments tell us that, within this inertial range, $x(k)$, α and the straining time scale $\tau(k)$ are related by

$$\alpha \sim kX(k)/\tau(k)$$

or in other words

$$X(k) \sim \alpha\tau(k)\ k^{-1} \ . \qquad (3)$$

For large values of k the integral in (1) converges, or at most diverges weakly, because of nonlocality; then $\tau(k)$ is asymptotically constant or increases logarithmically. It follows that the passive scalar spectrum follows a k^{-1} law (irrespective of the slope of the kinetic energy spectrum), with a possible logarithmic correction if nonlocality is just marginal. This phenomenological law has been verified in numerical calculations by Babiano et al (1984).

Numerical models of the passive scalar transport equation (2) must be able to simulate the cascade process, even though they do not resolve the physical dissipation range; if k_c is the cut-off wave number, we have to assume $k_I < k_c << k_D$. The right hand side of (2) must then be replaced by a turbulent diffusion term, which models the turbulent mixing due to sub-grid scales, extracting the right amount of variance from the resolved scales.

The simplest approach to turbulent diffusion is a spatial filtering operator, for example an iterated Laplacian (Basdevant et al, 1981); a numerical model of (2) then reads (from now on we omit the injection term)

$$\frac{\partial \xi_c}{\partial t} + \frac{\partial(\psi_c,\xi_c)}{(x,y)} = - \tau_c^{-1} L_c \xi_c \tag{4}$$

where the subscript c refers to truncated fields, and τ_c is a time scale of the order of magnitude of $\tau(k_c)$. The main defect of (4) is that turbulent diffusion is modelled independently of the velocity field. L_c is a high-pass filtering operator designed to concentrate the turbulent diffusion effects near the cut-off scale: for example, $L_c = (-k_c^2 \Delta)^n$.

In fact, a truly physical approach to turbulent diffusion should take into account the structure of the velocity field as well as the pollutant gradients. Intuitively, it is more natural to apply the filtering operator L_c, which expresses the fit of subgrid scale turbulent diffusion to the cut-off location, to the time derivative of the pollutant - which describes the transport activity, rather than to the pollutant field itself. This can be realized in a straightforward way by using the following model:

$$\frac{\partial \xi_c}{\partial t} + \frac{\partial(\psi_c,\xi_c)}{\partial(x,y)} = \tau_c \frac{\partial}{\partial(x,y)} \left(\psi_c, L_c \frac{\partial(\psi_c,\xi_c)}{\partial(x,y)}\right)$$

or, using vector notation instead of streamfunction,

$$\frac{\partial \xi_c}{\partial t} + V_c \cdot \mathbf{grad}\ \xi_c = \tau_c V_c \cdot \mathbf{grad}\ (L_c V_c \cdot \mathbf{grad}\ \xi_c); \tag{5}$$

Note that we must impose that the filter L_C be local in physical space, since diffusion is a local process: this is obviously the case for the iterated Laplacian. L_C being positive definite, equation (5) is parabolic and induces a global dissipation

$$\frac{dX}{dt} = - \tau_C \int\int (\mathbf{V}_C \cdot \text{grad } \xi_C)\, L_C\, (\mathbf{V}_C \cdot \text{grad } \xi_C)\, dxdy\ . \tag{6}$$

The method used in model (5) can be described as a scale-selective upwind method: the correction in the right-hand side corresponds to an upwind correction of the left-hand side advection term, with a scale-dependent time lag determined by the structure of L_C.

There is still one disadvantage in formulation (5): a uniform velocity field induces a residual diffusion, depending on the small-scale structure of the pollutant field. This is unphysical, since a uniform velocity field can be substracted from the equations by a Galilean transformation. In other words, model (5) is not Galilean-invariant. The break-down of Galilean invariance is related to the fact that turbulent diffusion of a passive scalar should depend on velocity shear rather than velocity itself. Model (5) can be readily modified to take this into account. Let us define a "shear" vector

$$\mathbf{S} = \text{rot } \mathbf{V} \frac{\mathbf{V}}{||\mathbf{V}||}$$

colinear with velocity, and whose magnitude is vorticity. Then the scale-selective upwind method can be replaced by the "scale-selective shear-gradient" method:

$$\frac{\partial \xi_C}{\partial t} + \mathbf{V}_C \cdot \text{grad } \xi_C = \tau_C k_C^{-2} \text{ div} \left[\mathbf{S}_C\, L_C\, (\mathbf{S}_C \cdot \text{grad } \xi_C) \right] . \tag{7}$$

Note the physical interpretation of (7). The product S_C · **grad** ξ_C is a differential advection of ξ_C due to the combined effect of lateral velocity shear and longitudinal pollutant gradient (see figure 1). Again L_C is a physically local operator which selects the scales of sub-grid influence. And the diffusion term is written in divergence form because its space integral must vanish - no net production of pollutant, S_C being no longer, like V_C in (5), non-divergent. Dissipation due to (7) has the same form as (6), replacing V by S. In model (7) like in model (5), turbulent diffusion is intimately related to the flow dynamics. In both cases the right-hand side vanishes whenever the isolines of ξ locally follow the streamlines (stationary conditions for the pollutant advection). Model (7) is not yet strictly Galilean-invariant because of residual directional effects, but the main defect of (5), the spurious diffusion by a constant velocity field, has been eliminated.

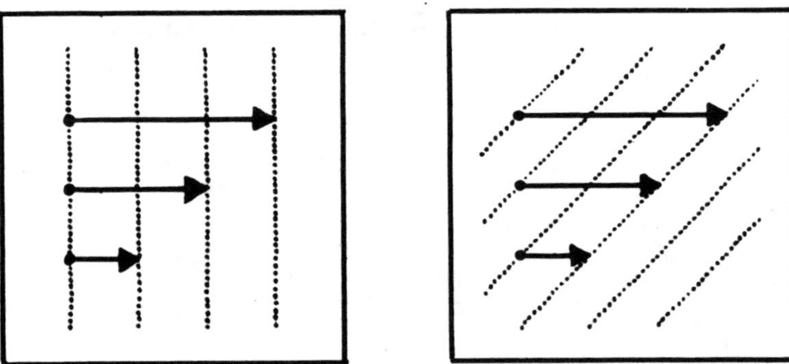

Figure 1 Schematic view of the combined effect of longitudinal gradient of pollutant concentration, and laterial velocity shear, on the pollutant diffusion. Isolines of concentration are indicated by dotted lines.

3. DIFFUSION OF VORTICITY IN TWO-DIMENSIONAL FLOW

We turn now our attention towards the dynamics of the flow itself, described by the barotropic vorticity equation

$$\frac{\partial \zeta}{\partial t} + \frac{\partial(\psi,\zeta)}{\partial(x,y)} = \nu \, \Delta \tag{8}$$

Equation (8) is the same as equation (2) for the passive scalar, except that now $\zeta = \Delta\psi$. We quickly recall here the classical phenomenology of two-dimensional turbulance, originally derived by Kraichnan (1967, 1971), Leith (1968) and Batchelor (1969). Like in the passive scalar problem, we consider an injection of variance at a given wavenumber k_I. Variance can be seen either as an energy variance K(k), or an enstrophy variance $Z(k) = k^2K(k)$. We suppose a constant rate of energy and enstrophy injection, which we denote respectively by ε and $\omega = k_I^2\ \varepsilon$.

For very large Reynolds numbers ($k_D >> k_I$) the ratio of enstrophy dissipation to energy dissipation, which is of the order of k_D^2 , is much larger than the injection ratio k_I^2: this means that, in the stationary régime, no energy can actually be dissipated by viscosity. From there it follows that enstrophy cascades at a constant rate ω from k_I to k_D, while energy cascades back to ever larger scales at a constant rate ε. The classical phenomenology of the enstrophy inertial range looks very much like the passive scalar case. The enstrophy cascade rate ω, the enstrophy spectrum Z(k) and the characteristic timescale of straining $\tau(k)$ are related by

$$\omega = k\ Z(k)/\tau(k) \ ,$$

or in other words

$$\omega = k^3\ K(k)/\tau(k) \ ,$$

which, together with (1), yields the classical (Kraichnan 1971) spectral law

$$K(k) \sim \omega^{2/3} k^{-3} (Lnk)^{-1/3} . \tag{9}$$

In this theory the passive scalar and vorticity have identical behaviours. Eliminating K(k) and $\tau(k)$ between (1, 3, 9) yields

$$X(k) \sim \alpha\omega^{-1/3} k^{-1} (Lnk)^{-1/3} \tag{10}$$

which has the same spectral dependency as enstrophy. (We are in the case, already mentioned in section 2, of marginal non-locality, which brings logarithmic corrections to the k^{-1} law.)

This identity between vorticity and passive scalar dynamics has actually been proven wrong in numerical experiments by Babiano et al. (1984). The phenomenology described above seems to apply well to the passive scalar but not to vorticity. The link between ψ and ζ, by which the flow reacts to vorticity advection, is an essential aspect of the dynamics which is too subtle to be taken into account by phenomenological arguments. Numerical experiments on forced two-dimensional turbulence indeed show kinetic energy spectra much steeper than expected from (9) (Basdevant et al., 1981); numerical experiments on decaying two-dimensional turbulence also seem to yield steeper spectra if they are integrated long enough (McWilliams, 1984, to compare with Bracket and Sulem, 1984). In such experiments steep spectra are always associated to the emergence of strong coherent vortices, which resist the enstrophy cascade process. Concentrations of this kind do not occur in passive scalar dynamics: see for instance figure 2, extracted from Babiano et al., 1984. The formation of coherent vortices is in fact a reverse energy cascade process which occurs only in vorticity dynamics, since there is no equivalent to the energy invariant in the passive scalar advection.

In modelling the turbulent diffusion of vorticity, we must thus follow the passive scalar diffusion model, but be careful to impose the energy conservation constraint. It is readily seen that model (5), applied to vorticity, is indeed energy conserving. Energy conservation can be verified directly by taking the streamfunction formulation, multiplying by ψ_c and integrating over the domain -- which makes all Jacobian terms vanish. Another way is to take the equivalent of (5) for the momentum equation:

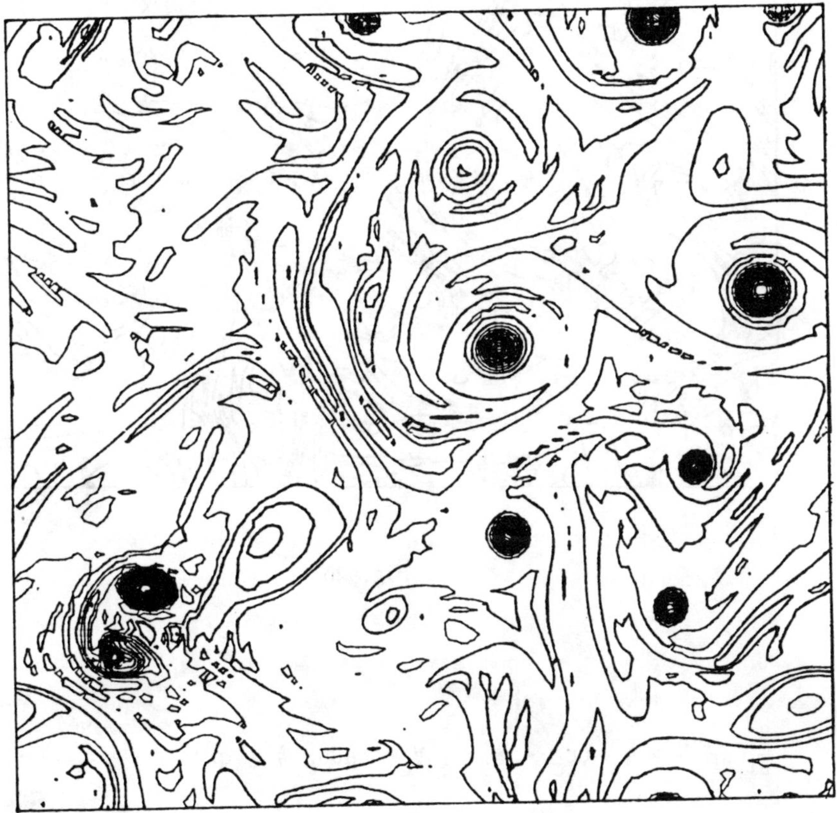

Figure 2 Vorticity (a) and pollutant (b) charts at coincident times. Note the close correspondance between structures, but the absence of strong concentrations in the pollutant field (from Babiano, Basdevant, Legras and Sadourny, 1984).

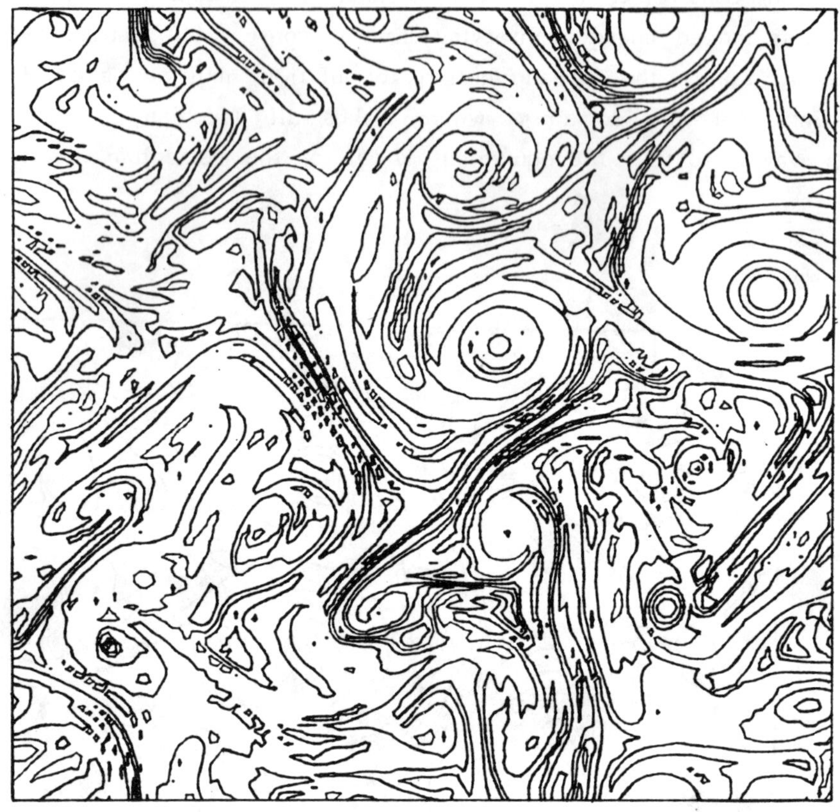

Fig. 2b

$$\frac{\partial \mathbf{V}_c}{\partial t} + \mathbf{N} \times \mathbf{V}_c \, (\zeta_c - \tau_c L_c \, \mathbf{V}_c \cdot \mathrm{grad} \, \zeta_c) + \mathrm{grad} \, (P + \frac{V_c^2}{2}) = 0 \, , \qquad (11)$$

where N is the unit vector normal to the flow domain. Here we see that the diffusion correction is normal to velocity, and therefore produces no work. Model (11), or (5) applied to vorticity, is the "anticipated vorticity method" of Sadourny and Basdevant (1981). Its advantage over the "superviscosity" (iterated Laplacian) approach (e.g.: Basdevant et al., 1981) is that it is based on the dynamics of the motion rather than on its spatial structure only. Its advantage over the spectral diffusion approach of Leith (1971) or Basdevant, Lesieur and Sadourny (1978) is that it is local in physical space. Finally, its advantage over Smagorinsky (1963)'s method is that it formally obeys the two basic constraints of two-dimensional flow: energy conservation and enstrophy dissipation. For a critical appraisal of these various approaches, the reader may refer to Basdevant and Sadourny (1983).

The defect of (5, 11), again, is that it violates the Galilean-invariance property. To get closer to Galilean invariance we may again shift to (7), replacing ξ by ζ, or its equivalent for the momentum equation:

$$\frac{\partial V_c}{\partial t} + N \times V_c \, \zeta_c + \text{grad}\,(P + \frac{V^2}{2}) = \tau_c k_c^{-2} \, N \times S \, L_c(S_c \cdot \text{grad}\, \zeta_c) \,. \tag{12}$$

Since S is colinear to V, the right-hand side again does no work and total energy is again formally conserved. Like in (7), the shear-gradient formulation appears quite physical because it is based on the differential advection of vorticity gradients by velocity shears.

4. DIFFUSION OF POTENTIAL VORTICITY IN QUASI-GEOSTROPHIC FLOW

To get closer to the actual dynamics of atmospheric and oceanic flows in mid-latitudes, we have now to include baroclinic effects and consider the general case of three-dimensional quasi-geostrophic flow. The phenomenological theory of quasi-geostrophic turbulence has been developed by Charney (1971), Salmon (1978, 1980), Hoyer and Sadourny (1982) and Sadourny (1985). In formal similarity to two-dimensional turbulence, it is characterized by a direct potential enstrophy cascade towards the smaller scales, associated with a reverse energy cascade towards the larger scales. Note that the scales here are three-dimensional, so that the larger scales are barotropic (barotropic modes having the largest vertical scales). Energy here refers to the sum of kinetic and potential energy:

$$E(k,l) = K(k,l) + P(k,l)$$

where l refers to the vertical wavenumber. It is related to potential anstrophy Q(k,l) via

$$Q(k,l) = \lambda^2(k,l)\ E(k,l)$$

with $\lambda(k,l)$ an increasing function of k and l, going infinity as k or l goes to infinity (see Sadourny 1985 for more details).

The key to the understanding of quasi-geostrophic dynamics is the combination of total energy conservation with potential enstrophy dissipation. This means that, like in barotropic flows, energy is trapped in the larger scales; but here the larger scales - in three-dimensional sense - are the barotropic scales. If energy is injected in thermal (baroclinic) form, it must eventually end up in kinetic (barotropic) form. The baroclinic instability, or conversion of potential (baroclinic) energy to kinetic (barotropic) energy, is really a reverse energy cascade process, associated with a dissipation of potential enstrophy without loss of energy.

Turning now to the formulation of turbulent diffusion, we see again that it will be essential to preserve these two basic integral constraints of quasi-geostrophic dynamics in order to model properly the energy conversions associated with baroclinic instability operating at the subgrid scales as well as the resolved scales. This will be done by using the same technique as before, this time on the quasi-geostrophic potential vorticity (q) equation. The anticipated potential vorticity method - analog to (5, 11) - reads

$$\frac{\partial q_c}{\partial t} + V_c \text{ grad } q_c = L_c V_c \text{ grad } (L_c\ V_c \text{ grad } q_c) \qquad (13)$$

where τ_c is now evaluated using the kinetic enstrophy at the level considered. This model has been successfully used by Sadourny and Basdevant (1985): they have shown that using such a turbulent diffusion scheme allows the use of very coarse resolution by adequately simulating the sub-grid scale energy conversion processes due to baroclinic instability. The "quasi"-Galilean-invariant from corresponding to (12) can also be used:

$$\frac{\partial q_c}{\partial t} + \frac{\partial(\psi_c, q_c)}{\partial(x,y)} = \tau_c k_c^{-2} \operatorname{div} \left[S_c L_c (S_c \operatorname{grad} q_c) \right] \tag{14}$$

The quasi-geostropic potential vorticity equation does not actually cover all the aspects of baroclinic dynamics. In addition to it, we have to consider the dynamics of frontogenesis, which can be also done, following Blumen (1978) in a quasi-geostrophic framework. Then we have to model, instead of potential vorticity advection, the advection of temperature or upper and lower boundaries. The problem is formally the same, and we may define the diffusion operator in the same manner, replacing q by ψ_c. It is worth noting that because of their "upstream" character, the turbulent diffusion schemes proposed here allow the formation of contact discontinuities like fronts, which would be strongly damped if ordinary space filters were used.

Finally we have to consider the general case of non-quasi-geostrophic flows. Equations (13) and (14) are still valid in this general case, provided q is defined as Ertel's (hydrostatic) potential vorticity, and the horizontal partial derivatives are taken along isentropic surfaces; all properties stated above are again valid. Frontogenesis dynamics reduces to the advection of potential temperature on free or rigid upper or lower boundaries, for which again we may repeat the above formulations with similar results.

REFERENCES

Babiano, A., C. Basdevant, B. Legras and R. Sadourny, 1984: Dynamiques comparées du tourbillon et d'un scalaire passif en turbulence bi-dimensionelle incompressible. C.R. Acad. Sci. Paris, 299 II, 601-604.

Basdevant, C., M. Lesieur and R. Sadourny, 1978: Subgrid scale modeling of enstrophy transfer in two-dimensional turbulence. J. Atmos. Sci., 35, 1019-1042.

Basdevant, C., B. Legras, R. Sadourny and M. Béland, 1981: A study of barotropic model flows: intermittency, waves and predictability. J. Atmos. Sci., 38, 2305-2326.

Basdevant, C. and R. Sadourny, 1983: Modélisation des échelles virtuelles dans la simulation numérique des écoulements turbulents bi-dimensionnels. J. Méca. Théor. Appl., No. spécial, 243-269.

Batchelor, G.K., 1969: High-speed computing in fluid dynamics. Phys. Fluids, 12 (Suppl. 2), 233-239.

Blumen, W., 1978: Uniform potential vorticity flow. Part I. Theory of wave interactions and two-dimensional turbulence. J. Atmos.Sci., 35, 774-783.

Brachet, M.E. and P.L. Sulem, 1984: Proc. Ninth Int. Comf. Numerical Methods in Fluid Dynamics, Beer Sheva, Dyn. Lecture Notes in Phys., 218, 103-108.

Charney, I.G., 1971: Quasi-geostrophic turbulence. J. Atmos. Sci., 28, 1087-1095.

Hoyer, J.M. and R. Sadourny, 1982: Closure modeling of fully developed baroclinic instability. J. Atmos. Sci., 39, 707-721.

Kraichnan, R.H., 1967: Inertial ranges in two-dimensional turbulence. Phys. Fluids, 10, 1417-1423.

Kraichnan, R.H., 1971: Inertial-range transfer in two and three-dimensional turbulence. J. Fluid Mech., 47, 525-535.

Leith, C.E., 1968: Diffusion approximation for two-dimensional turbulence. Phys. Fluids, 11, 671-673.

Leith, C.E., 1971: Atmospheric predictability and two-dimensional turbulence. J. Atmos. Sci., 28, 145-161.

Lesieur, M., J. Sommeria and G. Holloway, 1981: Zones inertielles du spectre d'un contaminant passif en turbulence bidimensionnelle. C. R. Acad. Sci. Paris, 292, II, 271-274.

Mc Williams, J.C., 1984: The emergence of isolated coherent vortices in turbulent flow. J. Fluid Mech., 146, 21-43.

Sadourny, R., 1985: Techniques for numerical simulation of large-scale eddies in geophysical fluid dynamics. Lectures in Applied Math., 22, 195-207.

Sadourny, R., and C. Basdevant, 1981: Une classe d'opérateurs adaptés à la modélisation de la diffusion turbulente en dimension deux. C. R. Acad. Sci. Paris, 292, II, 1061-1064.

Sadourny, R. and C. Basdevant, 1985: Parameterization of sub-grid scale barotropic and baroclinic eddies in quasi-geostrophic models: anticipated potential vorticity method. J. Atmos. Sci., 42, 1353-1363.

Salmon, R., 1978: Two-layer quasi-geostrophic turbulence in a simple special case. Geophys. Astrophys. Fluid Dynamics, 10, 25-31.

Salmon, R., 1980: Baroclinic instability and geostrophic turbulence. Geophys. Astrophys. Fluid Dynamics, 15, 167-212.

Smagorinsky, J., 1963: General circulation experiments with the primitive equations. I. The basic experiment. Mon. Wea. Rev., 91, 99-164.

INDEX